CRC

The Structure and Biological Function of Histones

Lubomir S. Hnilica
The University of Texas
M. D. Anderson Hospital and Tumor Institute
Texas Medical Center
Houston, Texas

published by:

CRC PRESS

A DIVISION OF
THE CHEMICAL RUBBER CO.
18901 Cranwood Parkway • Cleveland, Ohio 44128

CRC UNISCIENCE SERIES

Another information program recently initiated by CRC is the "Uniscience" Series. The function of "Uniscience" has been defined as the publication of highly authoritative scientific reference works dealing with those subjects for which there exists an urgent necessity for published information due to significant new developments.

To achieve this function, the "Uniscience" program utilizes the conventional reference book format, but with one, all-important difference: it has radically reduced the traditional lag between completion of the manuscript and publication of the printed book. Utilizing newly developed editorial procedures together with newly perfected computerized type-setting equipment, the time-span normally required from manuscript to printed book has been dramatically shortened to a matter of only a few months.

This very real "break-through" in production time makes it possible for the Uniscience Series to provide a new dimension in the distribution of scientific information.

B. J. Starkoff, President
THE CHEMICAL RUBBER Co.

Division of THE CHEMICAL RUBBER CO.

Editor-in-Chief
Robert C. Weast, Ph.D.
Vice President, Research, Consolidated Gas Service Company, Inc.

Editor-in-Chief, Mathematics
Samuel M. Selby, Ph.D., Sc.D.
Professor of Mathematics
Hiram College

Editor-in-Chief, Biosciences
Irving Sunshine, Ph.D.
Chief Toxicologist, Cuyahoga
County Coroner's Office, Cleveland, Ohio

HANDBOOK SERIES

BIOCHEMISTRY
Herbert A. Sober, Ph.D.
National Institutes of Health

BIOENGINEERING
David G. Fleming, Ph.D.
Case Western Reserve University
Lester Goodman, Ph.D.
National Institutes of Health

CHROMATOGRAPHY
Gunter Zweig, Ph.D.
Syracuse University Research Corp.

CLINICAL SCIENCES
Willard R. Faulkner, Ph.D.
Vanderbilt University Medical Center
John W. King, M.D., Ph.D.
Cleveland Clinic Foundation

ELECTRO-OPTICS
Robert J. Pressley, Ph.D.
Holobeam Corp.

ENGINEERING SCIENCES
Ray E. Bolz, D. Eng.
Case Western Reserve University
George L. Tuve, Sc.D.
THE Chemical Rubber CO.

ENVIRONMENTAL SCIENCES
Richard G. Bond, M.S., M.P.H.
University of Minnesota
Conrad P. Straub, Ph.D.
University of Minnesota

FOOD AND NUTRITION
Nicolo Bellanca, Ph.D.
CIBA-GEIGY Corp.
Giovanni Fenaroli, Ph.D.
University of Milano, Italy
Thomas E. Furia
CIBA-GEIGY Corp.

MARINE SCIENCES
F. G. Walton Smith, Ph.D.
University of Miami

MATERIALS SCIENCE
C. T. Lynch, Ph.D.
Wright-Patterson Air Force Base

MATHEMATICS AND STATISTICS
William H. Beyer, Ph.D.
University of Akron
Brian Girling, M.Sc., F.I.M.A.
The City University, London
Samuel M. Selby, Ph.D., Sc.D.
Hiram College

MICROBIOLOGY
Allen I. Laskin, Ph.D.
Esso Research and Engineering Co.
Hubert Lechevalier, Ph.D.
Rutgers University

ORGANIC CHEMISTRY
Saul Patai, Ph.D.
Hebrew University of Jerusalem
Zvi Rappoport, Ph.D.
Hebrew University of Jerusalem

THE AUTHOR

Lubomir S. Hnilica, Ph.D., is a biochemist and professor of biochemistry at the University of Texas at Houston, M. D. Anderson Hospital and Tumor Institute and at the Graduate School of Biomedical Sciences.

Dr. Hnilica received his Ph.D. in organic chemistry from Masaryk University, Brno, Czechoslovakia in 1952. He subsequently received postdoctoral training at the Czechoslovakian Academy of Science and the Czechoslovakian National Cancer Research Institute in Bratislava. Dr. Hnilica continued his research as a WHO research fellow at the Chester Beatty Research Institute in England. His present research concerns proteins of the cell nucleus; the biochemistry, structure, and metabolism of nucleohistones, the interaction of nuclear proteins with nucleic acids; and the biochemical mechanisms of gene regulations in eukaryotes.

PREFACE

Although the discovery of histones is approaching its first centennial, their exact biological role remains unknown. The explosion of interest in histones witnessed during the last decade has resulted in numerous publications in this area. Since the growing number of scientific papers concerning these proteins now amounts to several thousand, only the more recent references are discussed in detail, their selection reflecting the topics treated in the individual chapters. Because another group of chromatin proteins, the nonhistone or acidic nuclear proteins, is rapidly gaining interest among scientists, a brief discussion of them has been included in the last chapter.

In general, this book endeavors to present an analysis of the current knowledge of the structure, properties, and possible functions of histones. The broad scope of this topic prevents the discussion of certain individual aspects in any great detail. For this purpose, some of the more specialized review articles should be consulted. Hopefully, the present volume will contribute to a more general understanding of histone biochemistry and will provide stimulation, as well as a source of references, to the student of the cell nucleus and its functions.

L. S. Hnilica
Houston, Texas

TABLE OF CONTENTS

Chapter 1

HISTORIC BACKGROUND

Eighteenth century Germany was unquestionably the Mecca of natural sciences. Stimulated by the discoveries of Justus Liebig and his contemporaries, many places of higher learning, especially the medical schools, turned to investigations of the chemical principles of life. The century-old issues of the Biochemisches Zeitschrift and The Hoppe-Seyler's Zeitschrift für Physiologische Chemie are filled with discoveries of new natural substances, all found in living cells. A reader of these journals feels the excited pulse of science and scientists alike. The field of biochemistry was born with many prominent men of science standing at her cradle.

It was in the laboratory established and directed by Hoppe-Seyler at the University of Strasbourg that the essential components of the eukaryote cell nucleus, the DNA, protamines, and histones, were discovered. A gifted scientist and teacher, Hoppe-Seyler attracted a great many students. Among them were Miescher and Kossel, who are credited with the discoveries of the mentioned macromolecules.

Born of a professional family (his father was a professor of pathology and his uncle, Wilhelm His, was a professor of anatomy and physiology) Johan Friedrich Miescher[1] studied medicine in Basel and Göttingen. After completing his medical studies, Miescher became interested in the chemistry of natural products, especially in the cell nucleus. In Hoppe-Seyler's laboratory, Miescher succeeded in isolating pus cell nuclei from which he obtained, by alkaline extraction and acid precipitation, phosphorus-containing substances which he named soluble and insoluble nuclein.[2] The soluble nuclein was a crude preparation of DNA. After some interruptions, Miescher continued his studies on nuclear material in Basel where in 1872 he assumed the position of Professor of Physiology vacated by his uncle Wilhelm His, who interested Miescher in studies on spawning habits of Rhine salmon *Salmo salar*. Here Miescher isolated nuclein from salmon sperm and also a nitrogen-rich base, which he named protamine.[2] Although he was not aware that protamine was a primitive protein, as was later demonstrated by Kossel,[3] he probably realized before his death in 1895 that protamine was an arginine-rich polypeptide. From unripe testes of salmon, Miescher was unable to isolate protamine. He found a protein instead (histone) which was later referred to as Kernalbuminose. Miescher's scientific work was unfortunately cut short. After contracting tuberculosis, he died in the summer of 1895 shortly after his 51st birthday.

Although Miescher discovered DNA and protamines, premature death prevented him from further scientific exploits in this field. It was another student of Hoppe-Seyler, Albrecht Kossel, who laid a solid scientific base for the biochemical research on the cell nucleus, especially its proteins. Born in 1853, the oldest of seven children, Kossel studied natural science and medicine, first in Strassburg and later in Rostock.[4] After receiving his degree in medicine he spent several years with Hoppe-Seyler at his Physiologisch Chemisches Institut. It was here that young Kossel became involved in the studies on nuclear proteins, an interest which he retained for the rest of his life. After leaving Hoppe-Seyler in 1883, Kossel moved first to Berlin and then to Marburg. In 1886 he married Luise Holtzmann, daughter of a well-known Sanskrit researcher; they had three children. After the turn of the century, Kossel accepted a chair at the University in Heidelberg where he remained until his death. In 1908 to 1909 he was Prorector of the University and in 1910 his research was rewarded by the highest scientific award, the Nobel prize.

During his life, among other honors Kossel received honorary degrees from the Universities in Cambridge, Dublin, Genf, Greifswald, St. Andrews, and Edinburgh, and was a member of numerous Academies of Science in several countries. He visited the U.S. in 1911, spending most of his time at the Johns Hopkins University. In 1924 Kossel retired from his chair at the University of Heidelberg, remaining Professor Emeritus and director of the Institute for Protein Research, which was partially funded by the Behringer Co. He died on July 5, 1927, after several heart attacks.

During his very productive scientific life, Kossel published numerous papers describing his discoveries of histones, adenine, theophylline, histidine, etc. His main interest was with the nuclear proteins, their structure, composition, and relationship to Miescher's nuclein (DNA). Although some of his ideas on histone and protamine structure and interconversion are hardly acceptable at the present time, the thorough and careful work of Kossel and his students such as Huiskamp,[5-7] Lilienfeld,[8,9] Dakin,[10,11] Kutscher,[12] Edlbacher,[13] Staudt[14] and others collected a great wealth of information on histones and protamines which Kossel summarized in a monograph[3] finished shortly before his death. Several methods for the isolation of nucleohistone, histones, and protamines which were developed by Kossel, Lilienfeld, and others are still used with only small modifications in modern laboratories studying biochemistry of nuclear proteins.

After Kossel's death, research on nuclear proteins came virtually to a halt (probably due to the poor economic situation in postwar Germany) except for the laboratory of Kurt Felix, who devoted his life to studies on protamines.[15,16] During the late 1940's the interest in histones was revived, mostly due to the work of Edgar and Ellen Stedman in Edinburgh. It was here that the idea of histones functioning as gene inhibitors was first conceived. The Stedmans and Butler in Europe, Mirsky and Luck in the U.S., their associates, and a few others picked up the challenge of this difficult field and developed it into a scientific discipline with many followers. Presently, it is difficult to find a major university or research institute without one or more investigators studying the chemistry or biology of nuclear proteins. Hundreds of papers are published annually in this field which less than a half century ago was almost entirely the domain of one man, Albrecht Ludwig Karl Martin Leonhard Kossel.

Chapter 2

CHEMICAL PROPERTIES OF HISTONES

Isolation Procedures

The very first isolations of histones by Kossel and his associates took advantage of the solubility of histones in diluted strong acids such as HCl or H_2SO_4, a method still used for extraction of histones in many laboratories. As a rule, prior to the extraction of histones, it is necessary to remove cellular cytoplasm. It is not possible to obtain good histone preparations by extracting whole cells since many cytoplasmic proteins, mainly of ribosomal origin, are soluble in diluted acids and are frequently present in larger quantities than nuclear histones. Isolated nuclei or crude chromatin preparations are preferred by most investigators to prepare histones and their fractions.

Many animal and plant cells contain rather large quantities of cytoplasm compared to the nuclear mass. Using such cells (liver, kidney, embryonal cells, pea seedlings, etc.), nuclei are isolated from tissue homogenates employing one of several available techniques (centrifugation in isotonic or hypertonic sucrose, removal of the cytoplasm by diluted organic acids, detergents, isolation of lyophylized nuclei by flotation in nonaqueous media, etc.). It should be borne in mind that a technique yielding clean nuclei from a certain tissue may be utterly unsuitable for isolating nuclei from other cellular types. Since the procedures for isolation of nuclei have been discussed in several review articles, the reader is referred to them for detailed information.[17-23]

In brief, methods utilizing hypertonic sucrose are superior to those using isotonic sucrose; both are used frequently in studies on histones. Diluted citric or acetic acid (acetic acid was first used by Miescher for isolation of nuclei) as well as diluted nonionic detergents (Nonidet®, Triton X-100®, etc.) remove cytoplasm quite readily, resulting in clean nuclear preparations. However, since diluted organic acids are known to selectively extract certain histone fractions,[24,25] their use for isolation of nuclei is not recommended. Similarly, exposure of nuclei to detergents has been reported to result in a partial solubilization and loss of nuclear proteins, both basic and acidic.[26,27] Relatively little information is available in the literature on the suitability of nonaqueous nuclei (Behrens' procedure) for the extraction of histones.

Since the isolation of nuclei from tissue cultured cells and many neoplasms, both solid or growing in suspension (ascites) presents considerable difficulties in removing strongly adhering cytoplasmic reticulum, the cells are frequently exposed to a hypotonic shock for various time intervals. Although this treatment facilitates later removal of the cytoplasm, it also increases the chances for rupturing the lysosomes and enhancing proteolytic damage during the subsequent isolation of nuclei and histones.

If nuclei are used to isolate histones, homogenization and extraction of nuclear material first with isotonic (0.14 *M*) saline, usually buffered by sodium citrate (0.01 *M*), and then with Tris-HCl buffer (0.1 *M*, pH 7.4) is recommended.[28-30] The saline-citrate solution was more recently modified by Widholm and Bonner[31] (0.15 *M* NaCl, 0.015 *M* sodium citrate, pH 7.0), and is frequently referred to in the current literature as standard saline citrate (SSC). The homogenization and extraction of nuclei with buffered saline and with Tris-HCl buffer removes considerable quantities of proteins (nuclear sap) part of which are soluble in acid and would contaminate isolated histones. As pointed out by Johns and Forrester,[32] the saline wash removes cytoplasmic proteins which become adsorbed to the nuclei and chromatin during their isolation. A stronger (0.3 to 0.35 *M*) saline solution is recommended for better removal of these cytoplasmic contaminants. However, a further increase of the salt concentration to 0.4 *M* NaCl results in an altered chromatin structure, a rearrangement of chromatin proteins, and in partial extraction of histones.[33] After saline and buffer washes, the nuclear mass can be further extracted with 80% ethanol (aqueous) to decrease the contamination of histones by acid soluble proteins.[34]

Few tissues, such as thymus, nucleated erythrocytes, spleen, etc. can be extracted directly with buffered saline and Tris-HCl buffer solutions without prior isolation of nuclei. Typically, repeated homogenization of calf thymus tissue in SSC and buffer solutions produces crude chromatin preparations which can be used for the extraction of histones. It was perhaps the simplicity of this procedure together with the good purity of isolated histones that made calf thymus the starting material for most of the past studies on histone heterogeneity and physicochemical properties.

Using saline-washed nuclei or isolated chromatin, histones can be either dissociated from DNA by concentrated salt solutions or extracted by dilute acids. The acid extraction is preferred for it can handle large amounts of starting material and virtually eliminates possible proteolytic degradation of histones. Strong acids such as HCl or H_2SO_4 are necessary for the best extraction of histones since the use of weaker acids (H_3PO_4, HCOOH, citric, acetic, perchloric, etc.) or very dilute strong acids may result in a partial fractionation of these proteins.[24,25,35-38]

Due to their relatively simple conformation, histones fall an easy prey to various proteolytic enzymes. It was reported by several investigators[39-46] that histones can be considerably degraded during their isolation. Therefore, caution should be exercised when selecting a proper method for histone extraction. Although partially protected from proteolysis by their association with DNA, histones can be rapidly damaged during dissociation of the chromatin complex at neutral or slightly alkaline pH.[47] The presence of urea or guanidine-hydrochloride in the buffered NaCl solutions frequently used for dissociation of chromatin does not completely inhibit the proteolytic digestion of histones. A change to slightly acidic pH values (pH 5 or 6) decreases the proteolytic damage to histones considerably. Nevertheless, if the isolation of undegraded histone fractions from various tissues is desired (without preserving the nonhistone protein and DNA residue), extraction with diluted mineral acids (HCl or H_2SO_4) is recommended.

Another modification of histone isolation described in the literature is the extraction of nuclear preparations with distilled water.[48,49] This treatment solubilizes nucleohistone which can be isolated by centrifugation. Acidification of the aqueous chromatin extract precipitates the DNA and nonhistone proteins with histones remaining in solution. This method, however, results in considerable proteolytic damage to the histones and should be avoided.

To prevent proteolytic degradation of histones during their isolation, especially when dissociating nucleohistone with salt solutions of high ionic strength, several investigators included protease inhibitors in the solutions used for the isolation or dissociation of crude chromatin.[40,50,51] Unfortunately, these inhibitors are highly reactive compounds, frequently hydrolyzing rapidly in aqueous solutions. Panyim et al.[44] studied the presence of proteases attacking histones in various tissues. Low concentrations (5 mM) of sodium bisulfite or 1 mM diisopropyl fluorophosphate included in solutions used for the isolation of nucleohistone inhibited the proteolytic degradation of histones, especially the very lysine-rich fraction F1 (I). Although sodium bisulfite has been reported to interact with the pyrimidines in nucleic acids, forming sulfonates and deaminating cytosine,[52,53] its addition to the solutions for isolation of nuclei was found extremely useful in our laboratory and appears to be gaining general acceptance.[27]

After their solubilization, histones can be recovered by dialysis and lyophylization or by precipitation with an excess of ethanol or acetone. Since the molecular weights of most histone fractions are less than 20,000 daltons, there is a possibility of histone leakage across the cellophane membrane during dialysis. Although histone losses through the cellulose tubing were occasionally reported during dialysis,[54-58] such losses occurred only either at low pH values or during very prolonged dialysis intervals. It should be realized that during the final stages of dialysis, the pH of the dialysate becomes nearly neutral, thereby increasing the possibility of proteolytic degradation of histones by traces of acid soluble proteases. Such proteolytic damage may be responsible for the losses attributed to the passage of histone through cellulose membrane during prolonged dialysis. Nevertheless, the time of histone dialysis should be kept to the necessary minimum. If no excessive amounts of salt are present in the extracts, several changes of 25 to 50 volumes each of deionized water over a 24-hr period are sufficient. Numerous protein precipitants (phosphotungstic, picric, flavianic, trichloroacetic, or perchloric acid) will remove histones from aqueous solutions nearly quantitatively. In most instances, the collected precipitate can be converted to the hydrochloride or sulfate by suspending it in acetone containing 0.1% or less of hydrochloric or sulfuric acid.[36,59] Although complexes of histones with organic acids such as trichloroacetic or picric acid are partially soluble in acetone, the resulting histone hydrochlorides or sulfates are quite insoluble and precipitate readily. The precipitates can be washed several times with acetone and dried in vacuo or dissolved in distilled water and lyophilized after a short dialysis.

An extremely useful and quantitative technique for recovering histones even in small quantities was developed by Lindh and Brantmark.[60,61] In the presence of an excess of Reinecke salt, $NH_4[Cr(NH_3)_2(SCN)_4] \cdot H_2O$ the reineckate ions bind quantitatively and stoichiometrically to positively charged residues, resulting in the precipitation of protein from solution. For histones the precipitation is carried out at pH 1 to 2 with a cold saturated solution of Reinecke salt. As little as 10 μg of histone can be precipitated quantitatively. The histones are recovered from the complex with reineckate by suspending the precipitate in a mixture of acetone and 1 *N* HCl (98:2 v/v). The Reinecke acid is soluble in acetone and thereby is removed from the insoluble histone hydrochloride, which, after washing with acetone, can be dissolved in water, dialyzed, and lyophilized. The presence of acid in acetone is essential since practically all histone complexes with organic acids (picric, trichloroacetic, etc.) are partially or completely soluble in ethanol or acetone, and washing such precipitates with these solvents can result in a partial fractionation of the isolated histone samples (e.g., arginine-rich histone complexes with trichloroacetic acid are more soluble than similar complexes of lysine-rich histones). Similarly, histone hydrochlorides are quite soluble in ethanol and cannot be precipitated quantitatively with this solvent. Even the less soluble histone sulfates are differentially soluble in ethanol, which can be used for their fractionation.[62-64] On the other hand, in the presence of inorganic acid, acetone, in at least tenfold excess, precipitates histones almost quantitatively.

Histone Heterogeneity

General Occurrence and Characterization of Histones

Extensive studies on histones during the last two decades suggest a general presence of these proteins in the nuclei and chromatin of most eukaryotes. As supported by several reports,[65-71] histones have not been found in bacteria and appear also absent from primitive eukaryotic organisms, such as blue-green algae,[72] mold *Neurospora crassa,*[73] and other fungi.[74] Interestingly, the slime mold, *Physarum polycephalum*, was reported to contain several histone fractions quite similar to those common in higher organisms.[75]

Owing their solubility in dilute acids to their unique amino acid composition (approximately 25% of their amino acids are basic), protamines and histones were identified in cell nuclei of various organisms almost a century ago. Because of the primitive composition of protamines which probably resulted from a sequence of duplications of a simple gene,[76,77] there are no controversies in the characterization of these proteins. Unfortunately, this does not hold for histones which are larger and more complicated than protamines. Proteins similar in amino acid composition to the histones were isolated from ribosomes,[78-84] mitochondria,[85] nervous tissue,[86-89] and other sources. The amino acid composition of several such proteins shown in Table 1 can be compared with the composition of five main histone fractions from calf thymus in Table 2. In general, histones are relatively small proteins with molecular weights near 10,000 daltons [with the exception of the very lysine-rich histones F1 (I) and chicken erythrocyte specific histone F2c (V) which are about twice as large] with an excess of basic over dicarboxylic amino acids. As a result of this excess, histone molecules are positively charged and have their isoelectric points in the basic pH region. There are several histone species in higher organisms differing from each other in amino acid composition, electrophoretic mobility, etc. All typical histones lack tryptophan and, with the exception of the arginine-rich histone F3 (III), cystine or cysteine. The main cellular localization of histones, certain exceptions aside, is in the cell nucleus where they are associated with the DNA in the form of chromatin. Proteins more basic than histones are present in spermatozoa of many fish species and of some lower animals. These proteins, either protamines or special arginine-rich histones, will be discussed later.

It was recognized shortly after their discovery that histones are heterogeneous proteins. Although the chemical similarity of histone fractions frustrated many early attempts to fractionate these proteins, it was at least possible to distinguish two histone fractions, later designated as lysine-rich and arginine-rich. Due to its peculiar

TABLE 1

Some Basic Proteins Similar to the Histones

Origin of the fractions

	Ribosomes							Nervous tissue	
Amino acid	11(a)	12(a)	7(b)	11(b)	L20(c)	S20(c)	S21(c)	FB(d)	Brain(e)
Lysine	**11.5**	**18.3**	**7.7**	**14.0**	**11.0**	**14.3**	**12.9**	**6.5**	**8.1**
Histidine	1.0	2.5	1.2	3.0	1.1	3.2	1.5	8.1	5.9
Arginine	**15.8**	**7.4**	**13.1**	**8.5**	**13.1**	**10.4**	**20.6**	**13.8**	**9.9**
Aspartic acid	7.0	7.6	6.4	8.8	6.9	8.7	4.8	6.2	6.2
Threonine	5.5	3.2	4.3	3.5	1.7	3.3	4.4	5.4	4.3
Serine	5.8	**11.6**	5.4	6.3	4.3	4.9	3.4	**11.2**	8.3
Glutamic acid	6.1	8.7	11.9	9.5	8.5	9.4	11.3	5.9	6.6
Proline	3.9	3.9	3.1	2.3	0.5	1.9	4.8	6.0	7.8
Glycine	6.8	3.1	9.6	5.7	6.3	3.6	2.2	7.6	**15.1**
Alanine	**10.6**	**18.7**	8.6	**17.4**	**18.5**	**19.2**	**13.0**	**14.0**	9.2
Valine	4.8	2.3	8.1	4.6	7.7	2.8	7.8	0.1	1.7
Methionine	1.4	1.8	2.2	2.9	0.5	2.9	0.0	2.7	1.3
Isoleucine	7.3	**5.9**	5.5	6.8	7.5	7.0	1.0	0.1	1.9
Leucine	7.0	**6.8**	6.7	4.1	4.9	4.2	5.5	6.2	5.8
Tyrosine	2.9	2.6	3.0	0.2	3.9	1.9	2.7	3.0	2.9
Phenylalanine	2.3	2.1	3.0	2.3	3.1	2.4	4.0	3.0	4.9
Basic/acidic	2.2	1.7	1.2	1.4	1.6	1.5	2.2	2.3	1.9

All values represent percent of total moles of amino acids recovered. Features similar to histone fractions are printed in boldface type. Note the similarity of fraction 12(a) with the histone F2b (IIb2) and of fraction S21(c) with arginine-rich histones in sea urchin sperm.

(a) Proteins from *B. stearothermophylus*,[83] (b) proteins from *E. coli*,[82] (c) proteins from *E. coli*,[84] (d) protein from bovine spinal cord, Fraction B,[86] (e) protein from pig brain.[87] Although basic, the proteins from nervous tissues are quite different from the composition of histones.

amino acid composition (over 50% of all amino acid residues are lysine and alanine) the very lysine-rich histone F1 (I) was the first histone fraction isolated in good purity and described in detail.[24,25,62-64,90-93] Identification of other histones had to await the development of techniques permitting separation and analysis of closely related protein species.

Heterogeneity and Nomenclature

Although it had been suggested during Kossel's lifetime that histones are heterogeneous proteins containing at least two fractions, these suggestions were first supported experimentally almost half a century later by Edgar and Ellen Stedman[62,94] who separated histones from several vertebrate species into the main (arginine-rich) and subsidiary (lysine-rich) fractions. Histone heterogeneity was confirmed by other investigators in the late 1950's while several groups were busy fractionating histones and characterizing the fractions. To support their theory that histones act as gene regulators, Edgar and Ellen Stedman proposed that there must be several and possibly many kinds of histones.[62] After intensive research, six histone fractions were characterized in their laboratory according to their electrophoretic mobility and sedimentation properties ($\alpha 1$, $\alpha 2$, $\alpha 3$, β, $0.8s\gamma$ and $1.6s\gamma$ histones).[90] Experiments utilizing ion-exchange chromatography and selective extraction of nuclear material confirmed the existence of two main groups of histones designated as lysine-rich and arginine-rich because of the preponderance of one or the other amino acid. Each of these groups was found to consist of several subfractions.

The true heterogeneity of histones was realized,

TABLE 2

Amino Acid Composition of the Main Histone Fractions

Amino acid	Calf* thymus F1(1)	Calf thymus F2b(IIb2)	Calf thymus F2a2(IIb1)	Calf thymus F2a1(IV)	Calf thymus F3(III)
Lysine	**28.7**	**16.7**	**12.5**	9.8	10.1
Histidine	0.0	2.3	2.8	1.9	2.4
Arginine	1.7	6.4	9.3	**13.9**	**13.6**
ε-*N*-methyllysine	0.0	0.0	0.7	1.1	0.6
Aspartic acid	2.0	4.9	5.5	5.0	4.4
Threonine	5.4	6.2	4.9	6.6	6.5
Serine	6.7	**10.9**	5.0	2.5	3.8
Glutamic acid	3.4	7.6	8.7	6.2	10.2
Proline	10.1	4.8	4.1	1.3	4.4
Glycine	6.9	5.6	9.0	**15.9**	5.8
Alanine	**25.1**	10.2	**13.2**	7.5	**13.5**
Half cystine	0.0	0.0	0.0	0.0	0.3
Valine	4.1	7.1	6.0	7.8	4.6
Methionine	0.0	1.6	0.3	1.0	1.5
Isoleucine	0.8	**4.9**	4.2	5.6	4.8
Leucine	4.1	**4.9**	10.1	8.0	9.0
Tyrosine	0.5	3.8	2.3	3.5	2.1
Phenylalanine	0.5	1.6	1.0	2.2	2.0
Basic/acidic	5.6	2.0	1.8	2.4	1.8

*Unfractionated.
All values are expressed as percent of total moles of amino acids recovered. The serine values were corrected (10%) for hydrolytic losses. The main features distinguishing each histone fraction are printed in bold figures.

however, only after the starch gel electrophoresis technique was applied to histone preparations from various tissues. Neelin and Neelin[95] observed 16 to 22 protein bands in calf thymus histone preparations subjected to starch gel electrophoresis. Variation of the number of protein components in calf thymus histone samples depending on the conditions of electrophoretic analysis led to the conclusion that at least some of these protein bands were aggregates of relatively few basic histone fractions. Using a simplified version of horizontal starch gel electrophoresis and low pH to prevent aggregation, Johns et al.[96] separated calf thymus histone into 5 to 6 bands corresponding to the major histone fractions isolated by other fractionation procedures. Essentially, the amino acid composition of the three major bands matched that of the chromatographic fractions F1, F2, and F3. Similar studies carried out by MacPherson and Murray[97] where chromatographic fractions of calf thymus histone were further subfractionated by preparative electrophoresis again demonstrated the chemical heterogeneity of histones and the similarity in amino acid composition of chromatographically and electrophoretically fractionated histone components. Starch gel electrophoresis then became a valuable tool for studies on heterogeneity and tissue specificity of histones. In an attempt to adapt starch gel electrophoresis for quantitative studies, Hnilica et al.[29] found that including small amounts of $AlCl_3$ into the electrophoretic media greatly reduced the trailing of histones and improved their separation. Although starch gel electrophoresis of histones was almost entirely replaced by the more elegant and sensitive separation on acrylamide polymers, it still is the method of choice for metabolic studies on histones in vivo where low specific activities of histone fractions prohibit the evaluation of radioactivity of polyacrylamide gels (average sample size for an individual starch gel slot is 1 to 2 mg of protein as compared with 25 to 50 μg used in polyacrylamide electrophoresis). A typical histone electrophoretogram is shown in Figure 1.

The introduction of polyacrylamide gel electrophoresis to cationic protein analysis by Reisfeld et al.[98] and its application to histones by Cruft[99] further refined the studies on histone heterogeneity. This method, in various modifications, was employed by a great number of investigators to study the heterogeneity, tissue specificity, metabolism, and in vivo modifications of histones. The greatest advantages of this technique were the speed, sensitivity, and excellent resolution of individual proteins. The tendency of histones to aggregate at nearly neutral pH values first interfered with the identification of true histone fractions. Owing to the studies of Fambrough et al.,[100] this disadvantage was eliminated by the addition of urea to the system and the relatively low pH of the analytical conditions. As can be seen in Figure 1, a typical histone sample (rat liver) can be resolved by this technique into four major components. Further modification of this method by Panyim and Chalkley[101,102] consisting mainly of lengthening the poly-

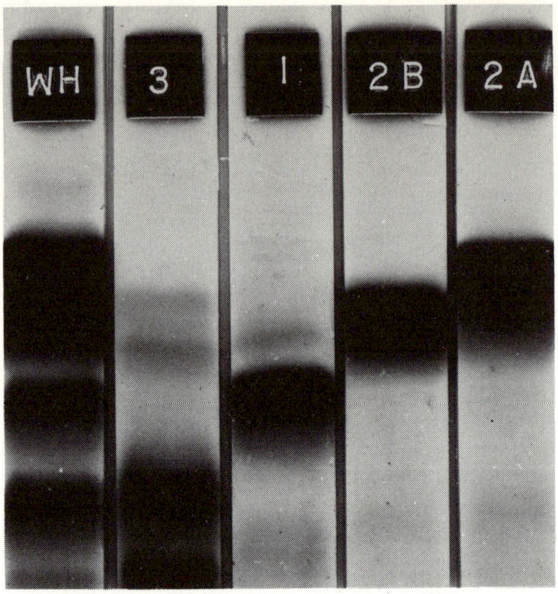

FIGURE 1A. Starch gel electrophoresis of calf thymus histones. WH : whole, unfractionated histone, 3 : fraction F3 (III), I : fraction F1 (I), 2B : fraction F2b (IIb2), and 2A : fractions F2a1 (IV) and F2a2 (IIb1). The electrophoresis was performed in 12% starch made in 0.01 *N* HCl containing 0.2 mM $AlCl_3$. The origin or migration is at the bottom of the figure.

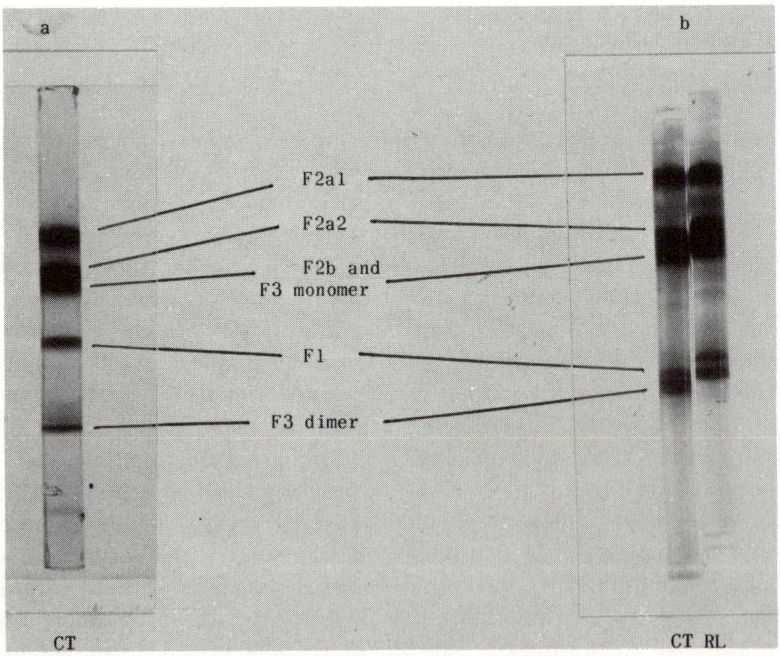

FIGURE 1B. Electrophoretic separation of calf thymus and rat liver histones in polyacrylamide gels containing a) 6 *M* urea and b) 0.1% sodiumdodecyl sulfate. CT : calf thymus, RL : rat liver. The fractions are identified by their symbols. The origin of migration is at the bottom of the gels.

acrylamide gel columns permits the separation of whole histone into the five individual bands corresponding to the fractions isolated by chemical procedures. It can be concluded both from electrophoretic and fractionation studies on histones that the chromatin of most eukaryotes contains only five main histones, each characteristic in amino acid composition and electrophoretic mobility.

Another, more recent, technique for polyacrylamide gel electrophoresis of histones consists of the separation of their complexes with sodium dodecyl sulfate (SDS).[103,104] Such complexes have equal charge densities imposed by this detergent and their electrophoretic mobility decreases proportionally with increasing molecular weights. Calf thymus and other histones can be separated into five bands. The main advantage of this method depends on the elimination of differences in charges of the various protein molecules thereby permitting simultaneous separation of positively and negatively charged proteins. Also the protein-protein interactions are greatly reduced or eliminated in the presence of SDS. The separation of a histone extract from rat liver chromatin in SDS containing gel is shown in Figure 1.

The contribution of methods for electrophoretic separation of histones to the studies on the properties and distribution of these proteins was enormous. Polyacrylamide gel and, to a lesser extent, starch gel electrophoresis permitted comparative investigations (both quantitative and qualitative) on histones from various sources. It can be said that the introduction of electrophoretic analysis made possible the considerable advances of our knowledge of histones witnessed in recent years.

As both the electrophoretic and chemical studies on histones demonstrate, there are five main histone fractions in most eukaryotic tissues. In addition to these main histones, several other fractions were detected either in minor quantities (probably due to chemical modification of the main histones) or in highly specialized tissues (chicken erythrocytes, spermatozoa, etc.). Amazingly, despite extensive studies on histones leading to the determination of total or partial amino acid sequences of the major fractions, no unified nomenclature exists for these proteins. Histone nomenclature remains confusing with variations depending on the imagination of individual investigators. The original terminology describing histones as arginine- or lysine-rich was complemented by introducing additional amino acids abundant in the individual fraction (e.g., VLR, SLR, AR, and AL histones).[105] Other authors refer to their fractions by letters or numbers indicating the sequence in which the fractions emerged from chromatographic columns (e.g., A, B, C;[57,106] Ia, Ib, IIa, IIb, III, IV, V;[23,56,107] F1, F2, F3;[34] etc.). Others characterize their histones according to their electrophoretic mobility or sedimentation properties (e.g., $\alpha 1$, $\alpha 2$, $\alpha 3$, β, 0.8sγ, 1.6sγ;[90] E1, E2, E3;[96] 1 to 5 and 1°;[102] etc.). Terminology based on turnover rates[108] or on the NH_2-terminal amino acid composition[109] is less usual. Obviously, there is a need for a unified histone nomenclature since the present situation can be a nightmare to an investigator unfamiliar with this field. Murray[110] attempted to introduce a system based on the molar ratios of arginine to lysine present in each histone fraction. Unfortunately, this ratio does not characterize each fraction sufficiently and additional parameters are necessary to avoid duplications. Hopefully the increasing attention paid by scientists to the histones and other nuclear proteins will lead to a generally accepted nomenclature for all the histones already known or yet to be discovered.

As was already mentioned, most tissues of higher animals contain five major histone fractions which can be separated chemically or by gel electrophoresis. There are three major histones in the lysine-rich group and two in the arginine-rich group. The heterogeneous very lysine-rich fraction F1 (I) contains several subfractions. Other histones with characteristic amino acid composition were isolated from the nuclei of specialized tissues such as spermatozoa and nucleated erythrocytes. The very lysine-rich histones, F1 or I, the moderately lysine-rich (lysine-serine-rich) histone, F2b or IIb2, and the slightly lysine-rich (lysine-arginine-rich) histone, F2a2 or IIb1, represent the lysine-rich group of histones in higher animals. The arginine-rich (glycine-arginine-rich) histone, F2a1 or IV, and the arginine-rich (alanine-arginine-rich) histone, F3 or III, are in the group of arginine-rich histones. Since histone nomenclature yet awaits standardization, various notations can be found in the literature. The more common ones are compared in Table 3. The two most widely accepted systems are the Roman numeral nomen-

TABLE 3

More Frequently Used Nomenclatures of the Five Common Histone Fractions

Fraction					Reference
Lysine-rich			Arginine-rich		62,94
F1	F2b	F2a2	F2a1	F3	36,59,111,164
Ia,Ib	IIa,IIb	IIb	IV	III	23,56,100,110
I	IIb2	IIb1	IV	III	97,100,110
a1,a2,a3	0.8sγ	1.6sγ	β	β	90,422
A	B	B	B	C	59
		6,7,8	9	5	166
		β6	β7	β5	264
E2	E1	E1	E1	E3	96
1,1′,1″	3	4	5,5′	2,2′,2″	101
VLR	SLR	AL	GAR	AR	105,166

clature of Luck et al.,[56] reflecting the elution sequence of the individual fractions from an Amberlite IRC-50® (GC-50) column and the F nomenclature of Butler et al.[111] which is based on the elution sequence of histones from carboxymethyl cellulose columns. This latter nomenclature with corresponding Roman numerals given in parentheses is used consistently in this book. The amino acid composition of the five major histone fractions from calf thymus is shown in Table 2.

Quantitative Assays

Lacking enzymatic activity or other characteristic features which would permit their selective identification in a mixture of proteins, histones cannot be quantitatively determined without their prior dissociation from the DNA or direct extraction from isolated chromatin or nuclei. Once extracted, histones can be quantitatively determined either by most of the usual protein assay techniques or some group chemical reactions. The in situ cytochemical staining procedures for histones fall largely in the latter category.

Although there are acid soluble proteins other than histones in both the cytoplasm and nuclei of most cells, their solubility in acid offers a definite advantage for quantitative histone assays since most acid soluble proteins other than histones can be removed by repeated washing of the nuclear material with buffered saline (SSC) or other solutions. Once extracted, the concentration of histones can be determined by the biuret assay or by the method described by Lowry et al.[112] If Lowry's procedure is used, it is advisable to use isolated histone standards since the extinction coefficients of histones are slightly lower than those of other proteins. Another, quite sensitive method for quantitative determination of histones is the turbidimetric measurement of their complexes with trichloroacetic acid developed by Luck et al.[56]

Among more specific reactions for histone determination is the Millon reaction modified by Mirsky and Pollister,[113] the precipitation with Reinecke salt,[60,61] and the fluorometric procedures employing interaction of 1-fluoro-2,4-dinitrobenzene[114] or 8-anilinonaphtalene-1-sulfonic acid[115,116] with these proteins. The Mirsky and Pollister[113] reaction is based on the solubility of histone complexes with $HgSO_4$ in the presence of H_2SO_4. Most other proteins form insoluble precipitates. In a typical assay procedure, suspensions of nuclei or chromatin are mixed with an equal volume of a mixture of 0.34 *M* $HgSO_4$ in 1.88 *M* H_2SO_4 (or 10% $HgSO_4$ in 10% H_2SO_4 w/v) and the mixture is heated to 60°C for 15 min. After this time, 0.2 ml of 1% $NaNO_2$ solution are added and the cherry-red color of the diazotized tyrosine complex is developed during 10 min heating at 60°C. The protein concentration is determined spectrophotometrically at 3,540 A.

There is only little interference of other proteins present in the suspension since their complexes with $HgSO_4$ are insoluble. The extinction coefficient of whole calf thymus histone is approximately 0.800 for 1 mg/ml/1cm. Other basic proteins such as calf globin form red precipitates, or at very low concentrations, reddish turbid solutions which can be sedimented by high speed centrifugation. The color contribution of these proteins is less than 5% of the histone absorbancy. Since the color intensity of this reaction depends on the amount of tyrosine present in histone fractions, it is obvious that various histones will differ in their extinction coefficients depending on their individual tyrosine content.

Another technique much less affected by the differences in amino acid composition of histone fractions is the procedure based on precipitation with Reinecke salt. This method, developed by Lindh and Brantmark,[60,61] utilizes the complexing of basic protein groups with Reinecke salt. The complexes are insoluble at acidic or neutral pH values and by manipulating the pH of precipitation, histones can be separated from other proteins. The actual protein content is determined by reading the absorbancy of Reinecke acid released with acidified acetone from the precipitated histone complexes. The Reinecke acid is recovered from acetone by evaporation, dissolved in water, and measured in a photometer at 3,020 or 5,200 A. Since the basic amino acid content of most histone fractions is similar, minor stoichiometric differences in histone complexes with Reinecke salt can be disregarded.

Less specific methods for measuring histone concentration are based on the stainability of these proteins, especially in the cell nucleus. Quantitative cytochemical measurements using the alkaline fast green stain were introduced by Alfert and Geschwindt[117] and discussed in detail by Alfert.[118,119] This method offers the advantage of studies on quantitative distribution of basic proteins (not necessarily only histones) in the cell nucleus and has been frequently employed with various modifications.[120-127] Other techniques employ the Sakaguchi reaction or fluorescent staining with phenanthrenequinone for the detection of arginine,[128] staining of histones with ammoniacal silver[129] or acridine orange,[130] and staining of the nuclear nonhistone proteins with toluidine blue,[131-133] etc. A comprehensive review of cytochemical staining procedures was presented by Smetana.[134] It should be realized, however, that the majority of staining procedures for histones in situ can only detect positively charged proteins and are not specific for histones.

Histone Fractionation

The differences in the amino acid composition between the very lysine-rich histone F1 (I) and the rest of the fractions proved sufficient to afford their separation. Stedman and Stedman,[62,94] using differential solubility of histone sulfates in ethanol, were able to fractionate histones from various tissues into the main (arginine-rich) and subsidiary (lysine-rich) fractions, both of which were still heterogeneous. Using free zone electrophoresis (Tisselius) or ultracentrifugation, it was possible to separate histones into two bands, the first one corresponding to the main (arginine-rich) and the other to the subsidiary (lysine-rich) histones.[28,54,63,64,135-140] Although the discovery of histone heterogeneity[62,94] stimulated considerable work on histone fractionation and characterization of the two fractions,[4,25,28,35,54,91,135-139] it was soon realized that the lysine-rich and arginine-rich histones were heterogeneous. The unfortunate property of histones to form aggregates at neutral pH values interfered with attempts to further fractionate these proteins for a long time and led to erroneous conclusions that there may perhaps be several hundred histone fractions in a single cellular species. Using preparative zone electrophoresis with starch powder as the supporting medium, Cruft et al.[90] obtained six fractions from calf thymus histone extract. Three of the fractions ($\alpha 1$, $\alpha 2$, and $\alpha 3$) were either subfractions of degradation products[141] of the F1 (I) histones and the remaining three histones were arginine-rich (β) or moderately lysine-rich (0.8sγ and 1.6sγ) histone fractions.

Soon after their discovery, ion-exchange resins were employed for fractionation of histones adding more sophistication to the attempts of various investigators to isolate homogenous histone fractions. The sulfonic acid based cation-exchange resins proved unsatisfactory in retaining substantial portions of the chromatographed histone samples. Initial trials with a less acidic (carboxylic) resin Amberlite IRC-50 (or CG 50) were also disappointing since histones were bound quite strongly to the resin matrix.[106] Using an increasing ionic strength of buffers containing calcium or barium acetate, Crampton et al.[106]

separated calf thymus histone into three fractions (A, B, and C), the first being the very lysine-rich histone F1 (I), the second arginine-rich histone, and the third a mixture of moderately lysine-rich histones. This technique was modified by several investigators[142,143] with the best modification being introduced by Luck et al.,[56] who used an increasing gradient of guanidine hydrochloride to elute histone fractions. This technique, which was adopted in many laboratories,[23] separates histones into the very lysine-rich group I (F1), moderately lysine-rich group II (F2b and F2a2), and arginine-rich groups III (F3) and IV (F2a1).[100] A typical elution pattern of calf thymus histones chromatographed on Amberlite IRC-50 column using this procedure is shown in Figure 2. A slight disadvantage of this method is its poor separation of arginine-rich histones which are frequently contaminated by nonhistone proteins eluted by the abrupt increase in the concentration of the guanidine hydrochloride buffer.[144] The chromatography of histone on Amberlite IRC-50 columns proved extremely useful in subfractionation of the very low lysine-rich histones F1 (I) from various tissues.[144-148] The Roman numeral nomenclature first used by Luck et al.[56] and modified by Fambrough et al.[100] is widely used in the histone literature.

Other types of chromatographic media are cellulose based cation-exchange resins introduced by Peterson and Sober.[149] While phosphocellulose retained histones quite strongly preventing complete elution of these proteins from chromatographic columns, a less acidic carboxymethyl cellulose proved suitable for the fractionation of histones. Using an increasing gradient of NaCl, Davison[150,151] was able to separate histones from several tissues on carboxymethyl cellulose columns into three fractions. Fractions F2 and F3 were the lysine-rich and arginine-rich histone groups, respectively. A modification of this technique by Johns et al.[34] showed that whole histone from calf thymus can be separated into three fractions: F1, F2, and F3. Instead of an increasing NaCl gradient, these authors employed steps consisting of buffered NaCl, 0.01 *N* HCl, and 0.02 *N* HCl. Amino acid analysis of the fractions and other studies demonstrated that the Fl fraction consisted of the very lysine-rich histone F1 (I), the F2 was a mixture of histones F2a1 (IV), F2a2 (IIb1), and F2b (IIb2), and F3 was the arginine-rich histone F3 (III). It was this elution sequence

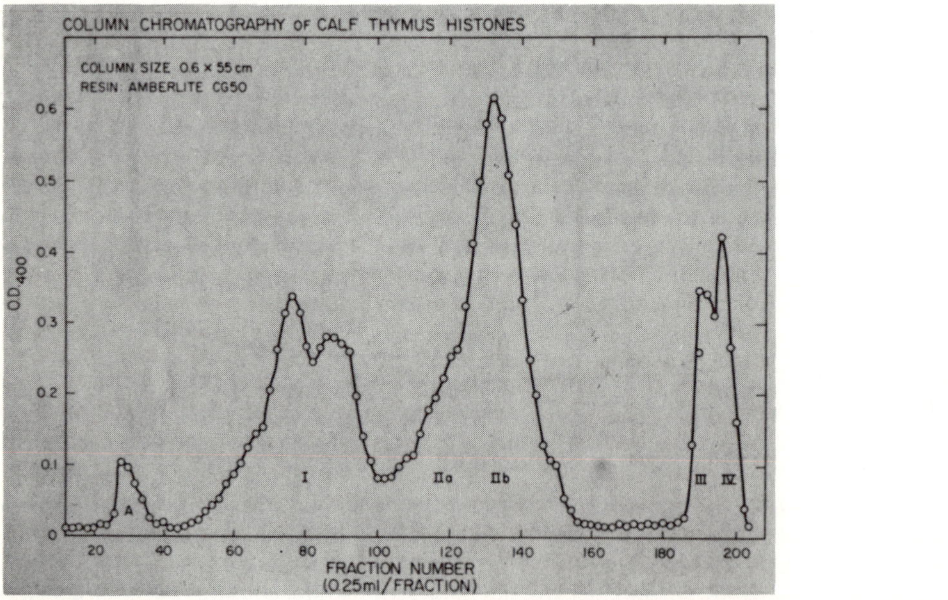

FIGURE 2. Fractionation of calf thymus histones by ion-exchange chromatography on Amberlite CG-50 (IRC-50)® column. A total of 3.0 mg of unfractionated histone were used. Protein concentration in the effluent fractions was determined turbidimetrically at 4,000 A by mixing 0.26 ml of each fraction with 1.1 *M* trichloroacetic acid in a total volume of 1.56 ml. ○ – ○ Turbidity, □ – □ concentration of guanidine HCl. A : nonhistone proteins, I : F1, IIa and IIb : mixture of F2a2 and F2b, III : F3, and IV : F2a1.[23]

which originated the F nomenclature for histones. This nomenclature, together with the Roman numeral system of Luck et al.,[56] as modified by Fambrough et al.,[100] represents the two most widely accepted nomenclatures for enumerating histone fractions (Table 3).

Chromatography of histones on carboxymethyl cellulose columns has been used by many investigators who found that although this technique satisfactorily separated histones from calf thymus, histones from other tissues were not well separated.[59] Various modifications, including the use of sodium formate gradients,[108,152] a stepwise increase in HCl concentration,[153] use of potassium salts,[50] urea,[154-156] formic acid and ethanol,[157,158] and formic acid and urea on sulfoethyl cellulose,[159,160] were introduced to improve the resolution of this method. Similar to the chromatography on Amberlite IRC-50, it was very difficult to resolve all the five main histones present in whole histone extracts by chromatography on carboxymethyl cellulose. Partial fractionations of the histones, e. g., by selective extraction procedures, were beneficial for chromatographic purification of histone fractions.

Since histones [with the exception of F1 (I) and F2c (V) fractions] have similar molecular weights, certain conformational differences probably make possible their fractionation by gel filtration. The feasibility of this approach was first demonstrated by Cruft[161] who separated whole calf thymus histone into several peaks on Sephadex G 75®. To prevent the aggregation of histones and their partial retention by Sephadex, Cruft used 0.02 *N* HCl as eluent. Similar to the separation of whole histones on carboxymethyl cellulose, gel filtration cannot fully resolve a mixture of all the histone fractions.[153] It was found by Hnilica and Bess,[162-164] however, that once prefractionated by extraction with ethanol-HCl mixture into the arginine-rich and lysine-rich groups,[34,59] histones from a variety of tissues can be easily fractionated into fractions F1 (I), F2b (IIb2), F2a1 (IV), F2a2 (IIb1), and F3 (III) by gel filtration on Sephadex G 75, G 200®, and Biogel P 60®. The gel filtration technique was applied to the studies of the heterogeneity of arginine-rich histones,[162-168] lysine-rich histones,[160,168-170] and whole histone extracts[171,172] by various investigators. As a rule, the arginine-rich histones are separated into the F3 (III) and F2a (IIb1 and IV) fractions on Sephadex G 75 or G 100. Further fractionation of the F2a histones into the fractions F2a1 (IV) and F2a2 (IIb1) can be accomplished on Biogel P 60.[164] For the separation of lysine-rich histones F1 (I) and F2b (IIb2) the resolution can be improved by using 2.5 *M* guanidine hydrochloride in 0.01 *N* HCl for elution.[160] Fractionation of arginine-rich and lysine-rich histones of Sephadex G 75 is shown in Figures 3 and 4, respectively.

Since the histone binding to the DNA is realized mostly through salt-like bonds between positively charged amino groups of histones and negatively charged phosphate groups of DNA, exposure of chromatin to diluted acids or to solutions of various ionic strengths will result in either complete or partial dissociation of such bonds depending on the pH and molarity of the solution. Although the dissociation of individual histone fractions is gradual with the lysine-rich histones dissociating at higher pH values and lower ionic strength than those required for the dissociation of arginine-rich histones,[23-25,35,36,38,173-184] crude fractionation of histones can be obtained by this technique. In general, NaCl or

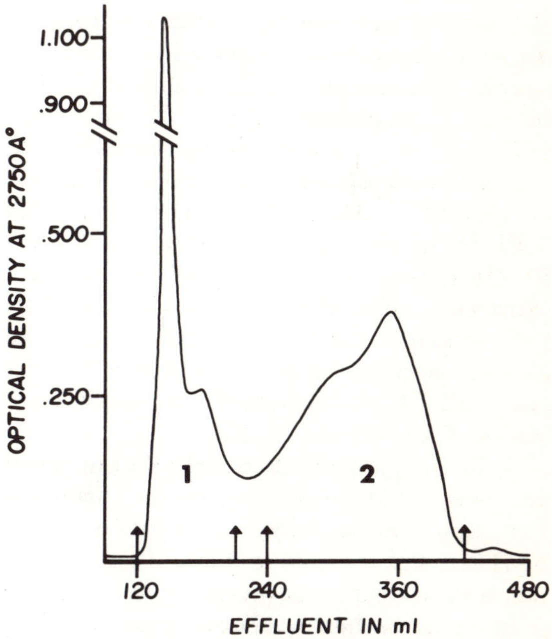

FIGURE 3. Separation on Sephadex G 75® of calf thymus histones soluble in ethanol-1.25 *N* HCl (4:1 v/v) mixture. The solvent is 0.01 *N* HCl saturated with chloroform. Peak 1 : F3 (III) oxidized form, Peak 2 : a mixture of F2a1 (IV) and F2a2 (IIb1) histones. The reduced form (monomer) of the F3 (III) histone will also elute in this peak.

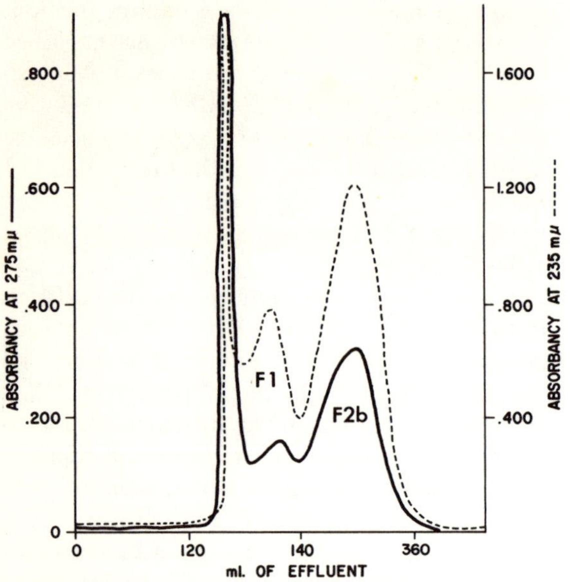

FIGURE 4. Separation on Sephadex G 75® of calf thymus histones obtained from the residue after ethanol-HCl extraction. The solvent is 0.01 *N* HCl saturated with chloroform. First peak contains aggregated proteins; the second peak consists of the F1 (I) fraction and the third peak contains the F2b (IIb2) histone.

KCl are the least selective; better selectivity in removing specific histone fractions was found for sodium perchlorate,[23,176] guanidine hydrochloride,[23] magnesium chloride,[181] ammonium sulfate,[182] and sodium deoxycholate.[183,184] Weak acids such as perchloric,[36] trichloroacetic,[173] or citric,[24,25] alone or in combination with NaCl,[185] remove selectively the very lysine-rich F1 (I) histone, which can also be selectively extracted by dilute (0.01 *N*) HCl. After the F1 (I) histones were extracted, a further lowering of the pH dissociates the lysine-rich fractions F2b (IIb2) and F2a2 (IIb1), followed by the arginine-rich histones F2a1 (IV) and F3 (III).[186]

While incrementally decreased pH or increased salt concentration usually removes the fractions in the approximate sequence F1→F2b→F2a2→F3→F2a1, this sequence is altered if the dissociation is carried out in the presence of urea. Histone fractions can be dissociated under these conditions with much lower salt concentrations (0.2 to 0.3 *M* NaCl). The slightly lysine-rich F2b (IIb2) and F2a2 (IIb1) histones are removed first, with the arginine-rich fractions F2a1 (IV) and F3 (III) solubilized at higher salt concentrations. The very lysine-rich histones F1 (I) and the chicken erythrocyte specific histone F2c (V) are extracted last.[187,188] Essentially the same sequence of histone fractions can be obtained by using increasing (0.01 to 0.1 *M*) concentrations of sodium deoxycholate for the dissociation of histones.[183,184] Since the dissociation of histones from DNA is not proportional to the ionic strength of the solution (only about half the concentration of sodium perchlorate is required as sodium chloride and sodium deoxycholate is about ten times as effective as sodium chloride), Smart and Bonner[184] suggested that the effectiveness of a given salt in dissociating histones depends considerably upon the binding specificity of the anion binding sites and hydrophobic regions on individual histones. Although the fractionation of histones by titration of chromatin with acid or by extraction with various salt solutions is possible, the fractions are usually mixtures of at least two histones (with the exception of F1) since there are no sharp transitions in the extractibility of individual histones from chromatin.

Another technique for histone fractionation, superior to the direct dissociation method, is the extraction of arginine-rich histones F3 (III), F2a1 (IV), and F2a2 (IIb1) with ethanolic HCl (absolute ethanol – 1.25 *N* HCl mixture 4:1 v/v) followed by acid extraction (0.25 *N* HCl) of the lysine-rich F1 (I) and F2b (IIb2) fractions. This method reverses the sequence in which the histone fractions are removed from chromatin as compared with the direct salt or acid dissociation and it was first introduced by Johns et al.[34] After some modifications,[189] this technique was perfected to yield each of the five histone fractions in large quantities and relatively good purity. The outline of this procedure is shown in Figure 5.

In a typical procedure, each histone fraction isolated according to the schedule in Figure 5 should be analyzed by polyacrylamide gel electrophoresis in a system preventing the aggregation of histones.[100-102] If contaminated, the fractions can be further purified by ion-exchange chromatography, gel filtration, and other methods. According to experiences from this laboratory, the selective extraction procedure, if accompanied by final purification, is the best technique available for the preparation of large quantities of pure histone fractions. For smaller amounts, as well as for analytical experiments, either chromatography on Amberlite IRC-50, as described by Luck et al.[56] and modified by Bonner et al.,[23] or selective extraction of arginine-rich histones and lysine-rich

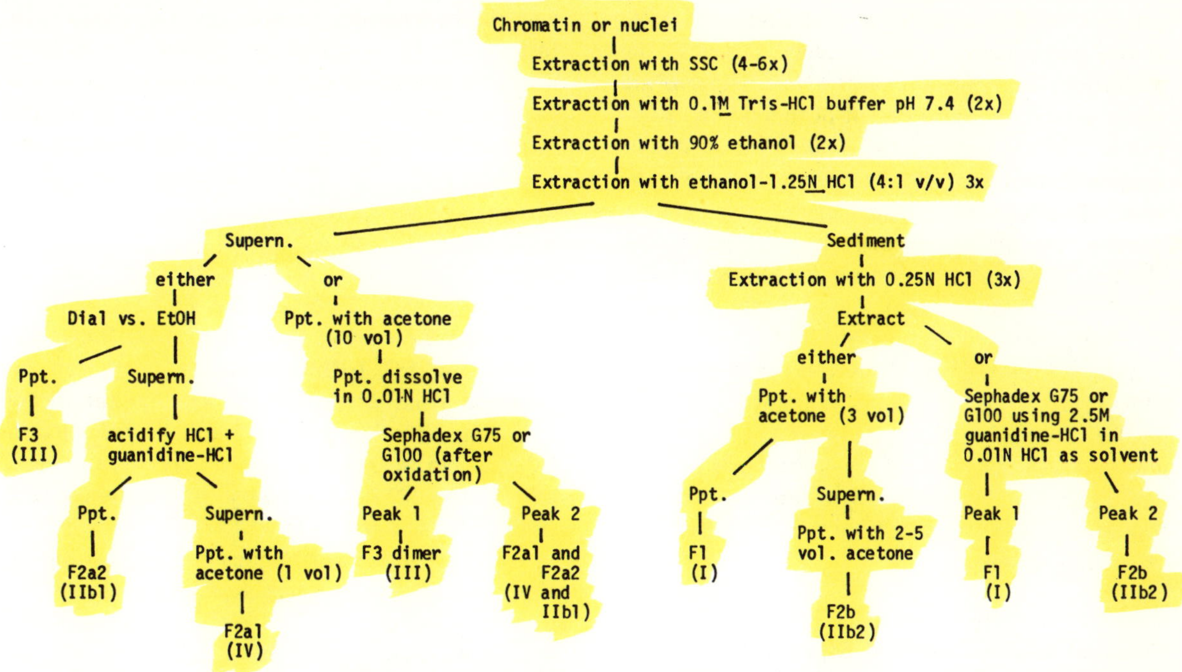

FIGURE 5. Isolation of the common histone fractions.

histones with ethanol-HCl and 0.25 *N* HCl, respectively (i.e., first part of the scheme in Figure 5), followed by gel filtration on Sephadex G 75 or G 100® can be recommended. The extractibility of histone fractions from isolated chromatin according to the scheme in Figure 5 may be influenced by the physiological conditions of the experimental animals or of the excised tissues. It was shown recently by Bartkowiak et al.[190] that relatively mild x-irradiation of excised lymph nodes alters the extraction of histone fractions. The most affected was the lysine-rich histone F2b (IIb2).

Among the less usual methods used for the fractionation of histones is the ethanol precipitation of histone sulfates first employed by Stedman and Stedman.[62,94] This method separates the very lysine-rich histones from the rest of the fractions.[54,64,191] The F1 (I) histones can also be isolated by precipitation of the histone solution with 5% trichloroacetic acid.[36,59] All the histone fractions precipitate except for the F1 (I) histones which remain in solution and can be recovered by increasing the trichloroacetic acid concentration to 20%.

The fractionation of histone sulfates with ethanol was further modified by taking advantage of the aggregation of the arginine-rich histones. Using a combination of both methods Cruft et al.[191-193] and Bellair and Mauritzen[194] separated histones into the fractions α (F1), β (F3), and γ (0.8sγ and 1.6sγ are mixtures of F2 histones). Countercurrent distribution has also been used for histone fractionation, but without much success.[195,196] Recently, a new procedure was reported for separating the F1 (I) and F2a2 (IIb1) histones from rat liver nuclei.[197] By rapidly lowering the pH of isolated nuclei to 2.8 to 2.9, these two histone fractions were solubilized. The remainder of the histones was obtained by extracting the nuclear residue with 0.25 *N* HCl. The method described by Van der Westhuyzen and Von Holt[198] was quite unique. It employed the dissociation of histones with an excess of protamine. The dissociated histones were then fractionated by precipitation with ammonium sulfate or by gel filtration on Sephadex G 100.

For the preparation and analysis of small amounts of histones, electrophoretic separation offers the advantage of rapid and reproducible separation, especially if the formation of aggregates is prevented by the addition of urea to the gels and electrophoretic medium. Electrophoresis in starch powder[90] or in starch[96,97,199-201] or polyacrylamide[202-204] gels yielded histone fractions, the analysis of which made possible rapid identification of chemically isolated fractions. While the recovery of histone fractions from starch gels is complicated by large quantities of soluble starch accompanying the protein samples, the limitations of polyacrylamide electrophoresis are those of the technical design of individual instruments. Gel electrophoresis has been an invaluable analytical tool in many studies on histone isolation, fractionation, tissue specificity, heterogeneity, etc. A new, promising procedure is the electrophoresis of histones on a cellulose acetate (Cellogel®) support matrix as recently described by Machicao and Sonnenbichler.[205] It clearly separates histones into the five main fractions (Figure 6), at pH 10.0 in 0.6 *M* ammonium borate buffer containing 10 mM EDTA, 10 mM β-mercaptoethanol, and 6 *M* urea. One of the advantages of this method is a rapid recovery of histone fractions by dissolving the cellogel support in organic solvents.

Chemical Composition and Structure of Histone Fractions

Although the definite number of histone fractions in various species of eukaryotes is unknown and may be quite large, especially when many unicellular and primitive organisms are analyzed, extensive analytical work on higher animals and some plants indicates that their chromatin contains only five major histone fractions. At least one of these fractions, the very lysine-rich histone F1 (I), is heterogeneous. The structure and chemical properties of these main histones, i.e., the very lysine-rich F1 (I), moderately lysine-rich F2b (IIb2), slightly lysine-rich F2a2 (IIb1), arginine-rich F3 (III), and glycine-arginine-rich F2a1 (IV), will be discussed in detail.

The Very Lysine-Rich Histones F1 (I)

Because of its unique amino acid composition and its hydrophilic nature, the very lysine-rich histone F1 (I) was first recognized as a definite histone fraction. It can be selectively dissociated from chromatin by dilute mineral acids in the pH interval 1.75 to 2.6[38,185,186] or by weak organic acids.[24,25,36,173,206–208] Salts in relatively low concentrations (0.4 to 0.5 *M* NaCl or KCl,[23,174–180,182,186,209] 0.1 to 0.3 *M* $NaClO_4$,[23,176] 0.2 *M* $(NH_4)_2SO_4$,[182] 0.1 *M* $MgCl_2$,[181] etc.) dissociate the F1 histones from chromatin selectivity. Since the F1 (I) fraction is soluble in 5% trichloroacetic acid, it can be prepared from whole histone solution simply by removing all other histone fractions precipitated by this acid.[59] Trichloroacetic acid concentration of 15 to 20% is necessary to precipitate the lysine-rich histone F1 (I).

In addition to their unique amino acid composition, the F1 (I) histones are the largest of the common histone family found in most higher organisms. Their molecular weights approximate 19,500 to 22,000 daltons[159,210,211] and their molecules consist of about 212 amino acid residues.[212] The F1 (I) histone fraction is heterogeneous and the number of its subfractions varies between individual tissues in the same animal as well as between various species. The heterogeneity of this fraction was first revealed by chromatography on Amberlite IRC-50, from which the calf thymus F1 (I) histone eluted in two peaks, Ia and Ib[56] and by starch[50] and polyacrylamide gel electrophoresis.[213] Murray,[214,215] who attempted to analyze this fraction, found over 60 tryptic peptides, i.e., more than would be expected from the molecular weight of F1 (I) histone, and decided this fraction was heterogeneous.

Recognizing that the heterogeneity of F1 (I) histone was not artifactual, Cole and his associates investigated this fraction in detail. Modification of the procedure for chromatography on Amberlite IRC-50 columns permitted the elution of four to five components in the tissues analyzed.[145–148,216,217] There are four or five F1 (I) histones in rabbit thymus, four in rabbit mammary gland, three or four in calf thymus, three in rabbit liver, two in chicken liver, five in rat and chicken spleen, four in cat spleen, and only three in calf spleen. Since careful analysis of these fractions indicates each to be an individual protein species,[146] it can be concluded that the F1 (I) histones are tissue and species specific.

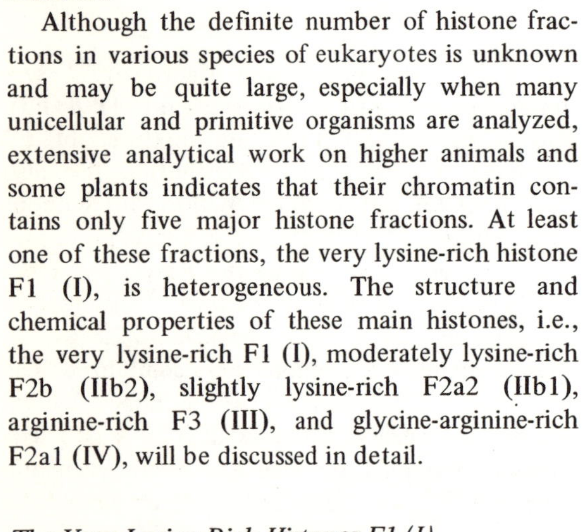

FIGURE 6. Electrophoresis of calf thymus histones on cellulose acetate (Cellogel®) sheets at pH 10.0 in 0.6 *M* ammonium borate.[205]

Based on their experience with fractionation of the F1 (I) histones, Kinkade and Cole[146] suggested that these proteins contain an invariant region of amino acid sequences common to all individual F1 (I) histones attached to a region which varies from one fraction to another. To support this claim, Rall and Cole[218,219] began to determine the amino acid sequences of F1 (I) histones from rabbit and calf thymus. The F1 (I) molecule was cleaved with *N*-bromosuccinimide at its single tryosyl residue. The carboxyterminal fragment (N_1), consisting of more than 140 amino acid residues, was found to be richer in lysine than the NH_2-terminal fragment (N_2), which represented 72 residues in the rabbit thymus component 3. It was this smaller NH_2-terminal fragment which appeared to contain most of the structural differences in the F1 (I) histone subfractions.[212] Subsequently, amino acid sequences of the NH_2-terminal fragments (N_2) from rabbit thymus F1 (I), histone components 3 and 4 (RTL3) and (RTL4), and from the calf thymus F1 (I) histone component 1 (CTL1) were determined.[219,220] They are shown in Figures 7 and 8. The tryptic peptides of components RTL4 and CTL1 were aligned with homologous regions of the RTL3 sequence either on the consecutive residue basis, which yielded minimal differences in corresponding amino acid residues (Figure 7), or by placing the lysine, arginine, and proline residues into the same positions in all three fragments (Figure 8).

As can be seen in Figures 7 and 8 all the components contain the same NH_2-terminal amino acid sequence as was reported by Fambrough[221] for unfractionated F1 (I) histones from calf thymus and pea seedlings. The NH_2-terminal residues of all components are blocked by acetyl groups, an observation previously reported by Phillips.[222]

The comparison of the amino acid sequence of N_2 fragments from rabbit thymus F1 (I) histone components RTL3, RTL4, and calf thymus component CTL1 shows that significant changes occurred during the evolution of these proteins in the two animal species. From Figure 7, there are 11 differences in amino acid residues when comparing the RTL3 to RTL4, 14 between RTL3 and CTL1, and 7 comparing RTL4 to CTL1. Using the arrangement in Figure 8, there are 12 differences between RTL3 and RTL4, 11 between RTL3 and CTL1, and 7 comparing RTL4 to CTL1. The sequential differences are accumulated between the residues 14 and 40 in all 3 com-

10
RTL-3 Ac -SER-GLU-ALA-PRO-ALA-GLU-THR-ALA-ALA-PRO-ALA-PRO-ALA-GLU-LYS-SER-PRO-
RTL-4 Ac -SER-GLU-ALA-PRO-ALA-GLU-THR-ALA-ALA-PRO-ALA-PRO-ALA-□ -LYS-SER-PRO-
CTL-1 Ac -SER-GLU-ALA-PRO-ALA-GLU-THR-ALA-ALA-PRO-ALA-PRO-ALA-PRO-LYS-SER-PRO-

20 30
ALA-LYS-LYS-LYS-□ -LYS-ALA-ALA-LYS-LYS-PRO-GLY-ALA-GLY-ALA-ALA-LYS-ARG-LYS-ALA-
ALA-LYS-THR-PRO-VAL-LYS-ALA-ARG-LYS-LYS-LYS-SER-ALA-GLY-ALA-ALA-LYS-ARG-LYS-ALA-
ALA-LYS-THR-PRO-VAL-LYS-ALA-ALA-LYS-LYS-LYS-LYS-PRO-ALA-GLY-ALA-ARG-ARG-LYS-ALA-

40 50
ALA-GLY-PRO-PRO-VAL-SER-GLU-LEU-ILE-THR-LYS-ALA-VAL-ALA-ALA-SER-LYS-GLU-ARG-ASN-
SER-GLY-PRO-PRO-VAL-SER-GLU-LEU-ILE-THR-LYS-ALA-VAL-ALA-ALA-SER-LYS-GLU-ARG-SER-
SER-GLY-PRO-PRO-VAL-SER-GLU-LEU-ILE-THR-LYS-ALA-VAL-ALA-ALA-SER-LYS-GLU-ARG-SER-

60 70 73
GLY-LEU-SER-LEU-ALA-ALA-LEU-LYS-LYS-ALA-LEU-ALA-ALA-GLY-GLY-TYR.
GLY-VAL-SER-LEU-ALA-ALA-LEU-LYS-LYS-ALA-LEU-ALA-ALA-ALA-GLY-TYR.
GLY-VAL-SER-LEU-ALA-ALA-LEU-LYS-LYS-ALA-LEU-ALA-ALA-ALA-GLY-TYR.

The amino acid sequences were determined and aligned for maximal ideality by Rall and Cole.[219] The gaps (—□—) allow for best alignment.

FIGURE 7. Amino acid sequences of NH_2 fragments from rabbit thymus F1 histone components RTL3 and RTL4 and from calf thymus component CTL1.

10

RTL-3 Ac -SER-GLU-ALA-PRO-ALA-GLU-THR-ALA-ALA-PRO-ALA-PRO-ALA-GLU-LYS-SER-PRO-

RTL-4 Ac -SER-GLU-ALA-PRO-ALA-GLU-THR-ALA-ALA-PRO-ALA-PRO-ALA-□ -LYS-SER-PRO-

CTL-1 Ac -SER-GLU-ALA-PRO-ALA-GLU-THR-ALA-ALA-PRO-ALA-PRO-ALA-PRO-LYS-SER-PRO-

20 30

ALA-LYS-LYS-LYS-□ - □ - □ -LYS-ALA-ALA-LYS-LYS-PRO-GLY-ALA-GLY-ALA-ALA-LYS-ARG-

ALA-LYS-LYS-LYS-THR-PRO-VAL-LYS-□ -ALA-ARG-LYS-□ -SER-ALA-GLY-ALA-ALA-LYS-ARG-

ALA-LYS-LYS-LYS-THR-PRO-VAL-LYS-ALA-ALA-LYS-LYS-PRO-□ -ALA-GLY-ALA-□ -ARG-ARG-

40 50

LYS-ALA-ALA-GLY-PRO-PRO-VAL-SER-GLU-LEU-ILE-THR-LYS-ALA-VAL-ALA-ALA-SER-LYS-GLU-

LYS-ALA-SER-GLY-PRO-PRO-VAL-SER-GLU-LEU-ILE-THR-LYS-ALA-VAL-ALA-ALA-SER-LYS-GLU-

LYS-ALA-SER-GLY-PRO-PRO-VAL-SER-GLU-LEU-ILE-THR-LYS-ALA-VAL-ALA-ALA-SER-LYS-GLU-

60 70 75

ARG-ASN-GLY-LEU-SER-LEU-ALA-ALA-LEU-LYS-LYS-ALA-LEU-ALA-ALA-GLY-GLY-TYR.

ARG-SER-GLY-VAL-SER-LEU-ALA-ALA-LEU-LYS-LYS-ALA-LEU-ALA-ALA-ALA-GLY-TYR.

ARG-SER-GLY-VAL-SER-LEU-ALA-ALA-LEU-LYS-LYS-ALA-LEU-ALA-ALA-ALA-GLY-TYR.

The amino acid sequences were determined and aligned by Rall and Cole.[219] The alignment respects the lysine, arginine and proline residues to occur at the same positions in all three proteins. The gaps (–□–) allow for best alignment.

FIGURE 8. Amino acid sequences of NH_2 fragments from rabbit thymus F1 histone components RTL3 and RTL4 and from calf thymus component CTL1.

ponents, while the sequences of the first 13 amino acids are identical. The region from residue 41 to the tyrosine is identical in RTL4 and CTL1 and differs from that in RTL3 only by 3 conservative replacements. Since the amino acid residues were aligned by Rall and Cole[219] to stress the maximal homologies between these 3 fragments, the actual dissimilarities may be even greater. The differences in the sequences of F1 (I) histones between various tissues and between animal species may further increase when the complete amino acid sequence of these proteins (including the much larger carboxy terminal fragment, N_1) is finally determined. Nevertheless, in contrast to the extremely invariable F2a1 (IV) histone, the very lysine-rich histones F1 (I) appear to have considerably changed during animal evolution. Whether this points to a less vital role of F1 (I) histones in chromatin, as compared with the F2a1 (IV) fraction, presently remains unknown. It can be argued, however, that the structural variability of sequences between the residues 14 to 40 indicates alterations of the molecule, essential for the F1 (I) histones to perform their functions in chromatin.

Each of the three F1 (I) components shown in Figures 7 and 8 contains two clusters of more than two basic amino acids (underlined). The NH_2-terminal *N*-bromosuccinimide fragments (N_2) have a relatively low content of basic amino acids; most proline residues and the majority of the lysine and alanine residues are accumulated in the larger carboxy terminal portion of the F1 (I) histones. The COOH-terminal sequence of all F1 (I) components from rabbit thymus,[148] as well as from the calf thymus,[222] is lysyl-lysine.

Most recently, Jones and Cole[223] completed sequencing additional amino acids in the RTL3 histone component corresponding to the sequence between residues 74 to 108. These sequences, together with the previous work of Cole and his associates, represent about one half of this histone molecule (Figure 9). Interestingly, this region does not contain any clusters of basic amino acids and is primarily hydrophobic. Based on the analytical data gathered by Jones and Cole,[223] which concern a number of peptides derived from the carboxy terminus of RTL3 component, it appears as if each molecule of the F1 (I) histones consisted of two highly dissimilar histone species joined together. The NH_2-terminal half resembles other histones (F2a1 or IV, F2b or IIb2, and F3 or III) in the accumulation of basic amino acid clusters within the first 50 residues. Similar to the other fractions, the central region between residues 50

RTL-3: Ac -SER-GLU-ALA-PRO-ALA-GLU-THR-ALA-ALA-PRO(10)-ALA-PRO-ALA-GLU-LYS-SER-PRO-ALA-

LYS(20)-LYS-LYS-LYS-ALA-ALA-LYS-LYS-PRO-GLY-ALA(30)-GLY-ALA-ALA-LYS-ARG-LYS-ALA-ALA-GLY-PRO(40)-PRO-VAL-

SER-GLU-LEU-ILE-THR-LYS-ALA-VAL(50)-ALA-ALA-SER-LYS-GLU-ARG-ASN-GLY-LEU-SER(60)-LEU-ALA-ALA-LEU-LYS-

LYS-ALA-LEU-ALA-ALA(70)-GLY-GLY-TYR-ASP-VAL-GLU-LYS-ASN-ASN-SER(80)-ARG-ILE-LYS-LEU-GLY-LEU-LYS-SER-

LEU(90)-VAL-SER-LYS-GLY-THR-LEU-VAL-GLU-THR-LYS-GLY(100)-THR-GLY-ALA-SER-GLY-SER-PHE-LYS

FIGURE 9. Amino acid sequence of the NH_2-terminal half of rabbit thymus F1 (I) histone component RTL-3 as determined by Jones and Cole.[223]

and 100 is not basic and contains a considerable number of hydrophobic amino acids. The carboxy terminal half of the F1 (I) histones appears, on the other hand, to consist mainly of alanine, lysine, and proline arranged in sequences punctuated mostly by glycine and a few other amino acids.[223] In this primitive arrangement, the carboxy terminal portion of F1 (I) histones resembles the amino acid sequence of protamines. The reason for this peculiar arrangement of alanine, lysine, and proline residues in F1 (I) histones is presently unknown. Bustin et al.[218] suggested that the more basic COOH-terminal region of F1 (I) interacts with the DNA in chromatin, while the less basic NH_2-terminal portion of the molecule may be engaged in more stereospecific interactions, perhaps with the nonhistone proteins or other molecules. The selective phosphorylation of the serine residue (corresponding to position 40 in Figure 8) in F1 (I) histones from rat liver described by Langan[224-226] may increase the specificity of such interactions.

The discovery that histones, especially the F1 (I) fraction, can be phosphorylated[227,228] complicates the problem of specificity and heterogeneity of these proteins. Selective phosphorylation can introduce reproducible multiplicity in chromatographic and electrophoretic patterns of histones, since both analytical techniques depend partially on the total positive charge of the proteins. Thus, varying degrees of F1 (I) histone phosphorylation may contribute heterogeneity to patterns showing tissue specificity. Another source of apparent heterogeneity of F1 (I) histones may result from the association of these proteins with other macromolecules, most likely nonhistone proteins or RNA. This possibility was recently investigated by Evans et al.[229] who found practically no binding of labeled RNA to the F1 (I) histones. Treatment of the rabbit thymus F1 (I) histones with alkaline phosphatase and with RNase did not change the fractionation patterns of these histones. These authors concluded that the tissue and species specificity of F1 (I) histones was mainly because of the differences in their primary structure and not because of the differences in phosphorylation of one basic molecular species.

A different conclusion was reached by Sherod et al.[230] who analyzed F1 (I) histones from several mouse tissues, including the Ehrlich ascites carcinoma, by high resolution electrophoresis on long polyacrylamide gels. In all tissues the F1 (I) histone was heterogeneous, as evidenced by at least three electrophoretic components in calf and mouse thymus and mouse liver and five components in Ehrlich ascites tumor. Two of the electrophoretic components, both only minor, disappeared after treatment with alkaline phosphatase, indicating that they were derived from at least one of the remaining three protein bands still present after the phosphatase treatment. This was interpreted to mean that the phosphorylation of F1 (I) histones contributes to their heterogeneity; however, the phosphorylated components were found only in the Ehrlich ascites cells.

Extensive studies of Langan[224-226,231] demonstrated the presence of a peptide in rat liver F1 (I) histones with a serine residue (serine residue 40 in Figure 8) which could be enzymatically phosphorylated. The amino acid sequence of this peptide correlated with the sequence of amino acids between the residues 36 - 48 of the rabbit thymus F1 (I) component 3 shown in Figure 9. However, in rabbit RTL3 histone, alanine replaced the serine residue present in the rat liver and in the rabbit RTL4 and CTL1 fractions and therefore the rabbit RTL3 histone cannot be phosphorylated at this site. A peptide similar to that in rat liver was also found in the F1 (I) histone from *Salmo gairdnerii.*[232] As in rat liver, this peptide was also phosphorylated. The structural and biological implications of histone phosphorylation will be discussed later in this book.

In invertebrates, the species and tissue specificity of the very lysine-rich histones is even more obvious. The F1 (I) fraction is reportedly absent in yeast,[233] fish erythrocytes,[234] and in early stages of embryonic development.[235-239] F1 (I) histones of quite variable amino acid composition and electrophoretic mobilities are found in plants,[157,203,240-247] insects,[240,248-251] mollusks and echinoderms,[159,202,252-254] *Tetrahymena,*[157,255] and numerous vertebrates.[46] The variability in the composition, quantities,[217,256] and other properties of the F1 (I) histone fractions in various plants and animals indicates considerable changes in the structure of these proteins during the evolutionary process. This feature sets the F1 (I) histones apart from other histone fractions (with the exception of chicken specific F2c histone), which are known to possess only limited tissue and species specificity. In the arginine-rich F2a1 (IV) histone only 2 amino acid residues out of the total of 102 changed during the time required for the pea plants and cattle to evolve.[257]

As was pointed out by several investigators and discussed here, the isolation and fractionation of very lysine-rich histones F1 (I) must be rigidly controlled. This fraction seemed to be more easily degraded by proteolytic enzymes present in the cell than the other histones. Partial degradation might result in electrophoretic multiplicity and changes in electrophoretic and chromatographic behavior of F1 (I) histones.[41-45,101,102,213,258] Another source of errors affecting especially the amino acid composition of F1 (I) histones was their contamination with a protein component rich in dicarboxylic amino acids.[141] Variable contamination of F1 (I) histones with this protein could significantly influence their amino acid composition. It is noteworthy that some F1 (I) histones reported specific for lower animals contain considerable quantities of dicarboxylic acids with a simultaneous decrease in their basic amino acid content. Therefore, experiments which do not control proteolysis during the isolation of nuclear material and histone extraction should be interpreted only with great caution with respect to F1 (I) histone heterogeneity and specificity.

The Moderately Lysine-Rich Histone F2b (IIb2)

Whereas the F1 (I) histones were the first fraction to be separated from the total mixture of all histones, the F2b (IIb2) was the first fraction to be prepared in sufficient purity to warrant the initiation of studies on its amino acid sequence and tissue and species specificity. The F2b (IIb2) histone can be prepared in large quantities by first removing the arginine-rich fractions by extraction of isolated nuclei or chromatin with a mixture of ethanol and 1.25 *N* HCl (4:1 v/v).[34,59] The residue is then extracted with 0.25 *N* HCl and the mixture of F1 (I) and F2b (IIb2) histones is fractionated by chromatography on carboxymethyl cellulose,[50,158,259] or sulfoethyl cellulose,[160] gel filtration on Sephadex G 75;[160] or by selective precipitation of the 0.25 *N* HCl extract with acetone.[206] This latter technique, although rapid and suitable for bulk preparations, yields F2b (IIb2) histone contaminated mainly with F1 (I) histones, and further purification by chromatography or gel filtration is required.

The complete amino acid sequence of F2b (IIb2) histone from calf thymus appears in a recent report by Iwai et al.[260] and is shown in Figure 10. A partial amino acid sequence of this fraction, also from calf thymus, established by Hnilica et al.,[160] is in general agreement with that shown in Figure 10. The F2b (IIb2) histone from calf thymus has 125 amino acid residues with a considerable asymmetry in the distribution of basic amino acids. The NH_2-terminal region is more basic than the COOH-terminal region and the basic amino acids are distributed irregularly; there are several doublets and one cluster of basic amino acids (underlined in Figure 10). Most of the neutral and hydrophobic amino acids are situated in the central portion of F2b (IIb2) sequence leaving both sides of the nonbasic portion enriched in lysine and arginine. During tryptic digestion this central, nonbasic portion of F2b (IIb2) histone precipitates, yielding a large peptide ('core') consisting of 15 amino acid residues. This represents the longest sequence of nonbasic residues in this protein. Similar 'core' peptides have also been reported in other histone fractions, except histone F1 (I).[111] This tryptic 'core' peptide also contains the bulk of nonhistone proteins occasionally contaminating histone fractions.

The F2b (IIb2) histonc was found in plants[203] and animals,[46,50,240,261,262] including invertebrates,[202,204,250] *Tetrahymena,*[255] and slime

10
NH_2-PRO-GLN-PRO-ALA-LYS-SER-ALA-PRO-ALA-PRO-LYS-LYS-GLY-SER-LYS-ALA-

20 30
VAL-THR-LYS-LYS-ALA-GLN-LYS-LYS-ASP-GLY-LYS-LYS-ARG-LYS-ARG-SER-ARG-

40 50
LYS-GLU-SER-TYR-SER-VAL-TYR-VAL-TYR-LYS-VAL-LEU-LYS-GLN-VAL-HIS-PRO-

60
ASP-THR-GLY-ILE-SER-SER-LYS-ALA-MET-GLY-ILE-MET-ASN-SER-PHE-VAL-ASN-

70 80
ASP-ILE-PHE-GLU-ARG-ILE-ALA-GLY-GLU-ALA-SER-ARG-LEU-ALA-HIS-TYR-ASN-

90 100
LYS-ARG-SER-THR-ILE-THR-SER-ARG-GLU-ILE-GLN-THR-ALA-VAL-ARG-LEU-LEU-

110
LEU-PRO-GLY-GLU-LEU-ALA-LYS-HIS-ALA-VAL-SER-GLU-GLY-THR-LYS-ALA-VAL-

120 125
THR-LYS-TYR-THR-SER-SER-LYS-OH

FIGURE 10. Amino acid sequence of the moderately lysine-rich histone F2b (IIb2) from calf thymus.[260]

mold.[75] In all of these species, especially in vertebrates, the amino acid composition of this histone is remarkably similar. Hnilica et al.[159,259,261,263] compared several corresponding peptides from the F2b (IIb2) histones of Walker carcinosarcoma 256 and calf thymus and found that they were identical in amino acid composition. Peptide maps (fingerprints) of this fraction prepared from several animal sources also proved to be very similar, if not identical. Minor differences in the position of several ninhydrin positive spots found by Laurence et al.,[264] who compared peptide maps (fingerprints) of F2b (IIb2) histones from calf thymus and rat osteogenic sarcoma, might be due to the contamination of their histones by other fractions. The total of 41 ninhydrin positive peptides found for the F2b (IIb2) histone by these authors was almost twice the amount which could be expected from the amino acid sequence of this fraction (shown in Figure 10). Since the F2b (IIb2) histone was reported to be heterogeneous,[159,265] the exceptionally high number of peptides found by Laurence et al.[264] might be due to an incomplete separation of the subfractions. The F2b (IIb2) histone has frequently been found to be contaminated by F1 (I) and F2a2 (IIb1) histones; contamination by the latter was usually rather difficult to remove and the heterogeneity of F2b (IIb2) histone from calf thymus reported by both Hnilica[159] and Johns[265] might be partly due to such contamination.

Although the amino acid sequences of F2b (IIb2) histones from tissues other than calf thymus remain to be determined, the close similarity of peptide maps of this fraction from several tissues of different animal species,[159,259,261] as well as some preliminary data on the sequence of chicken erythrocyte F2b (IIb2), indicates that there are only minor differences in the structure of this fraction in various animal species.

The work on the amino acid sequence of F2b (IIb2) histone from calf thymus did not reveal the presence of methylated or acetylated amino acids. In vivo phosphorylation of this fraction was reported by Gutierrez-Cernosek and Hnilica[266,267] in rat tissues and by Shepherd et al.[268] in Chinese hamster ovary cells (CHO) grown in tissue cultures. In vitro phosphorylation of F2b (IIb2) histone was carried out by Langan[224] who demonstrated that only one mole of phosphate can be incorporated per mole of histone. Preliminary results from this laboratory indicated that in calf thymus the P^{32} recipient was most likely the sequence between residues 47 and 57. However, only less than 10% of this peptide has been found phosphorylated in calf thymus F2b (IIb2) histone fraction.

The Slightly Lysine-Rich (Arginine-Lysine-Rich) Histone F2a2 (IIb1)

This histone fraction, which contains only a slight excess of lysine over arginine (Table 2), was among the last fractions to be recognized and purified. Most of the reported extraction and purification procedures resulted in a mixture of histones F2a2 (IIb1) and F2a1 (IV), which was for sometime regarded as fraction F2a. The F2a2 (IIb1) fraction was first isolated by preparative electrophoresis,[97] fractionation with acidified acetone,[206,269] chromatography on carboxymethyl cellulose,[155,158,269] aggregation in 1 *M* NaCl solution,[164] precipitation by dialysis against a mixture of ethanol and *n*-propanol,[163] and gel filtration on Sephadex G 75,[164,269] G 100,[166,270,271] or G 200.[167]

Initial information concerning the sequence of F2a2 (IIb1) histone has been accumulated by Phillips,[77,272] who found this fraction in calf thymus to consist approximately of 136 to 140 amino acid residues. This agreed with the 15,000 daltons molecular weight of F2a2 (IIb1) histone determined by Phillips and Clarke.[172] However, tryptic digestion of the F2a2 (IIb1) histone from calf thymus yielded peptides corresponding to 150 amino acid residues with perhaps a few amino acids still omitted due to the unknown number of basic amino acid clusters.[77] This discrepancy was interpreted by Phillips[77] as indicating a heterogeneity of F2a2 (IIb1) histone, which suggested that this fraction may be a mixture of two rather similar proteins. Since Leclerc et al.[273] found a homogeneous protein contaminating both the F2a2 (IIb1) and F2a1 (IV) fractions in calf thymus and other tissues, the heterogeneity of F2a2 (IIb1) histone as proposed by Phillips[77] might be due to this contamination. Preliminary reports on the amino acid sequence of the F2a2 (IIb1) histone from calf thymus showed the basic amino acids to occur in clusters, which appears to be a feature common to all histones. The basic amino acids were distributed irregularly and there were stretches of up to 11 nonbasic amino acids in the known portions of the F2a2 (IIb1) sequence. Similar to the F2b (IIb2) histone, the 'core' peptide resulting from tryptic digestion of F2a2 (IIb1) fraction appeared to be derived from the nonbasic part of the molecule.[77]

The nearly complete amino acid sequence of F2a2 (IIb1) histone from calf thymus as recently determined by Sautière et al.[273a] is shown in Figure 11. The 131 amino acid residues comprising this fraction are distributed irregularly with prominent accumulations of basic residues in both the NH_2- and COOH-terminal regions of this protein. There are two basic amino acid clusters (amino acid residues 29 to 32 and 74 to 77) and a basic region (residues 29 to 38) in the F2a2 (IIb1) histone. The central part of this fraction is represented by a sequence of 28 amino acid residues which are not basic (residues 43 to 70). A shorter nonbasic sequence of 18 amino acid residues is located near the COOH terminal (residues 106 to 123). The amino acid sequence of F2a2 (IIb1) calf thymus histone as determined by Sautière et al.[273a] is in good agreement with sequences determined by Phillips[77] and by Hayashi and Iwai[273b] for peptides derived from the F2a2 (IIb1) histones also from calf thymus.

Interestingly, the NH_2-terminal sequence of five amino acids (Ac-SER-GLY-ARG-GLY-LYS-) is the same as in the arginine-rich fraction F2a1 (IV) and there are additional similarities in the sequences of the fractions F2a1 (IV) and F2a2 (IIb1). Sung and Dixon[201] determined the sequence of the six NH_2-terminal amino acids in F2a2 (IIb1) histone from trout testis and found it identical to the NH_2-terminal sequence of trout and calf thymus histone F2a1 (IV), i.e., Ac-SER-GLY-ARG-LYS-GLY. Except for the sixth residue, which in calf thymus F2a2 (IIb1) histone is GLN and in trout testis GLY, the NH_2-terminal peptides are identical in these two species. Additional similarities between the sequences of histone F2a2 (IIb1) and

10
AcSER-GLY-ARG-GLY-LYS-GLN-GLY-GLY-LYS-ALA-ARG-ALA-LYS-ALA-LYS-THR-ARG-

20 30
SER-SER-ARG-ALA-GLY-LEU-GLN-PHE-PRO-VAL-GLY-ARG-VAL-HIS-ARG-LEU-LEU-ARG-

40 50
LYS-GLY-ASN-TYR-ALA-GLU-ARG-VAL-GLY-ALA-GLY-ALA-PRO-VAL-TYR-LEU-ALA-ALA-

60 70
VAL-LEU-GLU-TYR-LEU-THR-ALA-GLU-ILE-LEU-GLU-LEU-ALA-GLY-ASN-ALA-ALA-ARG-

80
ASP-ASN-LYS-LYS-THR-ARG-ILE-ILE-PRO-ARG-HIS-LEU-GLN-LEU-ALA-ILE-ARG-ASN-

90 100
ASP-GLU-GLU-LEU-ASN-LYS-LEU-LEU-GLY-LYS-ASN-LYS-ALA-LYS-GLY-LYS-(ASX,THR,

120
GLX$_2$,PRO,GLY$_2$,ALA$_2$,VAL$_2$,ILE$_2$,LEU)-VAL-LEU-LEU-PRO-LYS-LYS-THR-GLU-SER-

130
HIS-HIS-LYS(OH)

FIGURE 11. Amino acid sequence of calf thymus histone F2a2 (IIb1).[273a]

F3 (III) were discussed recently by DeLange and Smith.[274]

The F2a2 (IIb1) histone was reportedly acetylated in various tissues[222] but the site of the labile acetyl groups remains to be determined. In trout testis[201] and in regenerating rat liver,[200] this fraction was found both acetylated and phosphorylated at the NH_2-terminal serine residue. There has been very little information reported on the tissue or species specificity of the F2a2 (IIb1) histone. Its amino acid composition, as well as its electrophoretic mobility in polyacrylamide gels, was found to be virtually identical when F2a2 (IIb1) histones from various tissues of hog, calf, guinea pig, and rat were compared.[275] Similarly, the peptide maps of the F2a histones (unfractionated mixture of F2a1 or IV and F2a2 or IIb1 fractions) from calf thymus and rat osteogenic sarcoma were extremely similar, differing in the position of only several peptides.[264] Since the F2a1 (IV) histone is known to be evolutionarily conserved, all the observed differences might be attributable to the F2a2 (IIb1) histones from these two tissues. However, only a comparison of the complete amino acid sequences of these fractions from several species would determine the extent of specificity of F2a2 (IIb1) histones.

The Glycine and Arginine-Rich Histone F2a1 (IV)

Although the very first fractionation procedures were able to distinguish the two main histone groups, i.e., the lysine-rich and arginine-rich histones, it was only much later when the heterogeneity of the arginine-rich histones was first realized.[97,163,164,269] Two types of arginine-rich histones have been shown to occur in many plant and animal tissues. These are the glycine and arginine-rich fraction F2a1 (IV) and the alanine and arginine-rich fraction F3 (III). As seen in Table 2, both fractions are quite similar in amino acid composition, the main difference being the high glycine content in histone F2a1 (IV). The inverse is true for the alanine content of these two fractions. Perhaps the most significant difference is the presence of cysteine in F3 (III) histones as opposed to a complete absence of this amino acid in the other main histone fractions.[276-278] Since the F2a1 (IV) histone is usually extracted together with the F2a2 (IIb1) fraction, procedures applicable to the isolation of F2a2 (IIb1) also yield the F2a1 (IV) histone.[97,155,158,163,164,166,167,269-271] In addition to these techniques, the F2a1 (IV) and F3 (III) fractions can be extracted together from isolated chromatin with 2% NaCl in 75% ethanol or with 10% guanidine hydrochloride, also in 75% ethanol.[182,206] Although one of the last histones to be described in a pure form, the calf thymus and pea seedling histone F2a1 (IV) was the first fraction for which a complete amino acid sequence was established.[257,282-286] Furthermore, the existence of ϵ-*N*-acetylated lysine was documented in it and its amino acid sequence was compared to the homologous histone in pea seedlings.[257,279-283]

As can be seen in Figure 12, the F2a1 (IV) histone is the smallest of all histone fractions with only 102 residues forming its polypeptide chain. The basic amino acids are distributed irregularly, as in the other histone fractions. In addition to several pairs, there are two clusters of basic amino acids in the F2a1 (IV) molecule. The NH_2-terminal portion of this histone is enriched in basic amino acids as compared with the COOH-terminal portions. The neutral and hydrophobic amino acid residues are also accumulated in the COOH-terminal region. Methylated lysine is present at a single position (lysine residue 20) as a mixture of ϵ-*N*-monomethyllysine and ϵ-*N*-dimethyllysine.[257] In addition to the acetylated NH_2-terminal serine, the F2a1 (IV) histone also contains ϵ-*N*-acetyllysine which has been found, together with the methylated derivative, in amino acid hydrolysates of this fraction.[271,285,286] The ϵ-*N*-acetyllysine is located at the position 16 of the F2a1 (IV) amino acid sequence. Interestingly, only about 50% of the total lysine residues in position 16 are acetylated.[257] This affects the net positive charge of F2a1 (IV) histone and results in its electrophoretic heterogeneity. Using long poly-

```
                                                 10
Ac-SER-GLY-ARG-GLY-LYS-GLY-CLY-LYS-GLY-LEU-GLY-LYS-GLY-GLY-ALA-

16                    20
LYS(Ac)-ARG-HIS-ARG-LYS(Me)-VAL-LEU-ARG-ASP-ASN-ILE-GLN-GLY-ILE-

30                                            40
THR-LYS-PRO-ALA-ILE-ARG-ARG-LEU-ALA-ARG-ARG-GLY-GLY-VAL-LYS-ARG-

                50                                              60
ILE-SER-GLY-LEU-ILE-TYR-GLU-GLU-THR-ARG-GLY-VAL-LEU-LYS-VAL-PHE-

                                70                              77
LEU-GLU-ASN-VAL-ILE-ARG-ASP-ALA-VAL-THR-TYR-THR-GLU-HIS-ALA-LYS-

        80                                        90
ARG-LYS-THR-VAL-THR-ALA-MET-ASP-VAL-VAL-TYR-ALA-LEU-LYS-ARG-GLN-

                        100
GLY-ARG-THR-LEU-TYR-GLY-PHE-GLY-GLY-OH
```

FIGURE 12. Amino acid sequence of calf thymus histone F2a1(IV).[281]

acrylamide gels, Panyim and Chalkley[101] separated electrophoretically the F2a1 (IV) histone into the acetylated and nonacetylated fraction, each represented by a homogeneous protein band.

The tissue and species specificity of the F2a1 (IV) histone is exceptionally limited. As compared with the amino acid sequence of calf thymus F2a1 (IV), shown in Figure 12, this fraction differs in pea seedlings by only two amino acid residues;[257] valine residue 60 is replaced by isoleucine in the pea and lysine residue 77 is replaced by arginine. In addition to these conservative substitutions, lysine residue 20 is methylated in calf thymus F2a1 (IV) histone but not methylated in the pea. Also the distribution of acetylated amino acids is different between these two species. Lysine residue 16 is about 50% acetylated in calf thymus as compared to only about 6% in the pea seedlings, where at least one of the other three lysine residues in positions 5, 8, or 12 is also acetylated (most likely lysine in position 8).[257] The presence of cysteine in F2a1 (IV) histones from several species of Echinoderms has recently been reported by Subirana,[287] raising an interesting question of the possible origin of both the arginine-rich histone fractions (F3 or III and F2a1 or IV) from a common ancestral gene.

The complete amino acid sequences of F2a1 (IV) histone from porcine thymus,[288] rat Novikoff hepatoma,[289] and rat chloroleukemia[290] were recently found to be identical to the homologous fraction from calf thymus. Similarly, identical sequences of the 19 COOH-terminal amino acids in F2a1 (IV) histones from bovine lymphosarcoma, Novikoff hepatoma (rat), and fetal calf thymus were found by Desai et al.[291] and in rat chloroleukemia by Sautière et al.[292] The only compositional variations were found in the ϵ-*N*-monomethyllysine and ϵ-*N*-dimethyllysine content of this histone in several normal and neoplastic tissues.[293,294]

The amazing similarity of the amino acid sequences of F2a1 (IV) histones from calf thymus and pea seedlings, as well as other sources, indicates that the structure of this histone has been conserved practically unchanged during many years of evolution. According to Dayhoff,[295] as quoted in DeLange and Smith,[274] the F2a1 (IV) histone has a mutation rate of approximately 0.06 residues per 100 million years. This makes this fraction the most evolutionary stable protein known at the present time. How this remarkable evolutionary stability of F2a1 (IV) histones is related to their biological function in chromatin remains to be determined. Since it has been found that DNA cistrons coding for histones are highly reiterated,[296] a mechanism of suppression or elimination of mutated F2a1 (IV) histone cistrons, which is biologically required in order to conform to the observed evolutionarily "frozen" amino acid sequence of this histone, may be one of the most essential regulatory mechanisms in living cells. It appears that the highly conservative distribution of amino acids in this histone fraction in various species may be related to its fundamental biological role either as a genetic repressor or as a structural component of chromatin without which the genetic information cannot be transcribed properly to assure survival. It is of interest at this point to ponder the evolution of hypotheses concerning the biological function of histones.

The histones were first thought to be tissue and species specific in order to selectively suppress individual genes. Later, limited tissue and species specificity was discovered and led to the conclusion that histones are structural proteins which unspecifically associate with the DNA. This genetic "glue" theory, giving histones a position subservant to other components of chromatin, has now been challenged by the amazing lack of evolutionary changes of at least the F2a1 (IV) histone fraction. One could not reasonably expect a molecule useful only for the structure of chromatin to be as restrictive in its amino acid sequence as was found for the F2a1 (IV) histone. Since this fraction had at least the same chance of mutations during the developmental process as any other protein species, such mutations must have been incompatible with the biological function of this histone and resulted in the death of the mutant.

The variances in the extent and location of lysine methylation and acetylation may be related to structural modifications of chromatin as required by the DNA replicative and transcriptive processes. They seem to change with the functional state of the cell and will be discussed later.

The Alanine and Arginine-Rich Histone F3 (III)

The arginine-rich histone F3 (III) is the only histone fraction known to contain cysteine as part of its amino acid sequence[276-278] (barring the recent finding of cysteine in F2a1 (IV) histones from Echinoderms by Subirana[287]. Its isolation and purification follow methods used for the isolation of the histones F2a1 (IV) or F2a2 (IIb1). Chromatography on carboxymethyl cellulose,[50,59,158] Amberlite IRC-50,[144] gel filtration,[162-167,270,271] or selective extraction[34,189,206] followed by precipitation during dialysis against absolute ethanol,[34] etc. can be employed to obtain this histone in a variable degree of purity. According to Brandt and Von Holt,[297] dialysis against absolute ethanol yields F3 (III) histone contaminated by the F2a2 (IIb1) fraction (an observation confirmed in our laboratory). The contaminating F2a2 (IIb1) histone can be removed by gel filtration of the oxidized sample, thus separating the F3 (III) dimer from the nonaggregated F2a2 (IIb1) histone.[297] Since the F3 (III) histone frequently associates with other proteins of similar solubility, its purification, especially by ion-exchange chromatography or gel filtration, is difficult. The protein contamination of F3 (III) histones from a variety of tissues, including calf thymus, has been pointed out by several investigators[144,164,216] and frequently complicates metabolic studies on this fraction. Furthermore, because of its cysteine content, the F3 (III) histone forms polymer molecules joined together by interacting sulfur bridges. In a simple situation, F3 (III) histone preparation is composed of a mixture of monomers and oxidized dimers. In calf thymus, the F3 (III) contains two cysteine residues in each molecule[298] and the situation is more complicated. Furthermore, in addition to the chemical complexity, the F3 (III) histones are known to aggregate extensively at neutral pH values forming large macromolecular complexes[191-194] (a feature utilized for their isolation).

The F3 (III) histone from calf thymus is reported to be heterogeneous. Using countercurrent distribution, Yokotsuka et al.[299,300] obtained three subfractions of histone F3 (III) from calf thymus. A fraction containing two cysteine residues and having a molecular weight of approximately 16,000 daltons is being investigated in detail by this group.

The amino acid sequence of the F3 (III) calf thymus histone, as determined by DeLange et al.,[274,301,301A] is shown in Figure 13. Similar to the other fractions, histone F3 (III) contains a large region of nonbasic amino acids (sequence between residues 84 and 112). This part of F3 (III) molecule also contains both cysteine residues known to be present in F3 (III) histone from calf thymus. There are two lysine residues (14 and 23) in the F3 (III) histone which are acetylated and two lysine residues (9 and 27) which are methylated. The NH_2-terminal peptide NH_2-ALA-ARG has been isolated by Phillips and Simson,[302] who studied the composition of tryptic peptides from calf thymus F3 (III) histone, and has been reported to be present in the NH_2-terminus of the chicken erythrocyte F3 (III) histone.[297] Phillips and Simson[302] also found a relatively high frequency of alanine residues adjacent to lysine or arginine. This observation can be confirmed by inspecting the amino acid sequence of this fraction in Figure 13.

Although the complexity of F3 (III) histone

10
H_2N-ALA-ARG-THR-LYS-GLN-THR-ALA-ARG-LYS$(CH_3)_{0-2}$-SER-THR-GLY-GLY-LYS(Ac)-
20
ALA-PRO-ARG-LYS-GLN-LEU-ALA-THR-LYS(Ac)-ALA-ALA-ARG-LYS$(CH_3)_{0-2}$-SER-ALA-
30 40
PRO-ALA-THR-GLY-GLY-VAL-LYS-LYS-PRO-HIS-ARG-TYR-ARG-PRO-GLY-THR-VAL-ALA-
50 60
LEU-ARG-GLU-ILE-ARG-ARG-TYR-GLN-LYS-SER-THR-GLU-LEU-LEU-ILE-ARG-LYS-LEU-
70 80
PRO-PHE-GLN-ARG-LEU-VAL-ARG-GLU-ILE-ALA-GLN-ASP-PHE-LYS-THR-ASP-LEU-ARG-
90 100
PHE-GLN-SER-SER-ALA-VAL-MET-ALA-LEU-GLN-GLU-ALA-CYS-GLU-ALA-TYR-LEU-VAL-
110
GLY-LEU-PHE-GLU-ASP-THR-ASN-LEU-CYS-ALA-ILE-HIS-ALA-LYS-ARG-VAL-THR-ILE-
120 130
MET-PRO-LYS-ASP-ILE-GLN-LEU-ALA-ARG-ARG-ILE-ARG-GLY-GLU-ARG-ALA-COOH

FIGURE 13. Amino acid sequence of calf thymus histone F3 (III).[301a]

and its interactions with nonhistone proteins complicate studies on the tissue or species specificity of this histone, reports in the literature[264,274,298] suggest a conservation of its sequence similar to that found for the F2a1 (IV) fraction. The F3 (III) histones from calf thymus and pea embryos were found by Fambrough and Bonner[298] to be practically identical in their alanine COOH- and NH_2-terminal amino acids, amino acid composition, electrophoretic behavior, and tryptic peptide patterns (fingerprints). The chief difference was that calf thymus F3 (III) histone contains two cysteine residues per molecule while the pea histone has only one. The composition of the COOH-terminal cyanogen bromide peptide from calf thymus and chicken erythrocyte histone F3 (III) was recently reported to be essentially the same as in calf thymus.[297]

The formation of disulfide bridges between the individual F3 (III) molecules results in electrophoretic heterogeneity of this histone[298] which can be eliminated by complete oxidation or reduction. Taking such precautionary measures, the electrophoretic mobility of F3 (III) histones in a variety of tissues and animal species is very reproducible. Panyim et al.[46,240] found that in the histones from a great number of animal species only the histones F3 (III) and F2a1 (IV) had constant electrophoretic mobilities, again indicating the lack of extensive species and tissue specificity of these fractions.

It appears that sometime during the evolution of higher animals a second cysteine residue was introduced into the amino acid sequence of F3 (III) histones. Based on an extensive electrophoretic analysis of F3 (III) histones from a variety of plants and lower to higher animal species, Panyim et al.[303] reported that in rodents and other phylogenetically subordinate orders these proteins contain only one cysteine in their sequence, while in mammals more advanced than rodents, the F3 (III) histones have two such residues. Whether this mutation was retained by advanced mammalian orders (lagomorphs, carnivores, ungulates, and primates) because of its possible functional advantage is presently unknown.

Although the five main histone fractions represent products of individual genes, Phillips[77] suggested that histones may have evolved from a common ancestral peptide ("pre-histone") many millions of years ago. Indeed, a comparison of known amino acid sequences of the fractions F2a1 (IV), F2b (IIb2), and F3 (III) as well as of the partial sequences of histones F1 (I) and F2a2 (IIb1), showed that each individual histone most likely evolved from a primitive lysine and arginine containing peptide by a sequence of genetic duplications, mutations and conservative replacements. Finding further sequential homologies between the individual histone fractions, Phillips[77] concluded that the primitive "pre-histone" was most likely a pentapeptide composed mainly of lysine and arginine similar to the ancestral protamine. With evolution, this pentapeptide became less basic, perhaps containing only two or three basic residues. The small size of the primitive genome ("pre-nucleoprotein"), as well as the basicity of the pre-histone, probably satisfied the requirements for stability of the primitive genetic structures. Further evolutionary pressure increased the need for a more sophisticated chromatin conformation and, hence, led to a more complicated structure of the histones. Histone evolution progressed along with the evolution of species until, many millions of years ago, the process could not progress any further without endangering the functional stability of chromatin and chromosomes.

The sequential homologies reaching back to ancestral peptides are not unique to the basic proteins in chromatin. As shown recently by Bauer,[304] a systematic analysis of the amino acid sequence of the calf thymus histone F2a1 (IV) permits a postulation of an ancestral peptide common to a number of proteins (histones, ferredoxin, cytochrome c, the β-chain of haptoglobulin, chymotrypsinogen A, and two immunoglobulin H chains). It is difficult to conclude at this time whether the ancestral peptide, if rêal, represented certain functional and structural advantages over many other associations or whether it reflected only a more stable arrangement of nucleotides coding for its amino acids. It is also uncertain whether the small ancestral peptides were assembled by primitive ribosomes, or, more likely, by other interactions. While it is not difficult to imagine the biosynthesis of ancestral histones ("pre-histones") by primitive organisms (bacteria contain no histones but are able to biosynthesize many complicated proteins), there appears to be no known mechanism permitting all proteins to evolve simultaneously.

Tissue and Species Specificity of Histones

In an attempt to explain the presence of histones in the chromatin of eukaryotic cells, Stedman and Stedman[62,94] proposed that histones associate with the DNA and function as repressors of specific genes. To carry out this function, histones were expected to be tissue and species specific. In support of their theory, Stedman and his associates studied the chemical composition and other properties of histones from various sources. In addition to the chemical divergence between the somatic histones and sperm protamines which was already known to Kossel,[3] they were able to demonstrate considerable differences in arginine contents of histones from chicken erythrocytes and wheat germ as compared with similar fractions in calf thymus and other tissues.[62,94] Because the composition of histones could reflect their biological functions and lead to the discovery of genetic regulatory mechanisms, numerous reports comparing histones from various sources appeared in the literature. Although recent progress in the determination of amino acid sequences of several histone fractions answered some of the questions, the evolutionary history of histones and their specificity requires more investigation.

Similarities and Differences

In an extension of their initial work on chicken erythrocyte and wheat germ histones, Mauritzen and Stedman[305,306] reported a tissue specificity in the amino acid composition of the arginine-rich β histones from various tissues of domestic fowl and ox. However, these differences were only minor and they might easily have arisen from contamination of the β histones by other proteins. Crampton et al.[57] carried out similar studies with histones (fractions A and B which correspond to the very-lysine-rich histones and the remainder of the fractions, respectively) from calf and guinea pig tissues. These authors additionally supported their findings by comparing the chromatographic profiles of tryptic digests of histone fractions B from all the tissues and concluded that the homologous histones were remarkably similar. Practically identical amino acid compositions, starch gel electrophoretic patterns, and the distribution of NH_2-terminal amino acids in histone fractions F1 (I), F2 (mixture of F2a1 or IV, F2a2 or IIb1 and F2b or IIb2), and F3 (III) from normal and leukemic rat spleens and livers as well as calf thymus were found by Hnilica et al.[59] Similarly, no significant tissue specificity of histones was seen when the fractions F1 (I), F2a (mixture of F2a1 or IV and F2a2 or IIb1), F2b (IIb2), and F3 (III) were prepared from Walker carcinosarcoma and compared with those from calf thymus.[50] More detailed studies, including analysis of the distribution and composition of tryptic peptides from several F2b (IIb2) histones, further confirmed their essential identity and lack of tissue and species specificity.[259,261,263] Similarly, Laurence et al.,[264,307] who compared histone fractions F1 (I), F2a (mixture of F2a1 or IV and F2a2 or IIb1), F2b (IIb2), and F3 (III) from calf thymus, Crocker sarcoma, spontaneous mammary tumor of mice, and rat osteogenic sarcoma D177, also confirmed the lack of substantial differences between the homologous proteins.

Davis and Busch[108,152] analyzed acid extracts from crude nuclear preparations of various rat tissues by chromatography on carboxymethyl cellulose. The distribution of radioactivity was used as an additional parameter to study the tissue specificity of histones in cancerous tissues. The authors reported a chromatographic peak (RP2L) supposedly specific for neoplastic tissues. Isolation and analysis of the RP2L proteins revealed the presence of histones F1 (I), F2a (F2a1 or IV and F2a2 or IIb1), and F2b (IIb2).[308] Subsequent analyses of corresponding histone fractions from Walker carcinosarcoma and calf thymus demonstrated, however, no significant tissue and species specificity.[50,259,261] The peculiar chromatographic behavior of labeled histones from neoplastic tissues probably arises from the higher uptake of C^{14}-lysine by tumor cells and from contamination by nonhistone proteins and does not reflect true specificity of histones in neoplasms.

The similarity of homologous histones extended beyond mammalian species. Fractions similar to those in calf thymus were found in chicken liver and erythrocytes,[107,162,297,309,310] trout[262] and buffalo fish liver,[46] sea urchin embryos,[204,239,311] and sperm,[159,202,253,311] molluscs,[252,253] echinoderms,[254,287] plants,[75,203,233,241,245-247,312-315] insects,[248,250,251] *Tetrahymena pyriformis,*[244,255,316,317] and other species.[318-320] Arginine-rich histones appeared to be completely absent from wheat germ.[241] Many histones described in the cited references show compositional divergence from the stereotype pattern recognized for mammalian fractions. In general, the chief differences observed were in the contents of lysine, arginine, and dicarboxylic amino acids. Since at many occasions these latter values were significantly higher than in homologous histones from higher vertebrates, they might indicate a considerable contamination by acid soluble nonhistone proteins. Such contamination was expected to be more likely when working with lower living forms since the anatomical-histological relationship in these forms complicates the isolation purity of nuclei or chromatin. Ribosomal contamination could easily complicate starch or polyacrylamide gel patterns. Moreover, larger quantities of acid soluble nonhistone proteins and proteases might occur in the nuclei of such organisms. It is important, therefore, to base any claims of tissue or species specificity of histones on the analyses of carefully purified histone fractions (preferably by more than one technique).

The high resolving power of zone electrophoresis in starch or polyacrylamide gels was also extensively utilized in studies on tissue and species specificity of histones from various sources. Starch gel electrophoretic patterns of histones from domestic fowl revealed the tissue specificity of the erythrocyte and testis histones,[321] an observation which led to the isolation of chicken erythrocyte specific histone F2c (V).[107,309] Using starch gel electrophoresis, Vendrely et al.[322] observed a remarkable similarity in histone patterns from calf thymus, liver, and lung from rat thymus and liver. On the other hand, histone patterns from nucleated erythrocytes from chicken, duck, guinea chick, viper, tortoise, trout, carp, pike, pleurodele, and frog were distinctly different from mammalian histone patterns. Similarly, the electrophoretic analysis of histones from *Tetrahymena pyriformis*,[255,316,317] the mealy bug *Planococcus citri*,[248] cricket *Acheta domestica*,[248] mollusc *Mytilus edulis*,[252] *Drosophila melanogaster*,[251] broad bean (*Vicia faba*) roots,[247] embryonal tissues of bush beans (*Phaseolus vulgaris*),[323] maize seedlings,[314] wheat germ,[312,313] slime mold *Physarum polycephalum*,[75] and other species demonstrated variable differences in the mobility or number of components between the analyzed species and corresponding histones from calf thymus or other vertebrates.

On the other hand, only minor differences could be detected when the histones of several chicken tissues were compared with each other and with histone preparations from various developmental stages of chickens[324] or frogs.[320] Similarly, polyacrylamide gel electrophoresis of histones from rat spleen, thymus, liver, kidney, and testis;[325] rat heart, liver, kidney, spleen, brain, and lung;[318,326] or the brain, liver, and kidneys of newborn rats, adult rats, adult rabbits, and guinea pigs[327] showed only minor variations from the general pattern known for mammalian histones. No detectable changes were observed in the composition and electrophoretic patterns of histones during early embryogenesis,[328] and organogenesis[329] of chickens. Similarly, striking similarities were observed between plant and calf thymus histones.[203,246,315] The most extensive and careful study on the electrophoretic similarities of histones from numerous vertebrate species was recently performed by Panyim et al.[46,240] Using a system of long polyacrylamide gels,[101,102] the authors separated histones into protein bands corresponding to histone fractions F1 (I), F2a1 (IV), F2a2 (IIb1), F2b (IIb2), and F3 (III). In all species analyzed (Figures 14 and 15) the arginine-rich histones F3 (III) and F2a1 (IV) had a constant electrophoretic mobility indicating considerable conservation of these fractions during the evolution of vertebrates. Only small variations were seen for histones F2a2 (IIb1) and F2b (IIb2), while the very lysine-rich histones F1 (I) varied considerably in their electrophoretic heterogeneity and mobility (Figure 16), indicating substantial changes in the primary structure of this fraction during evolution. These findings were in excellent agreement with the chemical analyses of the arginine-rich and lysine-rich histone fractions discussed earlier, where the F1 (I) histones were found tissue and species specific.[145-148,217,253,254,274]

As was already mentioned, contamination or mild proteolysis of histones will change their electrophoretic behavior and amino acid composition. Additional care has to be exerted to prevent the formation of histone aggregates which could produce multiple protein bands in histone electrophoreograms and which could easily be interpreted as tissue or species specific components. Therefore, comparative studies on histones by electrophoretic techniques should be carried out at low pH values or in the presence of concentrated urea to prevent histone aggregation. As was pointed out

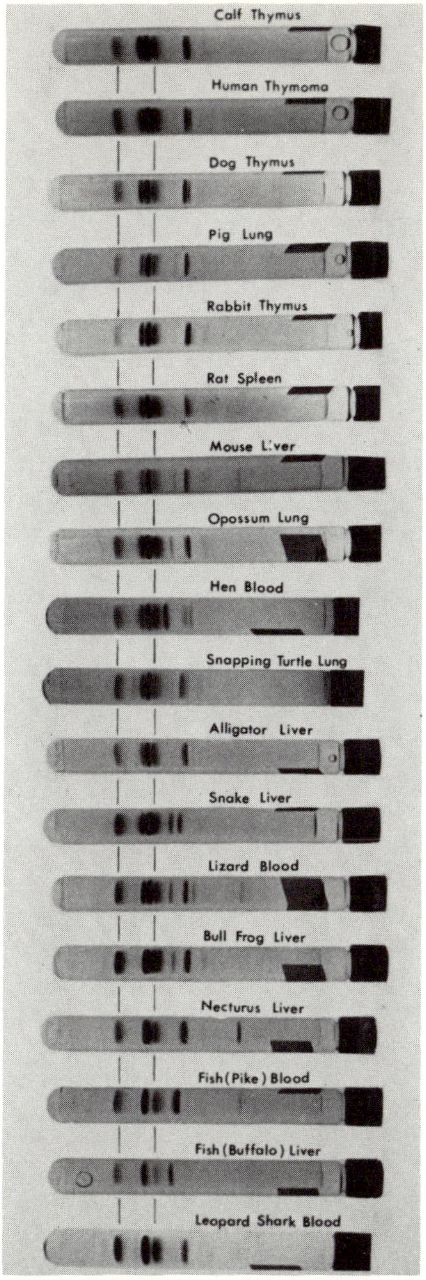

FIGURE 14. Electrophoresis of vertebrate histones. Electrophoresis was for 3½ hr at 130 V in the system of Panyim and Chalkley.[101] The direction of migration was from right to left.[46]

by Shepherd and Gurley,[330,331] minor electrophoretic differences between histones from calf thymus (or other homologous tissues) depend on the method by which the histones were isolated. Mild proteolytic degradation of histones increased their electrophoretic heterogeneity.[332] In certain extreme conditions, such as the exposure of fish to anoxia for only 3 min, the electrophoretic behavior of isolated histones changed profoundly, most likely due to active proteolysis of the histone fractions[46] (Figure 17). It is obvious that claims of histone specificity based solely on comparison of electrophoretic patterns and not supported by chemical analysis of the individual fractions should be considered only with great caution.

Of the various means for detecting tissue or species specificity of histones, perhaps the most prominent were those designed to demonstrate their immunospecificity. Because of their sensitivity to proteolysis and limited specificity, histones were found to be poor antigens. However, it was possible to immunize rabbits with covalently linked histone complexes with serum albumin[333] or with nucleic acid-histone complexes.[334,335] Stollar and Ward[334] obtained antibodies for each of the main histone fractions and, using histone antisera, they compared homologous histone fractions from human, calf, chicken, frog, and lobster tissues. Although it was possible to detect differences between the individual main types of histones, no differences were observed in the reactivity of homologous histone fractions isolated from any of these species. Similarly, Sluyser et al.[336] found no differences in antigenicity of the F1 (I) histones obtained from different tissues or species. On the other hand, Fukazawa and Shimura[337] claimed a species specificity of histones from the posterior silk glands of the silk moth (*Bombyx mori*) to be detectable by immunological techniques. However, since the silk gland histones were shown to exhibit considerably higher contents of dicarboxylic amino acids[250] than similar fractions from higher animals, the immunological specificity might have been caused by nonhistone proteins contaminating the silk gland histones. Similarly, the ammoniacal silver staining procedure reported by Black et al.[338-340] revealed a tissue specificity of histones which most likely detected differences in the binding of these proteins to chromatin DNA instead of reflecting the true differences in amino acid sequences of homologous histone fractions.

The lack of histone specificity seen among vertebrates and in some other species reflects an unusual phenomenon. Most proteins, both functional and structural, show in their amino acid sequences phylogenetic differences detectable by

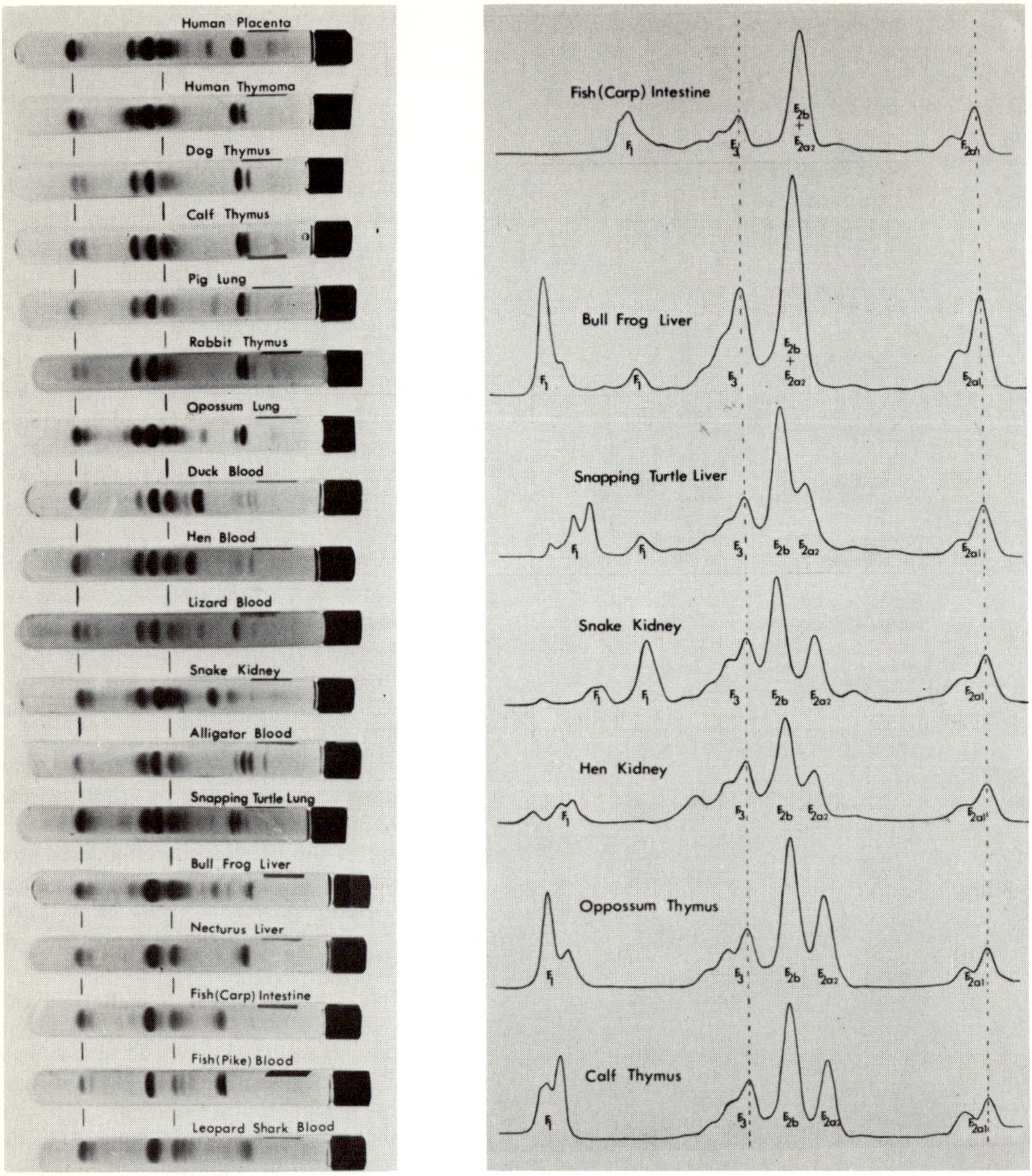

FIGURE 15. High resolution electrophoresis and microdensitometer scans of vertebrate histones. The length of migration was 20 to 25 cm.[101] Only the 25% of the gels containing the bands are shown. The migration was from the right to left.[46]

various analytical techniques. Yet, the histones, especially the arginine-rich fractions, are strikingly similar in most species. The biological significance of such similarity is not known. It appears that histones are part of a general and biologically essential mechanism and most likely participate in the highly specific structural and functional integrity of chromatin. The templates coding for histone m-RNA must then be a part of the primordial genome carried unchanged through many generations of evolution of the higher animals.

Histones in Nucleated Erythrocytes

In addition to the five histone fractions common to most vertebrate species, a limited

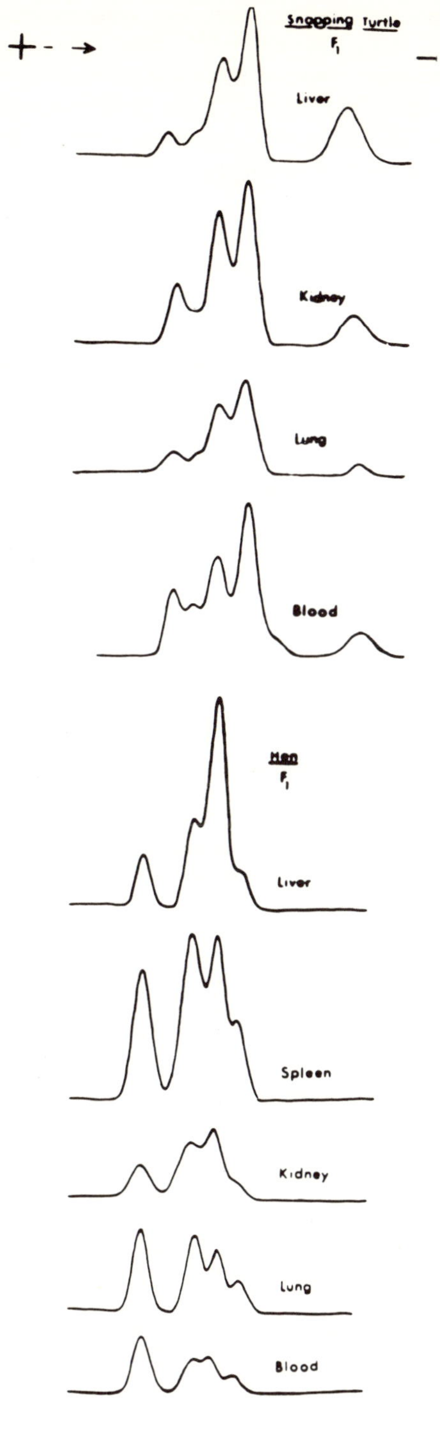

FIGURE 16. The microheterogeneity of very lysine-rich histones F1 (I) from different tissues of the turtle and the hen. High resolution electrophoresis was followed by microdensitometric analysis. The positions of various peaks indicate their relative mobilities under the conditions of the electrophoresis.[46]

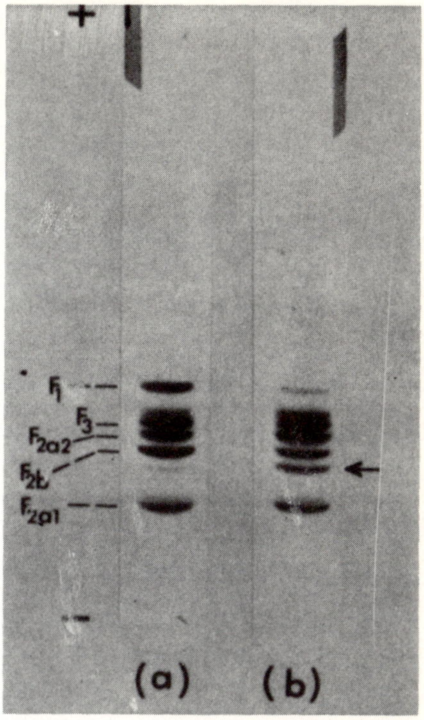

FIGURE 17. Effect of health on fish histones: (a) histones from erythrocytes of healthy Buffalo fish; (b) histones from erythrocytes of sick (anoxia) fish.[46]

number of histones specific for the tissues of their origin have been described in the literature. Most of these tissue specific histone fractions are confined to the male gametes and perhaps the best known example of extreme tissue specificity is the replacement of histones by protamines during sperm maturation in fish. The best analytically documented occurrence of a tissue specific histone in somatic cells is the presence of the serine- and lysine-rich histone F2c (V) in avian erythrocytes. The existence of this unique histone in chicken erythrocytes, as first suggested by the higher arginine content of the main histones (arginine-rich) from these cells as compared with a similar fraction from calf thymus,[62,94] is well-known. This finding has been confirmed by starch gel electrophoresis of chicken erythrocyte histones. Such gels show a characteristic electrophoretic band that is absent in other chicken tissues.[321] A histone corresponding to this electrophoretic band has been isolated and described by Neelin et al.[107,341] as fraction V, and by Hnilica[309] as fraction F2c. The amino acid composition of this histone differs from that of any of the five main histone fractions common to the vertebrates (Table 4).

The F2c (V) histone was initially isolated by chromatography on Amberlite IRC-50[107,341] or by selective extraction followed by chromatography on carboxymethyl cellulose and gel filtration on Sephadex G 75.[309] It was also obtained by dissociation of chicken erythrocyte chromatin with 0.7 *M* NaCl or diluted acid,[179,180,209] a combination of both,[185] or by extraction with 5% perchloric acid after previous removal of the F1 (I) histones with 0.5 *M* NaCl.[347] In most instances, a mixture of F1 (I) and F2c (V) histones was obtained by the selective extraction procedures. The mixture must be further resolved by ion-exchange chromatography.[343] A comprehensive fractionation procedure for avian erythrocyte histones was developed by Vidali and Neelin,[344] who demonstrated that the F2c (V) histone was not a degradation product of other proteins[345] or an artifact originating from the contamination of nuclei with ribosomes.[346] The comprehensive fractionation method was more recently modified by Adams and Neelin.[347] In polyacrylamide gel electrophoresis in the presence of urea,[101,102] the F2c (V) histone migrates with an intermediate mobility between the F1 (I) histone and the F3 (III) histone monomer (Figure 18). Because of its relatively large molecular weight, the F2c (V) histone migrates close to the F1 (I) fractions in the presence of SDS (Figure 18). The amino acid sequence of the first 31 NH_2-terminal amino acid residues of F2c (V) histone, as determined recently by Greenaway and Murray,[348,349] is shown in Figure 19.

In chicken erythrocytes, the F2c (V) histone showed an interesting polymorphism. Amino acid residue 15 was found to be either glutamine or arginine when F2c (V) histones from individual chickens were analyzed.[348] This substitution changed the overall positive charge of F2c (V)

TABLE 4

Amino Acid Composition of Some Tissue-specific Histones

Amino acid	**Nucleated Erythrocytes**					**Calf thymus**	
	Chick	Goose	Fish	Frog	Turtle	a	b
Lysine	24.9	22.6	23.9	28.7	30.3	27.8	31.3
Histidine	1.8	1.6	0.9	0.5	0.9	0.0	0.6
Arginine	11.4	11.5	7.1	9.9	3.3	2.0	2.6
Aspartic acid	1.7	2.3	3.9	3.1	3.0	2.7	3.3
Threonine	3.1	3.4	4.9	4.2	4.3	5.5	7.7
Serine	13.1	10.3	8.8	7.8	8.6	6.5	8.5
Glutamic acid	3.7	4.6	3.9	1.7	4.8	4.2	4.2
Proline	6.9	7.0	6.2	7.0	8.6	10.1	9.4
Glycine	4.7	5.4	5.6	4.7	3.6	6.9	4.3
Alanine	15.2	17.0	16.2	17.0	20.5	23.3	16.8
Valine	4.0	3.8	6.2	5.8	4.4	4.9	5.2
Methionine	0.5	0.6	0.7	0.1	0.4	-	-
Isoleucine	3.0	3.2	3.4	2.8	2.2	0.9	1.9
Leucine	4.0	4.3	5.8	5.3	2.7	4.1	2.1
Tyrosine	1.4	1.6	1.5	0.6	1.2	0.6	1.1
Phenylalanine	0.5	0.8	1.0	0.6	0.8	0.6	0.9

All values are expressed in mol %. The composition of goose erythrocyte F2c (V) histone is from Reference 345. The composition of chicken, fish (*Caranx hippos*), frog (*Rana catesbeiana*) and turtle (*Chelonia midas*) F2c (V) histones was determined in the author's laboratory.

Fraction (a) is the very lysine-rich histone F1 (I) and fraction (b) is the histone fraction found only in non-dividing tissues.[362] Both the a and b histones are from calf thymus.

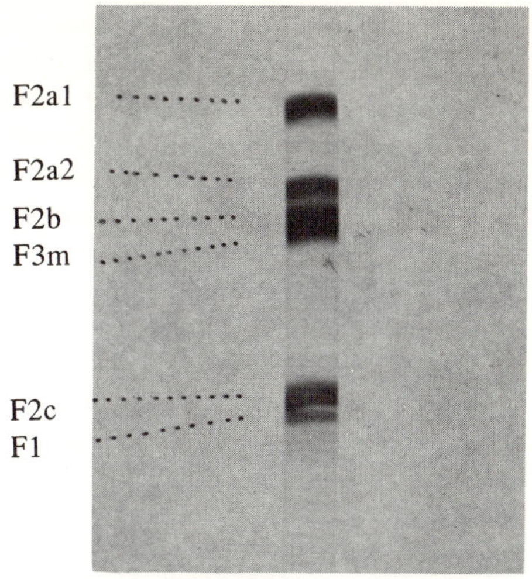

FIGURE 18A.

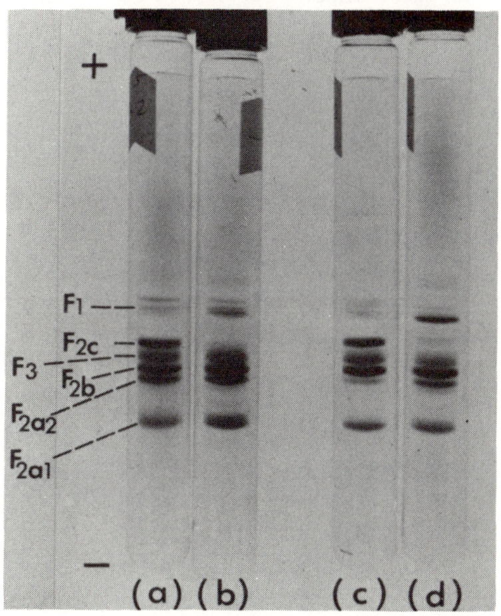

FIGURE 18B.

FIGURE 18. The erythrocyte-specific histone F2c of birds. A : polyacrylamide gel containing 0.1% of sodiumdodecyl sulfate (chicken erythrocytes) B : Urea-containing gel[101] (a) chicken erythrocytes, (b) chicken kidney, (c) duck erythrocytes, (d) duck liver. The gels in Figure 18B are from Panyim et al.[46]

histone permitting the separation of both types (histone Va and Vb) on Amberlite IRC-50 columns. Using this chromatographic screening procedure, Greenaway and Murray[348] selected a homozygous cock which they mated with a heterozygous hen. The progeny of such a mating were represented by one homozygote and five heterozygotes, indicating that there are allelic genes for the F2c (V) histone in chickens. Since the birds homozygous for F2c (V) histone appeared normal, the amino acid in position 15 of the F2c (V) histone was obviously not essential for its biological function. The authors suggested that the microheterogeneity of the F1 (I) histones might be caused by events similar to those described for the F2c (V) fraction. Indeed, the chemical similarities of both the F2c (V) and F1 (I) fractions might point to an ancestral gene common to these two histones.

The F2c (V) histone occurs in nucleated erythrocytes of several bird species, but not in other tissues from these animals.[46,107,159,345,350,351] It must undergo biosynthesis during the maturation of red blood cells, since it can be found in reticulocytes and mature erythrocytes but not in the bone marrow or spleen.[234,351,352] The unique amino acid composition of F2c (V) histone and its molecular weight of approximately 16,000 to 20,000 daltons[353,354] indicate that this fraction is not a proteolytic product of any of the five main histones.

Histones of a similar, but not identical, amino acid composition to chicken F2c (V) fraction were isolated from fish, frog,[234] and turtle[355] red blood cells and sea urchin sperm.[159,202,356] The presence of fractions similar to F2c (V) in vertebrate species with nucleated erythrocytes (fish, frogs, snakes, birds, etc.) was also detected electrophoretically by Vendrely et al.[322,357] and by Nelson and Yunis.[358] The latter authors also found this fraction in chicken liver and noted significant compositional variation of F2c (V) histones in chicken and frog, which were reflected

```
                                                    10
NH2-THR-GLU-SER-LEU-VAL-LEU-SER-PRO-ALA-PRO-ALA-LYS-PRO-LYS-

                          20
ARG/GLN-VAL-LYS-ALA-SER-ARG-ARG-SER-ALA-SER-HIS-PRO-THR-TYR-

   30
SER-GLU-HSR(MET)
```

FIGURE 19. Amino acid sequence of NH_2-terminal fragment resulting from cyanogen bromide cleavage of chicken erythrocyte F2c histone.[348,349]

in the differential solubility of this histone in dilute perchloric acid. Panyim et al.[46] made a detailed electrophoretic comparison of histones from nucleated erythrocytes and other tissues of birds, reptiles, amphibians, and fish. In accord with previous reports, the F2c (V) histone was detectable in electrophoretic patterns of histone extracts from chicken and duck erythrocytes and was absent in histone extracts from livers of the same animals. These authors were unable to detect the erythrocyte specific F2c (V) histone in snapping turtle, bull frog, and several species of fish. However, because of the chemical similarity of F2c (V) and F1 (I) histones, it is possible that one of the protein bands assigned to the latter may have represented a species specific variation of F2c (V) histone. Interestingly, the erythrocyte histones in fish were found to reflect the general well being of the animals. In buffalo fish, which were said to rapidly deteriorate in captivity, the intensity of an electrophoretic band corresponding to F2c (V) histone increased with the time the fish spent in captivity, i.e., with the decline of their health. In freshly captured specimens from which blood was isolated within seconds and rapidly frozen, the intensity of the F2c (V) histone band was dramatically reduced. In pike blood, which is not adversely affected by captivity, the F2c (V) histone was absent. This observation suggests that certain proteins with the electrophoretic mobility of F2c (V) histone may be proteolytic artifacts.

It was recently suggested by Greenaway and Murray[348] as well as Panyim et al.[46] that the F2c (V) histone present in erythrocytes may be related to the F1 (I) histones common to other somatic tissues. This is in accord with the findings of Dick and Johns[352] who reported that F1 (I) histones were completely replaced by the F2c (V) fraction during the process of erythrocyte maturation. As seen in Figure 18, chicken erythrocyte histones have the F1 (I) histone protein bands considerably reduced or absent. A similar replacement of F1 (I) histones in fish erythrocytes (common Jack) by a protein resembling the F2c (V) fraction in chicken was reported by Hnilica and Edwards[359] (Table 4). On the other hand, Panyim et al.[46] obtained F1 (I) histones of quite similar amino acid composition from calf thymus, chicken, bull frog, and carp blood. Additional experimental evidence seems warranted to show that the F2c (V) histone-like proteins in nucleated erythrocytes in reptiles, amphibians, and fish were either artifacts resulting from proteolytic degradation of other histone fractions or that they could be found only in some species while being absent in others. Although experiments reported by Panyim et al.[46] showed an increase of protein in bands corresponding to the F2c (V) histone in the buffalo fish during adverse conditions, this fraction cannot be derived by proteolysis from any of the common five histones. The F1 (I) histones are the only histones large enough to be degraded to a fragment similar in its size to the F2c (V) histone. However, the relatively high arginine content of the F2c (V) fraction excludes this possibility.

Presently, we do not understand the biological reasons for the existence of a histone fraction specific for nucleated erythrocytes. It has been suggested that this histone may accomplish the additional template restriction of DNA seen in chromatin from nucleated red blood cells. This resembles the appearance of protamines in trout sperm during its maturation which is also accompanied by a virtual inactivation of DNA transcription.[360] An increasing number of excursions into the biochemistry of nuclear proteins in primitive plants or animals show that early in evolution the fixed compositional pattern of histones found in the vertebrates had yet to become established. The erythroid cells are phylogenetically young, first appearing in chordates. Thus, the F2c (V) histones may represent the transcription of some ancestral genes normally repressed in most tissues and animals. Since our knowledge about the genetic template derives from studies on the products of operative genes (giving rise to the phenotype), it is possible for the repressed F2c (V) histone genes to go undetected until their restriction is released. The presence of protein similar to F2c (V) histones (assuming that these proteins are not proteolytic artifacts) in frog liver,[358] frog lens cells,[361] the posterior silk gland of *Bombyx mori L.*,[249,250] and sea urchin sperm,[159,202,204,310] in addition to the nucleated erythrocytes of birds and perhaps other vertebrate species, suggests such a possibility.

Another specific histone fraction detectable by polyacrylamide gel electrophoresis, the presence of which was substantiated by amino acid analysis, is a lysine-rich histone described by Panyim and Chalkley.[362] It was reported that this fraction is only present in mammalian tissues which do not replicate. In more rapidly dividing tissues such as thymus or Ehrlich ascites tumor, this histone could not be found by the authors. Although it can be classified as a typical very lysine-rich histone, its amino acid composition differs from that reported for the F1 (I) histone subfractions (Table 4). The main difference is the presence of histidine and the relatively low alanine content. [The lysine to alanine ratio is 1.86 as compared with 1.1 to 1.2 for the total F1 (I) histone.] However, very lysine-rich histones containing histidine with lysine to alanine ratios of 1.50 and 1.63 were reported in Walker carcinosarcoma[50] and in the sea urchin embryos,[239] i.e., rapidly dividing tissues. Because of its relatively small size, this new fraction may be a fragment resulting from a limited degradation of F1 (I) histones.

Although it is certain that there are species and tissue specific histones and basic proteins (both already discovered and still awaiting their discovery), it can be concluded that with the progress of evolution, the pattern of histones present in most animal tissues became more stable, reaching almost complete evolutional inertia in the arginine-rich F2a1 (IV) histone and perhaps in some of the other fractions as well.

Protamines

One of the most extreme examples of tissue specificity of basic proteins is the presence of protamines in the mature spermatozoa of a great many fish species. Although it is difficult to draw an exact line dividing histones from basic proteins in spermatozoa, typical protamines are easily characterized. The chemical difference between histones and protamines was sufficiently great to allow Kossel[3] to distinguish between these two protein groups despite the primitive analytical tools available almost a century ago. Protamines are frequently characterized as large basic polypeptides (average molecular weight 5,000 daltons), with arginine constituting more than 50% of all amino acid residues. Protamines are almost exclusively confined to the spermatozoa of a great variety of fish species. A protamine-like protein (galline) has, however, been described in rooster sperm.[363,364] Kossel,[3] who had an excellent knowledge about these proteins, classified protamines into three groups according to their content of basic amino acids, i.e., monoprotamines (arginine only), diprotamines (arginine and either lysine or histidine), and triprotamines (arginine, lysine, and histidine).

Chemically, protamines behave like large peptides, and spermatozoa of one fish species usually contain a mixture of closely related proteins categorized under one name derived from the species of origin (e.g., *Clupeus harengus*-clupeine, *Salmo salar*, or other species of salmon-salmine, *Scomber scombrus*-scombrine, etc.). Similar to the histone proteins, protamines can be extracted from spermatozoa nuclei by dilute inorganic acids and further purified or fractionated by chromatography, countercurrent distribution, etc.[15,16,214,365,366]

Typical protamines are composed of a relatively simple sequence of polyarginine interrupted by the occasional insertion of other amino acids. The NH_2-terminal amino acids of protamines are almost exclusively proline or alanine.[15,16] The NH_2-terminal amino acid of muguline is arginine while a mixture of alanine and glutamic acid has been found in sturine.[15] Present research on protamines has centered, especially after the death of Kurt Felix more than a decade ago, almost entirely in the laboratories of Toshio Ando in Japan and G. H. Dixon in Canada. Through the efforts of Ando and his associates, the complete amino acid sequence of seven protamines has been determined. These are clupeine components Y-I, Y-II, and Z from the sperm heads of herring *Clupea pallasii*, iridine components Ia, Ib, and II from the rainbow or steelhead trout (*Salmo gairdnerii* or *irideus*) and salmine component AI from the chum or keta salmon (*Oncorhynchus keta*). The amino acid sequence of these seven protamines as determined by Ando et al.[367-371] is shown in Figure 20.

The determination of the amino acid sequence of clupeine component Z by Ando et al.[368] permitted speculations on the possible phylogenetic origin of protamines. Using an analytical approach developed by Smithies et al.,[372] Black and Dixon[373] compared the incidence of maximal homology between the peptides of clupeine component Z. They arrived at the conclusion that since there was a fundamental subunit of four arginine residues with one nonbasic amino acid which occurred several times in the sequence of this protein, this sequence (ALA-ARG-ARG-ARG-ARG) represents the composition of an ancestral peptide from which clupeine component Z evolved.

Their evolutionary scheme, as accomplished by a sequence of partial or complete gene duplica-

Amino acid sequences of protamines as determined by Ando et al.[367-371]

Clupeine Y-I

10 20
ALA-ARG-ARG-ARG-ARG-SER-SER-SER-ARG-PRO-ILE-ARG-ARG-ARG-ARG-PRO-ARG-ARG-ARG-THR-
30
THR-ARG-ARG-ARG-ARG-ALA-GLY-ARG-ARG-ARG-ARG

Clupeine Y-II

10 20
PRO-ARG-ARG-ARG-THR-ARG-ARG-ALA-SER-ARG-PRO-VAL-ARG-ARG-ARG-ARG-PRO-ARG-ARG-VAL-
30
SER-ARG-ARG-ARG-ARG-ALA-ARG-ARG-ARG-ARG

Clupeine Z

10 20
ALA-ARG-ARG-ARG-ARG-SER-ARG-ARG-ALA-SER-ARG-PRO-VAL-ARG-ARG-ARG-ARG-PRO-ARG-ARG-
30
VAL-SER-ARG-ARG-ARG-ARG-ALA-ARG-ARG-ARG-ARG

Salmine A-I

10 20
PRO-ARG-ARG-ARG-ARG-SER-SER-SER-ARG-PRO-VAL-ARG-ARG-ARG-ARG-ARG-PRO-ARG-VAL-SER-
30
ARG-ARG-ARG-ARG-ARG-ARG-GLY-GLY-ARG-ARG-ARG-ARG

Iridine Ia

10 20
PRO-ARG-ARG-ARG-ARG-SER-SER-SER-ARG-PRO-VAL-ARG-ARG-ARG-ARG-ARG-PRO-ARG-ARG-VAL-
30
SER-ARG-ARG-ARG-ARG-ARG-ARG-GLY-GLY-ARG-ARG-ARG-ARG

Iridine Ib

10 20
PRO-ARG-ARG-ARG-ARG-ARG-ARG-SER-SER-SER-ARG-PRO-ILE-ARG-ARG-ARG-ARG-PRO-ARG-ARG-
30
VAL-SER-ARG-ARG-ARG-ARG-ARG-ARG-GLY-GLY-ARG-ARG-ARG-ARG

Iridine II

10 20
PRO-ARG-ARG-ARG-ARG-SER-SER-SER-ARG-PRO-VAL-ARG-ARG-ARG-ARG-ALA-ARG-ARG-VAL-SER-
30
ARG-ARG-ARG-ARG-ARG-ARG-GLY-GLY-ARG-ARG-ARG-ARG

FIGURE 20. Amino acid sequences of protamines as determined by Ando et al.[367-371]

tions and amino acid substitutions, is presented in Figure 21. Although this attractive scheme is hypothetical, the complete amino acid sequences of seven protamines (Figure 20) show that the X-ARG-ARG-ARG-ARG peptide indeed occurs several times in every one of the protamines. More information on the similarities in amino acid sequences of protamines and several histone fractions together with some speculations on their evolution can be found in the recent review of histone structure by Phillips.[77]

The amino acid composition of protamines remains remarkably constant for the same species regardless of the geographical location[374] or age[375,376] of the organism. Iridine isolated from three individual fish of the same species has been found to be heterogeneous[377] to a similar extent, thereby indicating that the heterogeneity of iridine is not artificially introduced by mixing spermatozoa of many fish of the same species. A similar situation has been found for clupeine from *Clupea pallasii.*[378]

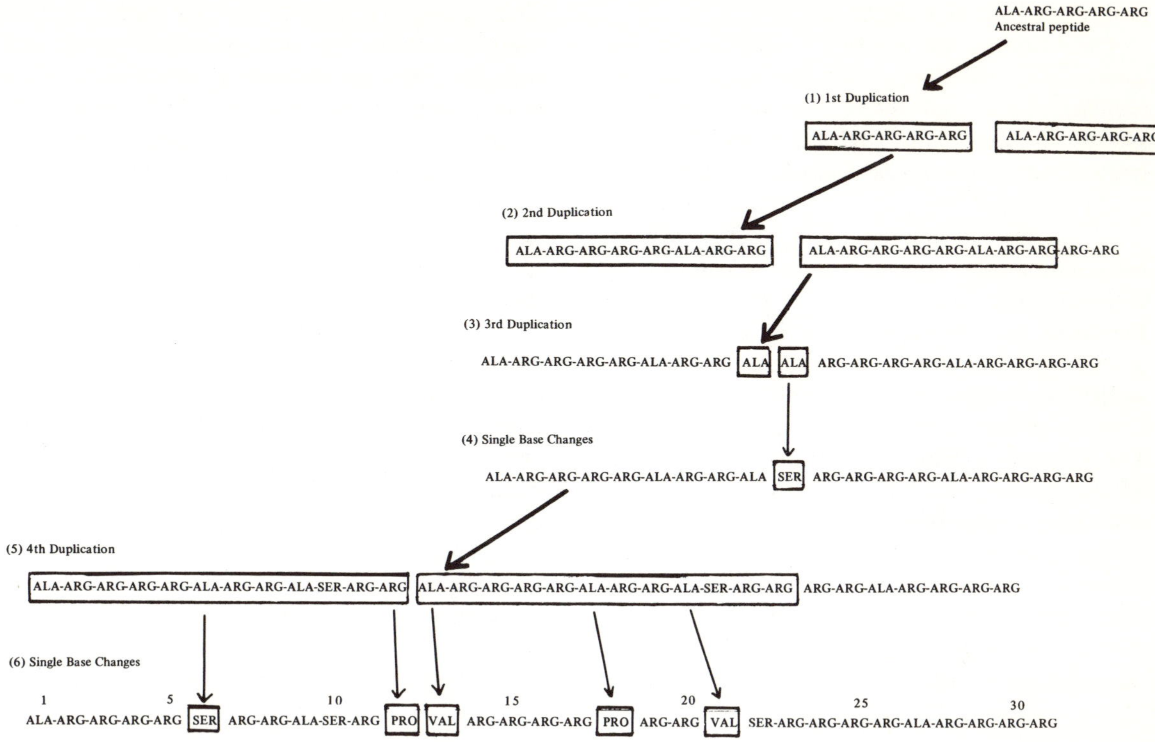

FIGURE 21. Possible evolutionary scheme for Clupeine component Z from *Clupea pallassii.*[373] The present amino acid sequence of Clupeine component Z may have evolved from an ancestral pentapeptide ALA–ARG–ARG–ARG–ARG by a series of partial gene duplications and single base mutations in the structural gene of this protein. The underlined segment of polypeptide corresponds to the portion of structural gene which was partially duplicated. The resulting new polypeptide sequence is enclosed in a box. Single arrows indicate single base mutations. Double arrows show partial gene duplications.

In the nucleus of fish spermatozoa, protamines are associated with DNA in the form of chromatin or nucleoprotamine. Before elucidation of the complete amino acid sequences of several protamines, it had been assumed that the arginine residues were regularly spaced along the protamine peptide opposing the DNA phosphates in nucleoprotamine. The occasional neutral amino acids were presumably bent outward. According to the early studies of Wilkins,[379] the polypeptide chain in nucleoprotamine has an extended β-form and follows the shallow groove of DNA. The number of basic amino acids in nucleoprotamines approximately equals the number of phosphate residues in DNA, indicating its almost complete electrostatic saturation.[15] The strong binding between the DNA and protamines supports the proposal of a periodic distribution of arginine residues along the protamine sequence[380] to achieve a parallel alignment of these polypeptides with the DNA. However, recent studies on the amino acid sequences of protamines supported by the measurements of affinities between the DNA and protamines carried out in concentrated salt solutions[381] speak against such simple conformation of nucleoprotamines.

Since protamines appear during the differentiation of male reproductive cells into metabolically inert spermatozoa, the much higher content of arginine in protamines as compared with that of the histones present in immature spermatocytes, was frequently correlated with the transcriptional inactivation of DNA. Also, it had been suggested that the arginine-rich protamines provided a means for the tight packing of the DNA in sperm heads as evidenced by many cytochemical studies. Unfortunately, protamines appear to be limited to fish and a wide spectrum of proteins consisting of protamine-like polypeptides, somatic histones, arginine-rich keratinoids, and perhaps nonhistone proteins was found in other animal species.[382]

Basic Proteins in Male Gametes

Early studies by Hultin[383] and Hultin and Herne[384] showed that in two species of molluscs (*Patella vulgata* and *Patella coerulea*) and three species of echinoderms (*Arbacia lixula, Brissopsis lyrifera*, and *Echinocardium cordatum*) the sperm nuclei contained proteins more similar to histones than to the protamines in fish. The average molecular weight of such proteins was estimated to be three or four times that of protamines and the amino acid composition of the basic proteins demonstrated an abundance of arginine and lysine in the two molluscs species and of lysine, arginine, alanine, glycine, and serine in the echinoderms. The presence of histone-like proteins in the nuclei of sea urchin sperm was confirmed by Hamer[385] and other investigators.[386,387]

Later, detailed analysis of the basic proteins isolated from the spermatozoa of sea urchins demonstrated the presence of typical histones in the sperm heads. In addition to the histones common for the tissues of advanced embryos and adult animals, a histone fraction specific for the spermatozoa was reported. In *Arbacia punctulata*, this sperm-specific histone (γ) is relatively rich in lysine, arginine, and alanine.[202] Its amino acid composition, together with the composition of the other basic proteins found in the spermatozoa of various animals, is shown in Table 5. A histone similar to the γ fraction was isolated from the sperm of *Arbacia lixula*[356] and *Strongylocentrotus purpuratus.*[159,204,310] The presence of the sperm-specific histone fraction (γ) may serve to additionally repress the DNA of sperm chromatin as compared with somatic tissues (average templating efficiency of sea urchin sperm chromatin is low, 0 to 2% of free DNA, as compared with 10 to 20% in embryonal chromatin[27,100,388]), since its removal with 0.7 to 1.0 *M* NaCl produced a spectacular increase in the templating efficiency of the treated chromatin accompanied by a considerable destabilization as reflected in the lowering of its thermal denaturation temperature.[199] Interestingly, the sea urchin

TABLE 5

Amino Acids Composition of Arginine-rich Sperm Proteins

	Origin						
Amino acid	(a) Clupeine	(b) Galline	(c) *Patella vulgata*	(d) *Patella coerulea*	(e) Boar	(f) Bull	(g) *Urechis caupo*
Lysine	–	–	**20.1**	8.0	3.6	–	**19.2**
Histidine	–	1.6	0.6	–	2.4	2.1	0.1
Arginine	**74.1**	**60.0**	**25.0**	**45.7**	**25.8**	**51.0**	**36.9**
Aspartic acid	–	0.8	–	–	5.1	–	0.9
Threonine	2.1	2.4	4.4	1.7	6.5	6.3	1.6
Serine	4.6	11.2	9.4	12.0	6.4	4.2	22.4
Glutamic acid	–	1.6	–	–	6.5	2.1	0.8
Proline	7.5	5.6	9.4	12.0	5.3	–	0.9
Glycine	–	8.0	9.4	8.0	5.0	4.2	2.7
Alanine	6.6	1.6	12.5	5.1	5.9	2.1	11.4
Half cystine	–	–	–	–	**6.5**	**12.6**	–
Valine	3.7	1.6	4.4	3.4	5.1	4.2	2.0
Methionine	–	–	–	–	1.1	–	–
Isoleucine	1.2	0.8	1.2	0.6	3.8	2.1	0.3
Leucine	–	–	3.7	3.4	4.9	2.1	0.4
Tyrosine	–	4.8	–	–	4.3	4.2	0.2
Phenylalanine	–	–	–	–	1.7	2.1	0.2

All values are expressed as mol %. The data were taken or recalculated from the following: (a) Reference 16, (b) References 16 and 369, (c) and (d) References 16 and 384, (e) Reference 409, (f) Reference 410, and (g) Reference 391. Distinguishing features are printed in boldface type.

sperm specific γ histone resembled in its amino acid composition the F2c (V) histone found in nucleated erythrocytes. The amino acid composition of several sea urchin sperm histones is presented in Table 8.

Subirana and Palau[386] analyzed the sperm proteins of ten species of echinoderms (three of *Echinoidea*, one of *Ophiuroidea*, four of *Asteroidea*, and two of *Holothuroidea*) and found all of them to contain typical histone-like proteins with only 8.0 to 13.5 mol % of arginine (a sample of calf thymus histone contained 9.5 mol % of arginine). In an extension of these studies, Subirana et al.[389,390] concluded that in echinoderms, most somatic histones are preserved in the sperm and not replaced by protamine-like proteins. In *Holothuroidea*, almost no changes occurred in the histones during spermiogenesis. In other echinoderm species, specific histone fractions appeared in addition to the somatic histones which were conserved except for the F1 (I) fraction which was replaced by the sperm-specific proteins. In molluscs (*Mytilus edulis*), the somatic histones exhibited considerable specificity and contained fractions rich in arginine and lysine (24.5 and 23.5 mol %, respectively), as well as histone similar to the F2c (V) fraction in nucleated erythrocytes (Table 6). During spermatogenesis, the arginine-lysine-rich histones appeared to be conserved (Table 7). In cephalopods and gastropods, the amino acid composition of sperm proteins varied from species to species, being either histone-like or protamine-like (Table 7).

In addition to the sperm-specific γ-histone fraction in sea urchins, histone-like proteins containing approximately 20 mol % of both arginine and lysine were found in the acrosomes of the spermatozoa from the marine mud snail *Urechis caupo*.[391] Their amino acid composition, shown in Table 8, resembled proteins found in the spermatozoa and muscle of the mollusc *Mytilus edulis*[390] and in the sperm of sea urchin *Strongylocentrotus purpuratus*.[392]

According to the recent findings from Dixon's laboratory,[393,394] testicular histones from the rainbow trout (*Salmo gairdnerii*) contained small amounts of highly specific proteins which can be classified as histones. One of them, named histone T, represented less than 1% of the total histone content and exhibited a very peculiar amino acid composition by lacking histidine, cysteine, methionine, isoleucine, tyrosine, and phenylalanine. It was relatively rich in lysine and proline (Table 8) and had proline as the NH_2-amino acid terminus. Another protein, also present in small quantities, was relatively rich in lysine and aspartic acid (Table 8). The biological significance of these testicular histone fractions is presently unknown. It would be of considerable interest to determine whether these specific histones can be linked to a particular cellular type comprising the salmon testes.

The appearance of arginine-rich proteins, partially or completely replacing histones during spermatogenesis, was reported in the snail (*Helix aspersa*),[395] sea snail (*Urechis caupo*),[391] squid (*Loligo opalescens*),[396] *Drosophila melanogaster*,[397,398] grasshopper (*Chortophaga viridifasciata de Geer*),[399] and other species.[382] Similar transitions occur during the sperm maturation in higher vertebrates such as mice,[400,401] rats,[402,403] and bull.[404] Since all

TABLE 6

Amino Acid Composition of Sperm Histones from *Mytilus edulis*[390]

Amino acid	Φ1	Φ3	Φ2a	Φ2b
Lysine	**23.5**	**27.6**	**11.1**	**12.9**
Histidine	0.4	1.6	1.8	2.2
Arginine	**24.5**	**9.2**	**11.6**	**9.2**
Aspartic acid	1.3	3.3	6.7	7.2
Threonine	4.0	4.0	4.5	6.1
Serine	**16.3**	**8.1**	3.4	7.7
Glutamic acid	1.3	5.2	8.1	8.1
Proline	5.1	7.1	2.6	4.1
Glycine	5.7	4.4	12.5	6.9
Alanine	**13.5**	**15.4**	10.1	10.7
Half cystine	Tr.	Tr.	Tr.	0.0
Valine	1.7	3.3	6.7	6.2
Methionine	Tr.	0.6	0.8	1.5
Isoleucine	0.8	3.3	5.0	4.2
Leucine	1.0	3.7	9.4	6.3
Tyrosine	0.3	1.3	3.3	3.4
Phenylalanine	0.3	1.5	2.1	1.7

All values are expressed in mol %.

The fraction Φ3 resembles the chicken-erythrocyte F2c (V) fractions. Φ2a and Φ2b are typical somatic histones resembling fractions F2a (IIa1 and IV) and F2b (IIb2). Distinguishing features are printed in boldface type.

TABLE 7

Amino Acid Composition of Basic Sperm Proteins from Various Molluscs[390]

Amino acid	Species					
	Cryptochiton stelleri	Mytilus edulis	Gibbula divaricata	Haliotis tuberculata	Loligo forbesi	Eledone cirrosa
Lysine	**15.5**	**21.3**	7.2	**10.7**	**2.6**	**5.1**
Histidine	2.8	0.6	0.2	1.2	1.8	7.0
Arginine	**34.4**	**19.3**	**53.6**	**25.0**	**61.7**	5.3
Aspartic acid	3.2	2.8	0.7	5.7	2.4	4.4
Threonine	4.1	4.7	2.2	4.2	1.3	3.6
Serine	4.2	**15.0**	**15.0**	**10.5**	**9.9**	4.9
Glutamic acid	2.0	2.8	0.9	5.6	2.9	6.2
Proline	4.4	5.7	0.2	3.2	3.1	**23.6**
Glycine	5.0	6.2	5.5	5.4	2.0	7.3
Alanine	10.3	12.8	9.9	8.8	1.7	4.7
Half cystine	ND	Tr.	0.1	0	0.5	**16.5**
Valine	6.8	2.6	2.6	5.7	1.2	4.0
Methionine	ND	0.5	0.2	1.7	0.2	0.2
Isoleucine	} 4.5	1.5	0.5	3.2	1.0	2.6
Leucine		2.4	0.7	4.3	1.4	3.0
Tyrosine	1.2	0.6	0.3	2.8	5.8	1.0
Phenylalanine	1.5	0.7	0.2	1.6	0.7	0.8

All values are expressed in mol %. Distinguishing features are printed in boldface type. Basic proteins in spermatozoa of the echinoderm *Holothuria tubulosa* are virtually indistinguishable from the somatic histones.[389]

TABLE 8

Amino Acid Composition of Specific Histone Fractions in Spermatozoa

Amino acid	(a) Sea urchin γ	(b) Sea urchin A	(c) *Urechis caupo* A	(d) Sea urchin H3	(e) Trout T	(f) Trout R
Lysine	**27.5**	**20.2**	**23.2**	**11.9**	**23.1**	**17.0**
Histidine	0.3	1.7	2.1	2.3	0.0	1.5
Arginine	**10.3**	**19.0**	**21.3**	**20.1**	**7.2**	4.5
Aspartic acid	2.8	3.4	4.7	4.6	6.7	16.0
Threonine	2.4	4.8	2.7	6.3	1.6	2.9
Serine	5.7	7.6	7.4	**11.2**	5.6	5.2
Glutamic acid	3.1	4.2	4.1	5.0	6.1	10.5
Proline	8.4	2.8	2.2	2.0	**12.3**	7.5
Glycine	7.1	4.2	4.1	10.0	7.4	7.9
Alanine	**19.3**	**18.9**	**15.2**	7.8	**25.4**	10.5
Half cystine	Tr.	0.0	0.4	–	–	1.2
Valine	3.3	4.5	4.8	5.8	3.4	3.5
Methionine	1.5	1.3	0.0	1.2	0.0	2.0
Isoleucine	3.0	2.5	2.1	4.2	0.0	1.5
Leucine	1.9	2.7	3.1	4.0	1.2	2.3
Tyrosine	0.7	1.1	1.1	2.3	0.0	2.3
Phenylalanine	0.6	0.8	1.4	1.1	0.0	3.8

All values are expressed in mol %.

(a) γ-fraction from the sperm of *Arbacia punctulata*[202]
(b) *Strongylocentrotus purpuratus* acrosomal histone
(c) *Urechis caupo* acrosomal histone[391]
(d) *Strongylocentrotus purpuratus* sperm histone fraction H3
(e) Trout testis histone T[394]
(f) Trout testis histone R[393]

these observations were based on cytochemical techniques, the chemical identity of the arginine-rich proteins replacing histones could not be determined from such studies. On the other hand, cytochemical studies on the nuclei during the spermatogenesis of the leopard frog (*Rana pipiens*) showed only the presence of a somatic kind of histone during each stage, including mature sperm.[405] This finding was recently confirmed by electrophoretic analysis of histones during the spermatogenesis of the toad *Bufo americanus.*[406] Although the sperm histones contained an additional, electrophoretically fast protein band, this protein was not rich in arginine.

Chemical studies using the extraction and analysis of proteins showed that in mammalian species, the sperm DNA appears to be enmeshed in a network of arginine-rich keratinoid resembling a sponge.[407] Released by oxidation with performic acid, the keratinoid protein in bull spermatozoa could be fractionated into three fractions with varying proportions of arginine, lysine, and cysteine. The ratio of total basic amino acids to the negative phosphate groups of DNA was unity indicating that the arginine-rich keratin-like protein serves a purpose similar to that of protamines in fish sperm.

A more detailed fractionation of the arginine-rich keratins was described by Henricks and Mayer,[408,409] who disrupted bovine or porcine spermatozoa with mercaptoethanol. By varying the mercaptoethanol concentration, the authors obtained several protein fractions subsequently soluble in 0.1 *N* HCl, differing from each other by their arginine and cysteine content. Representatives of these proteins are shown in Table 5. The mercaptoethanol fractions were electrophoretically heterogeneous on polyacrylamide gels. Diluted (0.3 *N*) HCl solubilized protein which was not basic in amino acid composition indicating the complete absence of histones. More recently, Coelingh et al.[410] purified a protein from bull sperm heads (Table 5) which was homogeneous by sedimentation analysis and had alanine and glutamine as NH_2- and COOH-terminal amino acids, respectively. From their data, the authors suggested this protein has approximate molecular weight of 6,200 daltons, is homogeneous, and has the following amino acid composition: ALA (ARG_{24}, CYS_6, THR_3, TYR_2, SER_2, GLY_2, VAL_2, LEU_1, ILE_1, PHE_1, HIS_1)GLN. Except for the six cysteine residues, this protein is similar in composition to the protamines. The existence of such a protein in bull spermatozoa supports the hypothesis advanced by Bril-Petersen and Westenbrink[407] that in the compact nucleus of bull spermatozoa, the DNA is enmeshed in a network of keratinoid fibers, rich in arginine. The network is most likely formed by identical subunits of the Coelingh's protein which are cross-linked by numerous disulfide bridges.

A basic nuclear protein of small molecular weight (5,600 ± 400 daltons) can be extracted with acid from the population of spermatids in mouse testes.[411] This protein is synthesized late during spermatogenesis and can be obtained from mature spermatozoa only after their disruption with mercaptoethanol. Its low molecular weight and relatively simple amino acid composition (ARG_{10}, LYS_{10}, HIS_2, ASP_2, THR_4, GLU_4, GLY_5, ALA_4, LEU_1) suggest its similarity to protamines. In contrast to the arginine-rich keratinoid from bull and boar spermatozoa, this protein in mice does not contain cysteine. A low molecular protein of similar amino acid composition has been recently isolated from rat testis in the author's laboratory.

A quite peculiar situation was found to exist in the sperm of various decapod crustaceans. Several investigators reported the presence of basic proteins in the cytoplasm.[382,412-417] The sperm nucleus of these species did not contain histones or protamines and the basic protein localized in the cytoplasm (decapodine) might represent sloughed nuclear histones. Das et al.,[391] who described arginine- and lysine-rich proteins in *Urechis caupo* acrosomes (Table 8), suggested that the cytoplasmic capsule in crustaceans, which contains basic proteins in large quantity, may be homologous with the *Urechis* acrosome. Indeed, according to Vaughn et al.,[412,413] cytoplasmic capsules in the sperm of *Emerita analoga* contain acid extractable proteins rich in both arginine and lysine. The cytoplasmic capsules, as well as sperm acrosomes, also contain considerable quantities of polysaccharides, which may bind the basic proteins. Polyacrylamide gel electrophoresis of acid soluble proteins from cytoplasmic capsules revealed their similarity to somatic histones. The authors proposed that during spermatogenesis, the histones were discarded from maturing spermatocytes and enclosed in the cytoplasmic capsules. Mature crab spermatozoa contain no cytochemically detectable histones or protamines and perhaps no DNA-bound protein at all.[414-417] The nucleoprotein isolated from mature sperm of the spider crab *Libinia emarginata* exhibited a thermal denaturation behavior essentially identical to that of free DNA.[418] This indicates that in the sperm of certain crustaceans, the DNA is not stabilized by basic proteins. It would be of considerable interest to study the biochemical and templating properties of crab sperm chromatin.

The changes in the chemical composition of the DNA-associated basic proteins during spermatid maturation in various animals present a serious challenge to the students of biological functions of histones and similar proteins. Disregarding the very important exception of the absence of basic proteins in sperm chromatin from various crustaceans, there is a general trend toward the replacement of somatic histones by relatively simple and quite basic arginine-rich proteins. The high arginine content appears to be, at least at the present time, the only feature common to the spermatogenesis of the many species analyzed. This change occurs late in spermatogenesis, i.e., during the absence of DNA replication. In contrast to the phylogenetically conservative histones, the genes coding for arginine-rich proteins in spermatozoa exhibit exceptional variations even within small taxonomic groups. The reasons for this phylogenetic variability are unknown. Assuming that every somatic cell within the organism is endowed with exactly identical DNA, the arginine-rich protein genes must be held inactive in all cells except a few undergoing terminal differentiation into spermatozoa. Since the biosynthesis of protamines, which will be discussed later, follows patterns known for the biosynthesis of other cellular proteins, it must be assumed that it is also ruled by similar regulatory mechanisms. Because of the unique product resulting from the transcription of arginine-rich protein genes, studies on the restriction of such genes may contribute considerably to the understanding of DNA transcription in eukaryotes.

The most favored explanation for the late appearance of protamines and other arginine-rich proteins during the schedule of spermatogenesis is to provide a better packing of DNA in spermatozoa, perhaps together with its transcriptional inactivation. This hypothesis has been recently challenged by Subirana et al.,[389,390] who found virtually no changes in the pattern of somatic histones during sperm maturation of two species of *Holothuroidea* (sea cucumbers). In spite of these small changes in the composition of somatic histones during spermatogenesis, the nucleus undergoes condensation typical for other species.[419] A similar situation has been described in plants (*Lilium longiflorum* and *Tulipa gesneriana*) where no arginine-rich proteins appear during meiosis, microsporogenesis, and pollen maturation. A specific protein, termed meiotic histone, is similar in its composition to the lysine-rich histone F1 (I).[420] In addition to this meiotic histone, a specific DNA-binding protein appears in meiotic cells of *Lilium*.[421] This protein exhibits a very high affinity for single-stranded DNA and it can be formed only during the prophase stages of meiotic cells. It occurs in the nuclei, most likely in association with lipids and it catalyzes rapid renaturation of the DNA, perhaps aiding the meiotic recombination of chromosomes.

The differentiation of chromatin in histone-containing spermatozoa does not advance to the extreme degree of condensation seen in spermatozoa of species with arginine-rich proteins since it stops at chromatin fibrils of about 200 A in diameter.[418] According to Zirkin,[422] the chromatin fibers in histone-containing sperm nuclei of the leopard frog, sea urchin, and goldfish average only 160 to 180 A in diameter as compared with about 250 A diameter fibers of somatic nuclei, i.e., exhibiting a lesser degree of supercoiling (Figure 22). It can be concluded that the arginine-rich proteins in sperm chromatin may not affect the condensation of DNA. Pladellorens and Subirana[418] proposed that arginine-rich sperm proteins may be involved in the removal of nonhistone proteins and other macromolecules associated with the DNA in metabolically active cells and perhaps not required in spermatozoa. During the sperm maturation of cancer crabs, the DNA becomes free of any detectable basic proteins and resembles, in electron microscope, the DNA in bacterial chromosomes.[414]

Basic Proteins in Female Gametes

The existence of histones or protamines in maturing oocytes or in mature eggs has not been demonstrated unequivocally. Since the eggs of higher animals are not available in quantities sufficient for chemical investigations, most of the studies on nuclear proteins in female gametes have been performed on echinoderms, molluscs, amphibians, or fish. Unfortunately, in these

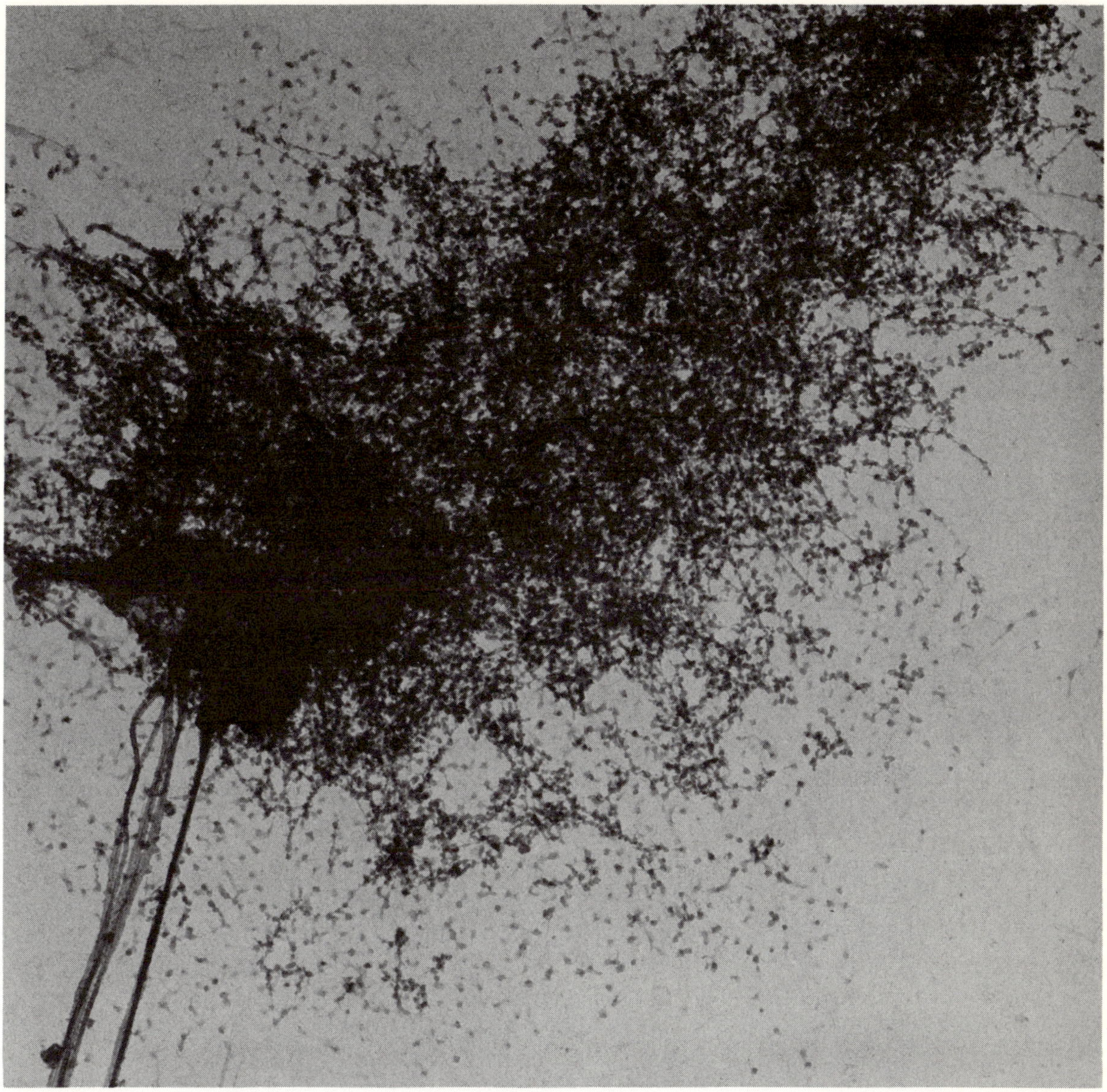

FIGURE 22A.

FIGURE 22. Chromatin from histone-containing sperm nuclei. (a) Isolated sea urchin sperm (X 26,000); (b) A portion of an isolated sea urchin sperm nucleus. The nucleus is composed of fibers whose average diameter is 164 A (X 100,000) (c) For comparison, a portion of an isolated bovine kidney chromosome is shown. The fibers average 264 A in diameter (X 100,000).[422]

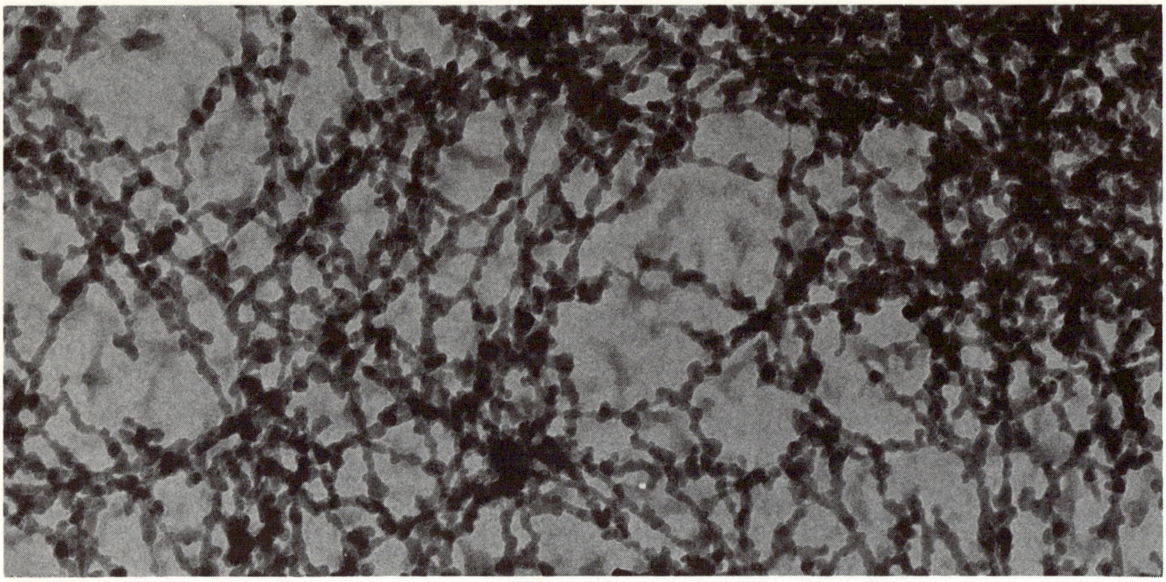

FIGURE 22B.

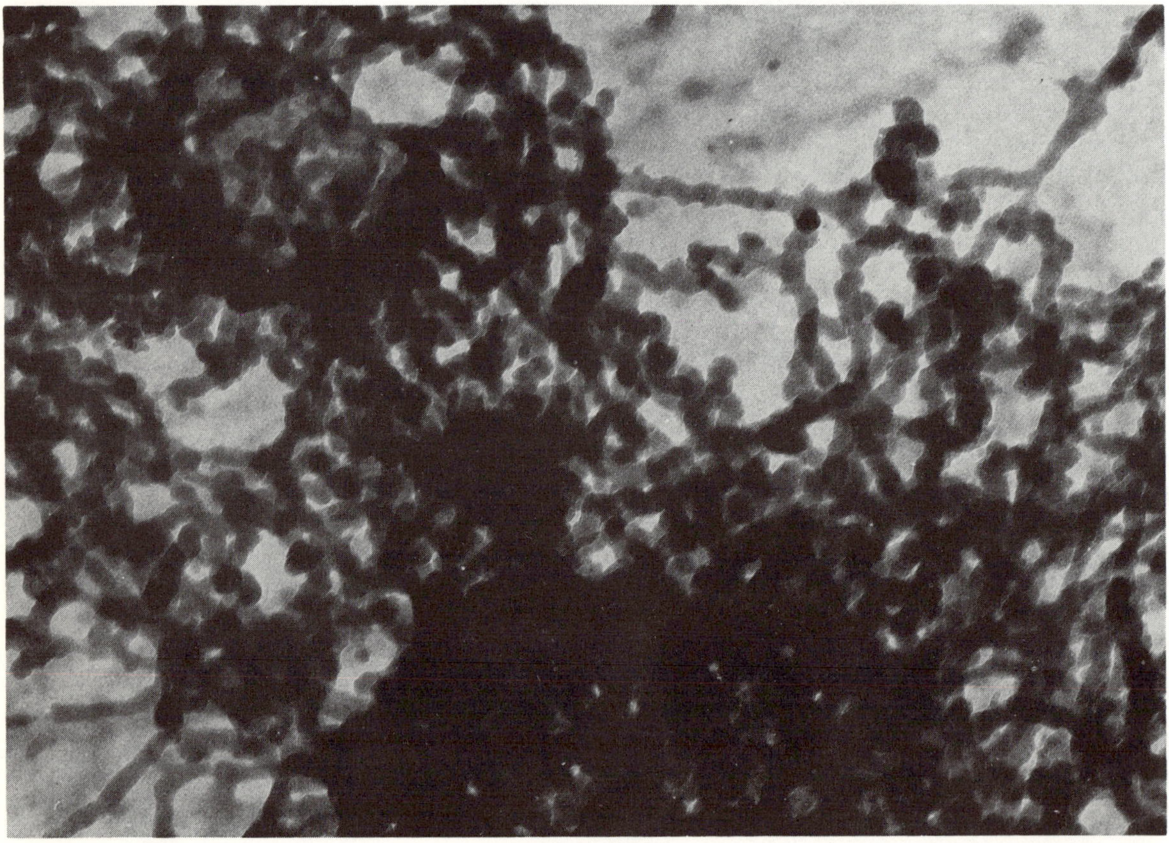

FIGURE 22C.

animals, the oocytes abound with basic nonhistone proteins present in both the cytoplasm and the nucleus.[423-431] These proteins may be ribosomal since they are associated with RNA[424-428,431] and decrease with the oocyte maturation. A comprehensive discussion of the cytochemical properties of basic proteins in oocytes has been presented by Cowden,[431] who has also reported the presence of an arginine-rich basic protein not associated with nucleic acid. The arginine-rich protein appears in the nucleoplasm of several echinoderm species and in mice; its function is presently unknown. In frog oocytes, the amino acid composition of the basic proteins resembles that of ribosomes.[159,429]

More detailed information concerning the composition of acid soluble nuclear proteins is available for sea urchin eggs. Because of their availability in large quantities and the relatively simple method for isolation of their nuclei, female gametes of sea urchins have been studied in several laboratories. Although some of the findings are controversial, there are proteins similar to the histones present in unfertilized sea urchin egg nuclei. Their electrophoretic patterns are different from mammalian histones.[311,431-434] The amino acid composition of the total egg histones, as well as of the fractions corresponding in solubility properties to the mammalian lysine-rich and arginine-rich fractions, is shown in Table 9.

TABLE 9

Amino Acid Composition of Basic Proteins from Unfertilized Eggs

Amino acid	(a)	(b)	(c)	(d)
Lysine	**9.1**	**13.5**	**7.6**	**11.2**
Histidine	5.7	1.7	1.7	0.7
Arginine	**9.1**	2.6	**6.0**	9.8
Aspartic acid	7.0	7.9	9.7	**9.5**
Threonine	5.1	4.4	5.0	5.0
Serine	5.5	**15.5**	6.7	6.6
Glutamic acid	8.9	**11.7**	**11.4**	**10.0**
Proline	3.6	4.3	4.2	5.9
Glycine	7.4	**13.9**	7.8	10.5
Alanine	9.9	11.1	8.7	7.7
Half cystine	0.6	–	0.4	0.5
Valine	6.4	4.0	6.3	5.5
Methionine	1.3	0.5	2.5	2.3
Isoleucine	4.7	1.9	5.5	3.9
Leucine	9.1	3.3	8.8	6.0
Tyrosine	3.1	1.3	3.1	2.5
Phenylalanine	3.2	2.2	4.2	2.3

All values are expressed in mol %.

(a) Frog oocyte nuclei, acid soluble proteins.

(b), (c), and (d) are nuclear proteins from *Strongylocentrotus purpuratus* eggs extracted sequentially with 5% perchloric acid (very lysine-rich), ethanol-HCl (arginine-rich) mixture, and 0.25 *N* HCl (lysine-rich).[311]

Chapter 3

MOLECULAR PROPERTIES OF HISTONES

The heterogeneity and aggregation of histones at nearly neutral pH values made studies of their molecular properties difficult. However, the advent of comprehensive fractionation techniques in conjunction with the determination of the complete or partial amino acid sequence of several histone fractions allowed new approaches to the physicochemical studies on histones and permitted the elucidation of their conformation both free in solutions and bound to the DNA in chromatin.

Molecular Weights of Histones

Histone Aggregation

Although measurements of the sedimentation, diffusion, osmotic, and light scattering properties of histones offered exact estimates of the extent to which histones aggregate in various solutions, the most elegant demonstration of this phenomenon was the analysis of histones by zone electrophoresis in starch or polyacrylamide gel. Using starch gel electrophoresis Neelin et al.[95,321,435] separated chicken erythrocyte, calf thymus, and other histones into more than 20 distinct bands. The high heterogeneity of histones in starch gel electrophoresis was confirmed by other investigators.[436,437] However, decreasing the pH of the electrolytes used for the electrophoresis reduced the number of histone components in calf thymus to only 5 to 6, indicating that the highly heterogeneous patterns observed earlier were due to the formation of aggregates.[96] The identity of the electrophoretic fractions was confirmed by chemical analysis. Including small amounts of $AlCl_3$ further decreased the self-aggregation of histones.[29]

Another refinement in the electrophoretic studies on histone heterogeneity, both natural and caused by aggregation, occurred with the introduction of polyacrylamide gel electrophoresis to this field by Reisfeld et al.[98] The superior resolution of this new technique led to an initial conclusion that there may be many, perhaps several hundred, histones in the nuclei of eukaryotes. Depending on the pH and chemical composition of the buffers used for polyacrylamide gel electrophoresis of histones, these proteins yielded a highly variable number of bands.[325,330,331,438,439] Decreasing the pH of the electrolyte medium[440] or including concentrated urea in the gel and sample buffer[100-102,246,441] proved again that the extensive heterogeneity of histones was due to their aggregation and could be completely abolished under proper experimental conditions. Presently, barring the aggregation of arginine-rich histone F3 (III) which is produced by chemical interaction of the cysteine residues present in this fraction,[298] only 6 to 9 electrophoretic components have been detected in most histone preparations. These correspond to the fraction F3 (III) – one or two major bands, depending on whether the sample is totally or partially reduced or oxidized; F1 (I) – one or more bands depending on tissue; F2b (IIb2) – one band; F2a2 (IIb1) – one band; and F2a1 (IV) – one or two bands, depending on the extent of its acetylation.

Although gel electrophoresis offers an extremely sensitive tool for the detection of histone aggregates, the information about their size and molecular weight was derived, in most instances, from physicochemical measurements. The earliest recognized aggregation was of arginine-rich F2a1 (IV) and F3 (III) histones (or the β fraction) by Mauritzen and Stedman,[305] who used this property of these histones for their preparation. In more detailed studies, Cruft et al.[191,192,442] found that the aggregation of arginine-rich histones could be promoted by increasing the pH to alkaline values, increasing the temperature of aggregation, or increasing the ionic strength of the solution. Divalent or trivalent cations, especially $AlCl_3$, repressed the aggregation, while divalent anions, especially sulfates, increased it. The lysine-rich α and γ histones were not considerably aggregated.[193]

An extensive aggregation of histones, as determined by ultracentrifugation, was also reported by other investigators.[58,63,91,135,142,443-447] Urea in high concentration, guanidine hydrochloride, or low pH decreased or completely eliminated aggregate formation with the exception of the chemical aggregates of the F3 (III) histones caused by the formation of disulfide bonds between the cysteine residues in individual molecules.[298] The interaction of histone molecules with each other in chromatin and its effects on the chromatin structure are not known.

Physicochemical Determinations

As previously described, exact measurements of the molecular weight of histones were complicated by their heterogeneity and exceptional tendency to aggregate. When the molecular interactions were prevented or at least decreased by dissolving the samples in urea or guanidine hydrochloride, the molecular weights of the histones decreased to a minimum between 8,000 to 24,000 daltons.[159,210,448] Similarly, initial estimates of the molecular weights of purified histone fractions using the quantitation of NH_2- or COOH-terminal amino acids gave variable values, partly because of the poor recoveries of the amino acid terminals and partly because of the acetylation of NH_2-termini in several histone fractions. Taking the acetyl content of the individual histone fractions into account, Phillips[222] arrived at average values for each fraction which were in the vicinity of 10,000 daltons. With the exception of the F1 (I) histones, Phillips results were in fair agreement with more recent information derived from ultracentrifugation in guanidine hydrochloride solutions[159] and from the determination of the complete amino acid sequences of several histones.

Probably because of their high positive charge and peculiar conformation, the hydrodynamic behavior of histones is anomalous[172] and cannot be used for accurate measurements of their molecular weights by gel filtration. Better estimates can be obtained from the electrophoresis of histones in polyacrylamide gels in the presence of sodium dodecyl sulfate.[103,104,449,450] When histones from calf thymus are electrophoresed in such a system (Figure 1) the lysine-rich histone F1 (I) and the F3 (III) dimer migrate considerably slower than the rest of the fractions. The F3 (III) monomer and F2b (IIb2) migrate much faster and form one band indicating a similarity of their molecular weights. This is closely followed by a slightly faster F2a2 (IIb1) histone. The well-separated and fast-moving F2a1 (IV) histone migrates well ahead of all the other histones. Since the positive charges of the histones are neutralized by sodium dodecyl sulfate, the electrophoretic mobilities of the histones are proportional to their molecular weights. This would then indicate that in calf thymus, the F3 (III) histone has an almost identical molecular weight to that of the F2b (IIb2) fraction, while the F2a2 (IIb1) histone is slightly smaller. The lysine-rich F1 (I) histones are the heaviest of all the fractions.

The molecular weights of the individual histone fractions determined physicochemically were in good agreement with the electrophoretic data. *The very lysine-rich histones F1 (I)* are the heaviest of all the fractions. Although Ui[63,444] and, later, Trautman and Crampton[448] reported its molecular weight to be about 10,000 daltons (a value similar to that based on the recovery of NH_2- terminal acetyl by Phillips,[222]) the detailed studies of Teller et al.[210] established that the F1 (I) histones were much heavier and had molecular weights of approximately 22,000 daltons. A similar molecular weight (21,000 daltons) was found by Ansevin (quoted in Reference 159) when the F1 (I) histones were analyzed by sedimentation equilibrium in 5 *M* guanidine hydrochloride. Haydon and Peacocke[211] obtained a value of 19,500 daltons and found a considerable reduction of the apparent molecular weight of F1 (I) histones with increasing concentration of this protein. This may explain the low values reported earlier by Ui[63,444] and Trautman and Crampton.[448] The F1 (I) histones have been shown to be heterogeneous (containing several subfractions). As was reviewed earlier in this book, the rabbit thymus F1 (I) histone component 3 contained approximately 212 to 216 residues, which was well compatible with the 20,000 to 22,000 mol wt of the F1 (I) mixture found by ultracentrifugation. Electrophoretic analysis of F1 (I) histones in polyacrylamide gels containing sodium dodecyl sulfate (Figure 1) indicated that the sizes of the F1 (I) histone subfractions were extremely similar.

The molecular weight of *the lysine-rich F2b (IIb2) histone* was initially said to be about 22,000 daltons, based on sedimentation analysis,[259] and between 15,000 to 25,000 daltons based on the recovery of NH_2-terminal proline.[222,259] Lower molecular weights were suggested (7,000 to 11,000 daltons) by its content of COOH-terminal lysine[259] and from the acetyl contents.[222] Ansevin (quoted in Reference 159) found the molecular weight of the F2b (IIb2) histone in 5 *M* guanidine hydrochloride to be close to 13,000 daltons, i.e., a value similar to that calculated from the amino acid sequence of this histone fraction (13,770).[447] Edwards and Shooter[170,451] have shown that the F2b (IIb2) histone undergoes progressive aggregation with an increasing salt concentration of the solvent. It existed in the form of single molecules only in NaCl solutions of 33 mM or less. In such solutions (0.01 *M* NaCl) the molecular weight of the calf thymus F2b (IIb2) histone was established as 15,200 daltons, a value slightly higher than calculated from the amino acid sequence of this protein. The authors also suggested that the F2b (IIb2) histone might be heterogeneous. In 1.0 *M* NaCl the molecular weight increased to about 100,000 daltons and a further rise of the NaCl concentration occasioned the formation of large aggregates. It was speculated that the aggregation of F2b (IIb2) histone might be due to the gradual salting out of its hydrophobic central part. In chicken erythrocytes, the molecular weight of F2b (IIb2) histone was determined to be 14,400 ± 500 daltons.[354]

Most of the available data on the F2a histones, which was a mixture of the F2a1 (IV) and F2a2 (IIb1) fractions concerns the arginine-rich F2a1 (IV) fraction. Its complete sequence contained 102 amino acid residues corresponding to a molecular weight of 11,300 daltons.[447] This fraction underwent considerable aggregation in concentrated salt solutions[451] – a property used for its isolation.[164] The molecular weight of the F2a mixture, as determined by Ansevin (quoted in Reference 159) at 12,000 daltons, well accomodated the average of the values calculated from the acetyl and NH_2-terminal amino acid content (9,000 to 12,500)[222] and from the amino acid sequences of the F2a1 (IV) and F2a2 (IIb1) histones (11,300 daltons[447] and about 15,000 daltons,[77,447] respectively). The electrophoretic behavior of the F2a2 (IIb1) histone in polyacrylamide gels containing sodium dodecyl sulfate (Figure 1) also suggested the molecular weight of this fraction to be larger than that of the F2a1 (IV) histone and perhaps only a little smaller than those of the F2b (IIb2) and F3 (III) fractions. Diggle and Peacocke[354] found the molecular weight of chicken erythrocyte F2a2 (IIb1). histone to be 16,800 ± 400 daltons. The molecular weight of calf thymus F2a2 (IIb1) histone alculated from its amino acid sequence (Figure 11) is 14,231 daltons.

The molecular weight of F3 (III) histones in 5 *M* guanidine hydrochloride was determined by Ansevin (quoted in Reference 159) to be 14,000 daltons. A very similar value of 14,800 daltons was reported by Edwards and Shooter,[451] who found this fraction quite homogeneous. This agreed well with the electrophoretic mobility of the F3 (III) histones in the presence of sodium dodecyl sulfate (Figure 1), which was practically the same as that of the F2b (IIb2) fraction. The molecular weight of the calf thymus F3 (III) histone calculated from its amino acid sequence (disregarding its acetyl and methyl residues) is 15,229 daltons. The F3 (III) histones form large aggregates in concentrated salt solutions and in the presence of divalent anions,[192,451] which can explain the much higher molecular weight values reported earlier for these proteins. A considerably higher molecular weight (19,100 ± 500 daltons) of this histone was reported for the chicken erythrocyte histone.[354]

The erythrocyte specific histone F2c (V) has a molecular weight similar to that of the very lysine-rich F1 (I) fractions. It further resembles these proteins in that it does not aggregate in the presence of salts.[353,354] The molecular weight of the F2c (V) histone from chicken erythrocytes has been determined by Champagne et al.[353] and by Diggle and Peacocke.[354] The values of 16,700 daltons and 20,800 ± 500, respectively, are in agreement with the electrophoretic mobility of the F2c (V) histones which, in the presence of SDS, migrate between the F1 (I) and F2b (IIb2) + F3 (III) fractions. A value of 15,714 daltons has been calculated from the amino acid composition of this fraction.[353]

The Conformation of Histones

The first ideas about the conformation of histones associated with DNA in the nucleohistone were obtained from x-ray diffraction, infrared spectrophotometry, and deuteration studies. In denatured nucleohistone, the presence of polypeptide chains in the β conformation was indicated by x-ray diffraction measurements. The interpretation of the data was difficult, however.[452] The infrared spectra of isolated histones, as determined in dry films cast from aqueous solutions, supported by optical rotatory dispersion measurements, revealed the presence of the α-helix[452-454] in the unfractionated protein samples in amounts corresponding to 25 to 37% of the theoretical values. In films cast from dichloroethanol (an α-helix-promoting solvent), the α-helical content of histone increased to about 50 to 60%.[454-455] That lyophilized or aged histone samples became partly denatured was reflected in the increasing presence of a β-structure.[452] The rates of deuteration of histone films exposed to

D_2O also indicated the presence of a partial α-helical conformation.[447,452,453]

In aqueous solutions, the optical rotatory dispersion of water soluble calf thymus deoxyribonucleohistone suggested that about two thirds of the proteins were α-helical. On the other hand, the determination of Moffitt constants for the unfractionated histones in either water solutions or in 8 *M* urea did not show any appreciable α-helical conformation,[447,456-461] indicating that these proteins are fully disordered in the absence of ions. This was probably due to the electrostatic repulsion of their positive charges as could be expected from the behavior of whole histone. All the main histone fractions F1 (I), F2a1 (IV), F2a2 (IIb1), F2b (IIb2) and F3 (III) as well as the erythrocyte specific histone F2c (V) were found to be disordered and lacking in α-helical conformation in aqueous solutions.[156,456-461]

An entirely different situation is observed if unfractionated histone or the fractions are exposed to ions, aqueous *n*-propanol, 2-chloroethanol, trifluoroethanol, or anionic detergents. Their α-helical content increases considerably.[447,456-461] In salt solutions, the α-helical content of the very lysine-rich histone F1 (I) increases only to a small extent and most of this protein remains disordered or partially ordered in the β-conformation.[447] Champagne et al.[353] found the erythrocyte specific F2c (V) histone to behave similarly, increasing its α-helical content in concentrated salt solutions only to a small extent. While the F2c (V) histone is not α-helical in water, in aqueous methanol it becomes more ordered and exists partly in the β-conformation.[461] The α-helicity of the other histones increases even in dilute salt solutions, reaching values between 25% and 30% of the theoretical values. According to Jirgensons and Hnilica,[456,460] there is no substantial difference in the α-helix forming abilities of Na^+, K^+, Mg^{++}, and Al^{+++} ions in proper concentrations. Anions, however, differ considerably. While 0.1 to 0.2 *M* chloride is necessary to induce α-helical conformation in histones F2a (mixture of F2a1 or IV, and F2a2 or IIB1) and F2b (IIb2), only 0.01 *M* phosphate is required for a similar effect. Changing the pH values from basic or acidic also has been found to increase the amount of ordered structure in the histones,[447,458,459] but only to a small extent. Anionic detergents are very active in organizing the histones, much more so than salt solutions.[460,462-464] Tetradecyl- and dodecylsulfates induce some α-helical conformation even in the F1 (I) histones. When the F1 (I) histone is compared to the F2b (IIb2) histone, the same amount of detergent formed more helix in the latter fraction.

It is known that dilute solutions of flexible linear polyelectrolytes exhibit an abnormal increase of the reduced specific viscosity upon dilution. This behavior is nearly completely abolished by the increase of salt concentration in the solution of polyelectrolyte. If histones behave in the solution as flexible polypeptide chains, they should exhibit such anomalous viscosimetric behavior. Jirgensons et al.[464a] determined the reduced specific viscosities of the main histone fractions in the presence of various salt concentrations. In water, the very lysine-rich histone F1 (I) showed nonlinear dependence of its reduced specific viscosity on protein concentration. This nonlinearity was alleviated in the presence of electrolytes such as 0.1 *M* NaCl, 0.1 *M* KCl, 0.02 *N* HCl, etc. A similar viscosimetric behavior was also seen for all the remaining histones (F2b or IIb2, F2a or IV + IIb1, and F3 or III). However, the initial nonlinearity between the reduced specific viscosity and histone concentration was much smaller than that of F1. The authors concluded that because of the neutralization of electrostatic charges on histones by salts, the conformation of these proteins changed becoming less asymetric. The F1 (I) histone was more affected by such changes than the other fractions. The conformational changes produced in histones by neutralization of their surface charges in the presence of salts include the formation of α-helix in all fractions except the F1 (I) histone. This latter fraction is not α-helical and its conformational change in the presence of salt must depend on the formation of structures other than the α-helix.

In the α-helix promoting solvents such as *n*-propanol, 2-chloroethanol, or trifluoroethanol the histones considerably increase their ordered conformation. The Moffit constant (b_o) and the mean residual specific rotation at the wave lengths $[m']\lambda$, corresponding to the Cotton effect extremes, are proportional to the degree of ordered structure. The values are $b_o = -630$ for the α-helix and $b_o = 0$ for random coil. The Cotton effect minima (trough) values $[m']_{233}$ are approximately −2,500° for random coil and −16,000 for the α-helix. To facilitate comparison of the α-helical contents of individual histones, as well as the effects of various solvents, the b_0 and $[m']_{233}$ values of the main histone fractions are compiled in Tables 10 and 11.

Since it has been reported that the α-helical content of histones increases in the presence of DNA or polyanions and becomes similar to that in diluted salt solutions,[456,465] it can be expected that the histones are partially α-helical in chromatin. This assumption is confirmed by the work of Tuan and Bonner,[466] who found in chromatin that the α-helical conformation of histone in association with DNA approaches that of the histone fractions in solutions of high ionic strength.

Optical rotatory dispersion measurements on histones indicated that the maximum amount of α-helix which could be formed in each fraction was proportional to its proline content. It was shown by Jirgensons[460] that in the acid solutions of sodium tetradecyl-sulfate the maximum α-helical content of the F2b (IIb2), which contains 3.4% proline, was 42% (67% in 2-chloroethanol). The proline-rich (9.4%) F1 (I) histones contained only 31% of α-helix in the detergent and 50% in 2-chloroethanol. It was also found that aliphatic alcohols (*n*-propanol, ethanol, methanol) increased the amount of ordered structure in histones similar to the more widely used 2-chloroethanol.[460,464] The presence of some β-structure in the histones was also indicated.

As was pointed out by Bradbury and Crane-Robinson,[447] optical rotatory dispersion or circular dichroism measurements could only offer quantitative estimates of the ordered (mainly α-helical) structures in the histone fractions. These techniques could not exactly determine in which part of the histone sequences the ordered structures were localized. The analysis of proton magnetic resonance spectra, although applicable to the sequence determinations of peptides, was at its best when used on polypeptides of known amino acid sequences containing an irregular distribution of amino acids. The recent elucidation of the F2a1 (IV) and F2b (IIb2) histone sequences, as well as some work on the partial sequence of F1 (I) histones, prompted Bradbury and his associates[447,467-469] to investigate changes in the proton resonance spectra produced by changes in the conformation of histones. The resonance peaks for the specific proton groups were found to remain sharp as long as the peptide chain was disordered and flexible. Any changes in the molecule or conditions which reduced the flexibility and motion of the molecules and amino acid residues within (e.g., formation of α-helical regions of histones dissolved in diluted salt solutions) were expected to be reflected in the broadening of the individual resonance peaks and a loss in their area. Hence, the analysis of the nuclear magnetic resonance spectra revealed that in F1 (I) histones, the ordered segment of their molecule was restricted to the area between residues 51 and 100.[468] The reduced mobility of the area between 51 and 100 probably resulted from the formation of either intra- or interchain β-conformations.[468] It appeared unlikely, then, that the F1 (I) histones form appreciable amounts of α-helix in aqueous salt solutions.

Similar measurements performed on the F2a1 (IV) histone from calf thymus showed the ordered (α-helical) segment of the polypeptide chain to be located over a large part of the COOH-terminal half of the molecule. Since the F2a1 (IV) histone contained only about 15% of α-helix in 0.2 *M* NaCl, the extensive changes in resonance spectra for this half of the molecule were at least partially due to a specific aggregation involving the COOH-terminal half of the F2a1 (IV) histone molecules. The most likely location of an α-helical conformation was between residues 55 and 72 of this histone fraction.[468]

A different situation was reported by Bradbury et al.[447,469] for the F2b (IIb2) histone. In this protein, it was the central part of the molecule (residues 50 to 105) which became immobilized in dilute salt solutions. This left both ends of the F2b (IIb2) molecule largely unaffected. Again, since the α-helical content of this fraction probably accounted for only a part of this immobilization, the other part might be due to certain interactions (aggregations) between individual F2b (IIb2) molecules. Based on their work on the conformation of histones, Boublik et al.[467-469] proposed a sequence of conformational changes for the aqueous solutions of histones: (a) F2a1 (IV); (b) F1 (I); and (c) F2b (IIb2) with an increasing salt concentration. Their scheme is reproduced in Figure 23.

Knowing the complete amino acid sequence of histones F2a1 (IV) and F2b (IIb2), it is possible to analyze the probability of certain amino acid sequences occurring in an α-helical conformation. It appears that the localized accumulation of certain amino acids, such as alanine, valine, leucine, etc. creates conditions favorable for the

TABLE 10

Some b_o Values for Histones and Protamines

	Calf thymus						Chicken erythrocyte		Protamines				
										Clupeine			
Solvent	Whole	F1 (I)	F2b (IIb2)	F2a1 (IV)	F2a2 (IIb1)	F3 (III)	Whole	F2c (V)	Whole	YI	YII	Z	Salmine
Water	- 40	0	0	0	0	- 8			+ 28	+ 25	+ 31	+ 30	
	0	+ 13	- 28	- 47	- 47	- 70							
0.1 *M* NaCl	- 97	- 26	-147	-123	-123	- 87							
0.15 *M* NaCl								-386					
0.25 *M* NaCl	-118	- 43	- 27	- 77	-149	-135	-115	- 93	+ 20				+ 27
1.0 *M* NaCl	-146		-180			-155		-309					
	-147	- 71	-172		-185	-119	-146	- 99	0				+ 5
0.15 *M* $NaClO_4$								-426					
0.4 *M* $NaClO_4$								-539					
1.0 *M* $NaClO_4$								-457					
8 *M* urea	- 22	0	- 10	- 14	- 34	- 25	0	0	+ 16				
									+ 35				
Water, pH 1.5	- 50	0	- 40	- 50	- 35	- 90							
Water, pH 3.0	- 40	0	- 28	- 35	- 54	- 70							
pH 7.0	- 65	- 11	- 55	- 60	- 52	- 75	- 86	- 52					
pH 9.5	- 90	- 23	-110	- 90	- 57	-100							
98% propanol		-280		-395	-395	-330							
2-chloroethanol	-335	-291	-375	-390	-463	-298	-388	-396	-315	-262	-198	-243	-375
	-401	-300	-444			-408			-222				
Trifluoroethanol	-484	-380	-525	-557	-535	-519	-520	-530					

The b_o values were compiled from References 353, 447, 456, 457, 460, and 477.
It can be assumed that the fully helical protein has a b_o value of –630.
The b_o value of a random coil is 0.

TABLE 11

Some [R′] Minimum Values for Calf Thymus Histone and Salmine

Solvent	Whole	Fl (I)		F2b (IIb2)		F2al (IV)	F2a2 (IIb1)	F3 (III)		Salmine
	$[R']_{233}$	$[R']_{233}$	$[R']_{min}$	$[R']_{233}$	$[R']_{min}$	$[R']_{233}$	$[R']_{233}$	$[R']_{233}$	$[R']_{min}$	$[R']_{233}$
Water	−3,500	−3,500		−3,600		−2,900	−2,900	−3,600		
Water, pH 1.5				−3,300				−4,800		
pH 7.0	−3,400	−3,800		−3,600		−3,000	−3,300	−3,900		
pH 9.5				−4,300				−4,700		
0.1 *M* NaCl	−3,800	−3,500		−4,300		−3,800	−3,800	−3,600		
1.0 *M* NaCl	−4,200	−4,100		−4,300				−3,500		
	−5,000	−4,300		−5,300			−5,200	−5,100		−2,950
98% propanol			−6,900**						−7,100**	
2-chloroethanol						−10,700**	−10,700**			−6,600
trifluoroethanol	−10,300	−8,900		−11,360		−11,500	−11,450	−11,860		
8 *M* urea	−2,400	−3,000		−2,600		−2,300	−2,200	−2,600		−2,850
Sodium dodecyl sulfate at 50°C			−6,500*		−6,800**					

* $[R']_{231}$, ** $[R']_{232}$
The $[R']_{min}$ values were compiled from References 447, 456, 460, and 463.
$[R']_{233}$ values of −2,500° and −13,500° correspond to random coil and fully helical form, respectively.

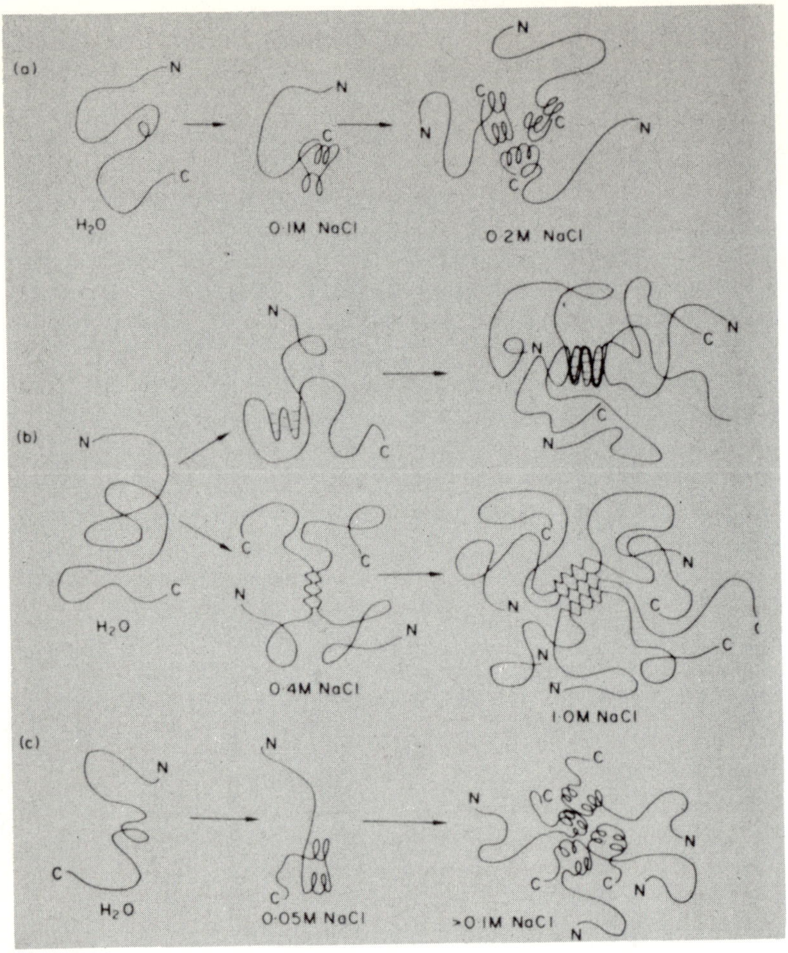

FIGURE 23. Proposed conformational changes in aqueous solution of histones (a) F2a1 (IV); (b) F1 (I); and (c) F2b (IIb2) with increasing salt concentration.[447]

α-helical conformation, while the accumulation of proline, serine, threonine, etc. is incompatible with an ordered protein structure.[470-473] In brief, it was proposed that any sequence of five amino acid residues will be helical if at least three include ALA, VAL, LEU, or GLU and any segment of seven residues will be helical if at least three are ALA, VAL, LEU, or GLU and if one more is ILE, THR, or GLN.[473] In addition, other hydrophobic residues (TYR, PHE, MET, CYS) may also contribute to the stabilization of helices.[474] Using known amino acid sequences of several proteins, Schiffer and Edmundson[474] arranged three-dimensional α-helices into two-dimensional "helical wheels" where 18 amino acids represent a complete repeat of the helix. There are 3.6 residues per turn and 5 turns per repeat. The adjacent residues are separated by an angle of 100° and side chains which are close to one another on one side of the α-helix are also close to one another on the wheel. Such a projection has the advantage of dividing the amino acid sequences into segments with either helical or nonhelical potential. Knowing the sequences of histones F2a1 (IV) and F2b (IIb2), Boublik et al.[468,469] have constructed helical wheels for these two fractions. As can be seen from Figures 24 and 25, there is an amazing agreement between the nuclear magnetic resonance data and the theoretical considerations concluded from the amino acid sequences of histones F2a1 (IV) and F2b (IIb2). Even more instructive are the two dimensional helical surfaces of these two proteins constructed by Bradbury and Crane-Robinson[447] and reproduced in Figures 26 and 27. As already suggested by the magnetic resonance experiments, histone molecules seem to have at least two interactive regions. One, perhaps less specific, shows a localized accumulation of

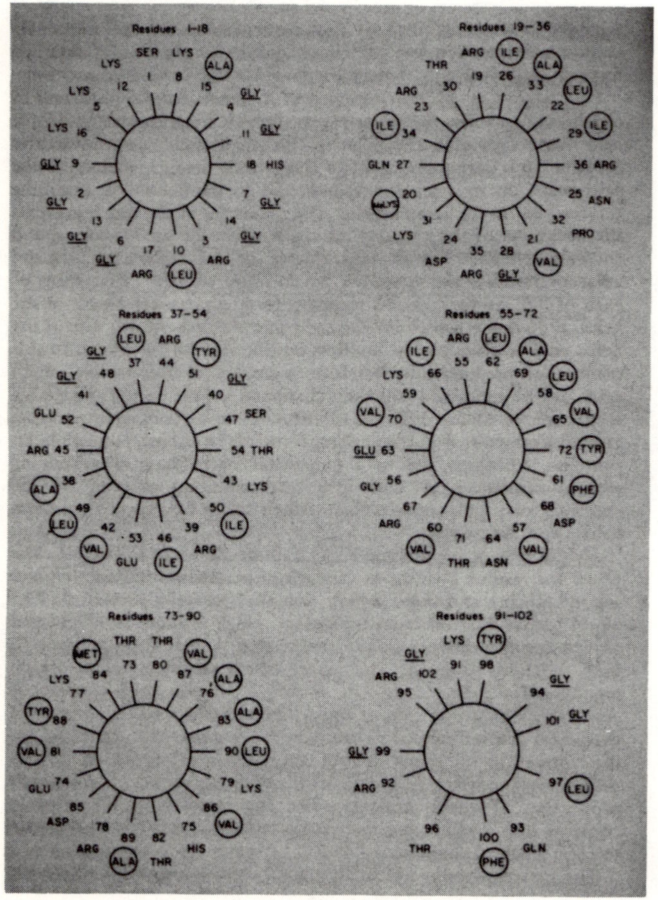

FIGURE 24. Helical wheels of arginine-rich histone F2a1 (IV) according to Bradbury and Crane-Robinson.[447]

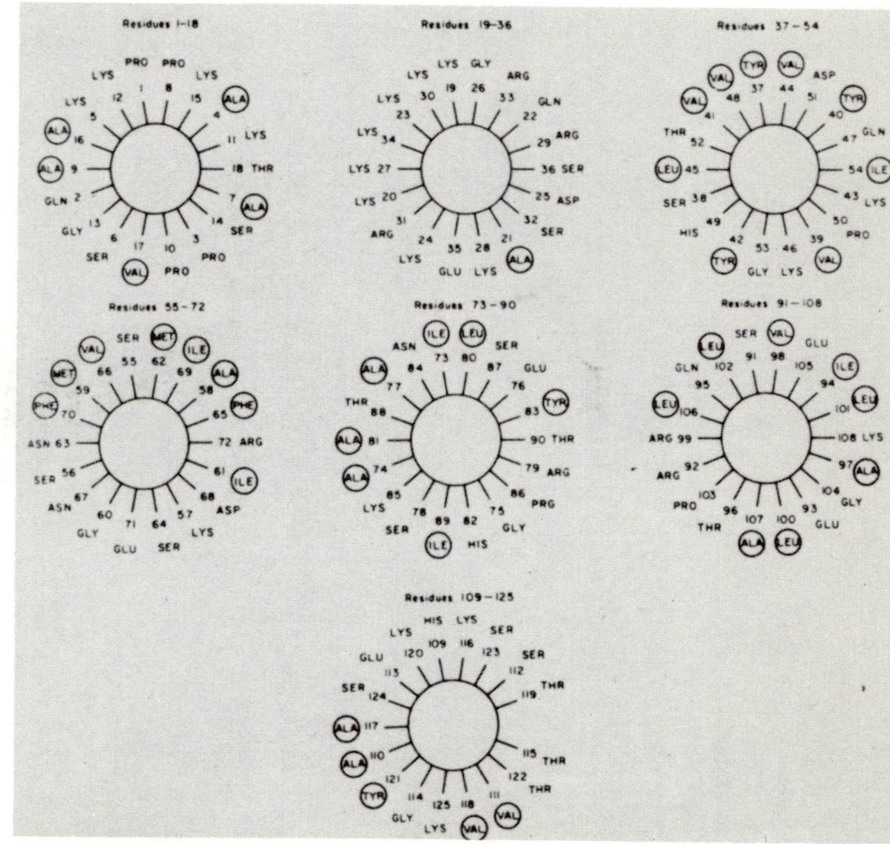

FIGURE 25. Helical wheels of the moderately lysine-rich histone F2b (IIb2) according to Bradbury and Crane-Robinson.[447]

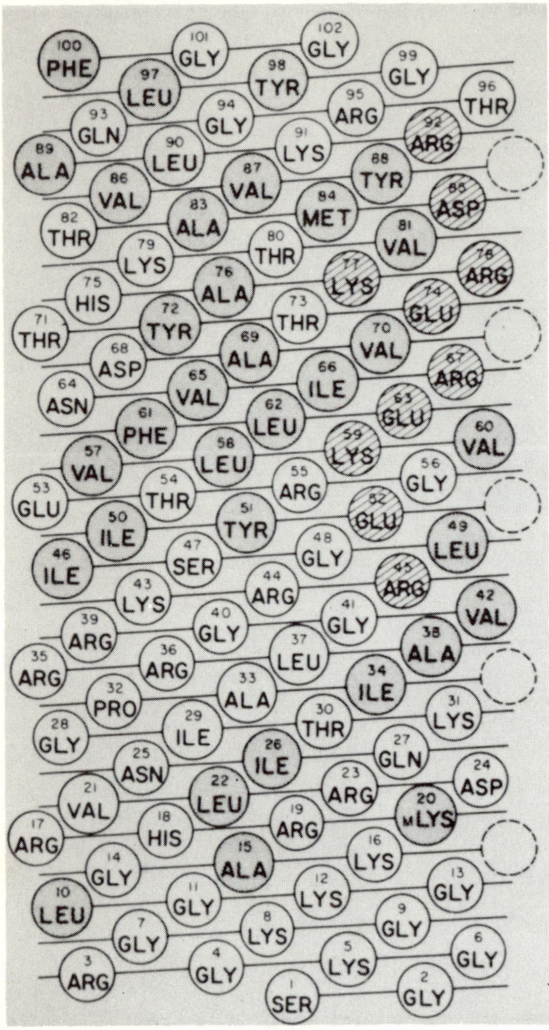

FIGURE 26. Two-dimensional helical surface of arginine-rich histone F2al (IV) according to Bradbury and Crane-Robinson.[447]

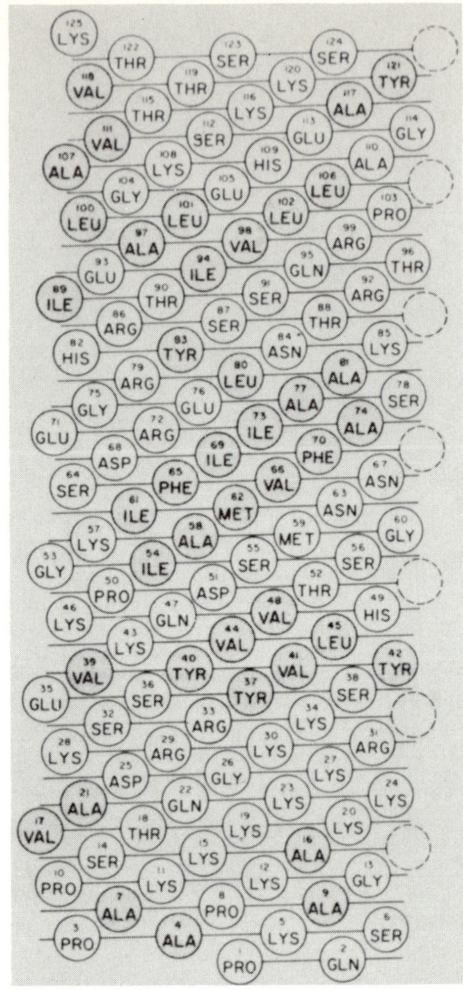

FIGURE 27. Two-dimensional helical surface of the moderately lysine-rich histone F2b (IIb2) according to Bradbury and Crane-Robinson.[447]

basic amino acids and represents the interaction sites with DNA.[475,476] The other, comprising the ordered part of the histone molecules, contains the hydrophobic (helix-forming) amino acid residues and may function in more specific protein-protein interactions.[475] In F2b (IIb2) histone, the sequences of residues 1 to 37 and 106 to 125 are most likely involved in the interactions with DNA, while the central part (sequences 38 to 105) is ordered and free for interactions with other molecules.[476] Treatment similar to that in Figures 23 to 27 of the remaining histone fractions has to await elucidation of their amino acid sequences. It can, however, be predicted from the incomplete sequence data discussed earlier that the distribution of basic amino acids, as well as that of the hydrophobic residues, will be irregular with clustering patterns similar to those seen in the fractions F1 (I), F2b (IIb2), and F2al (IV). The significance of this asymmetry for the biological function of histones is presently unknown and is the subject of many speculations.

Chapter 4

HISTONE METABOLISM

Since histones are essential components of eukaryotic chromatin and most likely regulate both its structure and function, ordered rates of histone biosynthesis closely paralleling the replication of other chromatin components appear to be necessary for the perpetuation of the differentiated state. Therefore, studies on histone biosynthesis and turnover represent an important aspect of the general knowledge of cellular biochemistry and physiology.

Histone Biosynthesis

Of all the macromolecules constituting chromatin, the biosynthesis and turnover of DNA and histones has attracted the most attention. Interest stimulated by the possibility that histones may function as gene repressors[62,94,478] has led to many publications concerning DNA and histone biosynthesis. Although there is no doubt that all the common histone fractions are biosynthesized at their maximum rates during the S period of cell cycle (the time of DNA synthesis), the problem of whether the synthesis of all the main types of histones is controlled centrally or whether the fractions turn over independently from each other remains unsettled. Similarly, it is not known, at the present time, how the newly made histone proteins reach the DNA segments with which they associate and whether there is a pool of unbound histones available in cells which do not undergo cell division for considerable time periods.

Localization of Histone Synthesis

Early studies on isolated nuclei based on cytochemical and autoradiographic findings present considerable evidence indicating active protein synthesis within the nuclei of many cell types. This evidence has been supported by the finding that nuclei contain all the elements necessary for protein synthesis. Indeed, as pointed out by Allfrey,[479] the cell nucleus is the source of the ribosomes and the ribonucleic acids essential for this process. There are many reports in the literature describing various aspects of nuclear protein synthesis. Since they are out of the scope of this monograph, the reader is referred to recent reviews concerned with this matter.[479,480]

Because of the close association of histones with DNA in the nucleus, the nuclear site of histone biosynthesis was frequently suggested. It appeared quite logical that such highly specialized proteins as histones should be made at the site of their utilization. The first evidence indicating the incorporation of labeled amino acids into the histones by isolated calf thymus nuclei in vitro was presented by Allfrey et al.[481,482] and confirmed by others. Since the incorporation of amino acids into histones by isolated nuclei was inhibited by puromycin[483–485] and RNase,[213] it was suggested that histones are made by biosynthetic mechanisms similar to those operating in cytoplasm. The incorporation of labeled amino acids into the proteins biosynthesized by isolated thymocyte nuclei was shown by radiation analysis of the tryptic peptides resulting from the digestion of the in vitro labeled F1 (I) histones.[213] Using an improved fractionation procedure, Reid et al.[485] separated the very lysine-rich histones F1 (I) from calf thymocyte nuclei into four fractions. All of these fractions were synthesized in isolated thymocyte nuclei in vitro and their biosynthetic rates did not differ significantly one from another. From the kinetic data on histone labeling in vitro the authors concluded that the F1 (I) histones are synthesized in the thymocyte nucleus.

Additional support for the nuclear localization of protein synthesis came from autoradiographic studies,[486] some of which revealed the most active incorporation of labeled amino acids in the nucleolus and perinucleolar chromatin.[487,488] Similarly, Birnstiel et al.[489,490] found high biosynthetic rates for the acid soluble proteins (histones) in the perinucleolar chromatin from isolated pea seedlings or tobacco cell nuclei incubated with labeled amino acids. The nucleolar proteins became labeled first. This was followed by a gradual loss of radioactivity in the nucleoli with a simultaneous increase of the specific activity of the extranucleolar proteins, indicating that perinucleolar chromatin may be the site of histone biosynthesis. Experiments on Novikoff hepatoma cells in vivo[491] revealed a similar situation, i.e., the nucleolar histones incorporated ^{14}C-lysine at approximately twice the rate of total nuclear histones. However, the incorporation kinetic did not indicate the nucleolus to be the primary site of histone synthesis.

The nuclear site of protein synthesis has been recently reinvestigated in several laboratories[492-498] and it has been confirmed that these cellular organelles can biosynthesize proteins. There is a general agreement, however, that the proteins biosynthesized in isolated nuclei are nuclear nonhistone proteins, perhaps ribosomal particle proteins, but not histones. Since the nonhistone nuclear proteins, including the nucleolar precursors of ribosomes, are partially acid soluble, experiments reporting the radioactivity incorporated into acid soluble proteins as histone biosynthesis should be evaluated carefully, considering the possibility of histone contamination by these proteins. This is especially true for the acid soluble ribosomal proteins accumulated in the nucleolus, which are known to be biosynthesized in vitro by isolated nuclei. It is possible that histone biosynthesis by isolated thymocyte nuclei[213,481-485] actually reflects the incorporated radioactivity of ribosomal proteins. Additionally, isolated nuclei frequently contain cytoplasmic ribosomes which adhere tenaciously to the outer nuclear membrane. These ribosomes can be removed with nonionic detergents such as Nonidet® or Triton X-100®. Studies on protein synthesis in nuclei isolated in the presence of nonionic detergents, which solubilizes the outer layer of the nuclear membrane together with the attached ribosomes, show that only the incorporation of labeled amino acids into the nonhistone proteins can be demonstrated.[494,495]

Even before the experiments on detergent-treated isolated nuclei, Bloch and Brack[399,499] pointed out that during spermatogenesis in the grasshopper *Chortophaga viridifasciata*, the arginine-rich sperm proteins were synthesized in the spermatid cytoplasm. Autoradiography of protein-bound labeled arginine has shown an initial incorporation into the RNA-containing layer of the cytoplasm surrounding the spermatid nucleus. There was no arginine incorporation in the nucleus. Later, the radioactivity accumulated on the DNA in the nucleus, indicating that the arginine-rich protein was first synthesized in the cytoplasm and then transferred into the nucleus. By electron microscopy, the RNA-rich cytoplasmic layer, peripheral to the nucleus, was shown to contain polysomes.[399,499]

Bloch's discovery was followed by another report of the cytoplasmic synthesis of a typical gamete cell nuclear protein, the protamine. Working with suspensions of salmon testis cells which are capable of in vitro protamine biosynthesis, Dixon and his associates[366,500,501] labeled such cells with very short pulses of ^{14}C-arginine. After an exposure of the cells to unlabeled arginine (cold chase), protein synthesis was arrested at various time intervals with cycloheximide. Labeled protamine was isolated and its specific activity was determined, together with that of cytoplasmic microsomes. The labeling kinetics, which clearly demonstrated the cytoplasmic synthesis of protamines and their rapid transfer into the nucleus, are reproduced in Figure 28. Sedimentation analysis of the isolated labeled ribosomes showed that the newly synthesized protamine was attached to small, slowly sedimenting polysomes (probably disomes).

In HeLa cells grown in tissue cultures, Robbins and Borun[502] and Borun et al.[503] showed that the cytoplasmic biosynthesis of nuclear proteins is not restricted only to the basic proteins of male gametes. In pulse-chase experiments with ^{14}C-tryptophan and ^{3}H-lysine, the lysine-labeled protein decreased rapidly in the cytoplasm with a simultaneous increase in the nucleus. The tryptophan label failed to exhibit a similar characteristic. Using the ratio of ^{14}C-tryptophan to ^{3}H-lysine as an indicator of lysine-rich histones biosynthesis (histones contain no tryptophan), the authors were able to demonstrate the biosynthesis of proteins rich in lysine and low in tryptophan. These proteins were assembled on small polysomes which could be separated by centrifugation in sucrose gradients. These small polysomes were found only in cells engaged in DNA synthesis and only during the S phase of the cell cycle. They were absent during the G 1 phase and, if the DNA synthesis was inhibited with cytosine arabinoside or 5-fluorodeoxyuridine, they disappeared rapidly. The small polysomes contained 7 to 9 s-RNA exhibiting many properties of m-RNA. This RNA disappeared with the cessation of histone-like polypeptide biosynthesis. The cytoplasmic biosynthesis of histones received further support when Gallwitz and Mueller[504,505] isolated the small polysomes and demonstrated that they can assemble histone-like polypeptides in vitro. The newly synthesized proteins coelectrophoresed with histones in polyacrylamide gels and their appear-

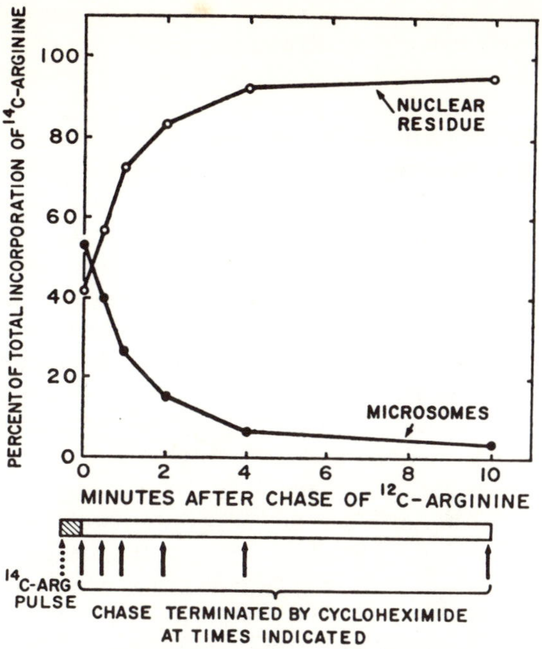

FIGURE 28. The kinetics of labeling of protamine in the ribosomal and nuclear fractions of *Salmo gairdnerii* testis cells following a pulse of L-arginine-^{14}C. Suspension of testis cells was exposed to a 0.5 min pulse of labeled arginine followed immediately by a chase of 10 mM L-arginine. The chase period was terminated in different aliquots by the addition of cycloheximide (to 0.05 mM) at the times indicated by arrows. The nuclei and ribosomes were isolated and extracted with 0.5 *N* HCl containing 4.2% trichloroacetic acid. Carrier protamine was added, the solutions were neutralized, and unlabeled arginine was added. The protamine was specifically precipitated by the addition of 0.25% sodium tungstate – 5% trichloroacetic acid (adjusted to pH 2.0) and the precipitate washed 5 times with tungstate-trichloroacetic acid containing 1 mM arginine (the last time in a boiling water bath). The precipitated protamine was finally collected on a Millipore filter, transferred to a scintillation vial and counted in Bray's solution.[500]

ance was strictly coupled with active DNA replication. When DNA synthesis was arrested by hydroxyurea or thymidine, the small polysomes from such inhibited cells terminated their in vitro histone synthesis. The inhibition of RNA synthesis with actinomycin D or 2-mercapto-1-(β-4-pyridethyl) benzimidazole did not, on the other hand, affect the in vitro histone biosynthesis for at least 90 min, indicating relatively long half-lives of the histone m-RNAs. Isolated nuclei were unable to synthesize histone-like proteins. The authors also confirmed the existence of 8 to 9s-RNA associated with the small polysomes only at the time of histone biosynthesis. This RNA, presumably the m-RNA for histones,[506] appeared to be a product of reiterated, closely clustered genes.[296] It was of interest, that several years before the experiments on histone biosynthesis by small polysomes in vitro, Griffin et al.[507] showed that ribosomes isolated from Novikoff hepatoma and incubated in vitro with labeled amino acids could biosynthesize polypeptides resembling histones in their amino acid composition.

The biosynthesis of histones in the cytoplasm appears to be rather a general phenomenon since a situation similar to that reported in HeLa cells was also reported for developing sea urchin embryos. In these animals, the appearance of typical histones coincided with blastulation, at which stage they could be demonstrated unequivocally. Nemer and Lindsay[508] isolated small polysomes (s-polysomes) from the cytoplasm of cleaving sea urchin embryos which incorporated considerable amounts of labeled arginine and only small amounts of labeled tryptophan into the newly assembled polypeptides. Maximum incorporation of arginine occured during early blastula, i. e., the time of definite appearance of histones in the nucleus. The cytoplasmic biosynthesis of histones in sea urchin embryos was also confirmed in other laboratories.[509-511] In addition to using the ^{14}C-lysine or ^{14}C-arginine to ^{3}H-tryptophan ratios to detect ribosomal histone biosynthesis, Moav and Nemer[510] showed that the basic proteins assembled on small polysomes behaved very much like histones both by polyacrylamide gel electrophoresis and chromatography on Amberlite CG-50®. Although the presence of histones in the nuclei of early embryos of sea urchins (before blastulation) could not be demonstrated unequivocally, in 32 to 64 cell embryos, histones were actively biosynthesized in the cytoplasm.[511] This observation might suggest a cytoplamic storage of histones which are not released into the nucleus until the start of blastulation.

The mechanism for the initiation of histone synthesis appears different from the *N*-formylmethionyl-t-RNA initiation known in most protein biosynthesizing systems. According to Liew et al.,[512] the chain initiation of F2a histones (a mixture of F2a1, or IV, and F2a2 or IIb1) involves the participation of *N*-acetyl-seryl-t-RNA. Although assembled on polysomes, the biosynthesis of histones is not regulated at the translational level.[513] As was discussed earlier, histone biosynthesis appears to depend intimately on DNA replication. Presently, the evidence supporting these observations on the cytoplasmic biosynthesis of histones followed by their subsequent transport into the nucleus is only circumstantial and is based on the electrophoretic similarities and chromatographic behavior of the newly synthesized proteins. Unequivocal demonstration of cytoplasmic histone synthesis awaits the actual isolation and chemical characterization of the basic proteins made on small polysomes.

Another problem born to the cytoplasmic biosynthesis of histones is their transfer to the nucleus. Experimental evidence on protamines from salmon testes[500,501] and histones from HeLa cells[502,503] indicates that these proteins are transferred immediately following their synthesis. Since histones are quite vulnerable to proteolysis, a question arises as to whether and how they are protected during their transfer. Labeled histones can be actively incorporated into cells and find their way to the nucleus. Whether their transport is merely due to the positive charge of the histones affecting the permeability of the cellular membranes or whether they use a readily available transport mechanism ordinarily used by the newly synthesized histones within the cell is not known at the present time. Nuclear transplant experiments on *Amoeba*[514-520] have shown the existence of proteins which are lost to the cytoplasm during mitosis and rapidly reenter the nucleus after cell divison. These "rapidly migrating proteins," which are not histones, appear to be synthesized in the cytoplasm and migrate back and forth by diffusion.[518] They represent about 40% of the nuclear protein content, their concentration in the nucleus being about 25 to 50 times higher than in the cytoplasm.[517] A similar migration has been observed for chromosomal proteins by Kroeger et al.,[521] who transplanted protein-labeled salivary gland chromosomes with their nucleoli into the cytoplasm of nonlabeled salivary gland cells of *Chironomus thummi* and found that the protein bound radioactivity rapidly accumulated in the host cell nucleus. The chemical nature and biological significance of these "shuttle" proteins in *amoeba* and *Chironomus* is not known. Although there is a remote possibility that they serve as carrier molecules for the transport of other proteins, it is more likely that they may be involved in the coordination of nucleocytoplasmic activities.

Rates of Histone Biosynthesis

Information about the biosynthesis of the common histone fractions represents one of the most controversial aspects of histone biochemistry. In the past, studies on the biosynthetic rates of histones have been complicated by the procedural shortcomings of the fractionation of these proteins. For this reason, some older papers report values for mixtures of histones which were further resolved only later. If histones play a role in genetic restriction of the DNA, it is reasonable to expect the biosynthesis of the individual fractions to reflect the immediate need of each fraction for the always changing transcriptional activity of each cell. Therefore, if histones participate in the variable restriction of DNA, their biosynthetic rates should respond readily to the changes in the physiological state of the cells with equilibria obtained only in metabolically inert cells. On the other hand, if histones function mainly as permanent repressor molecules which inactivate genes designated during cytodifferentiation to remain silent, their biosynthesis should be invariable and parallel to DNA replication. Unfortunately, although the latter possibility appears to be indicated by many experiments, the situation is far from clear. Because of the vulnerability of histones to proteolytic degradation,[39-47] histones in living cells may be constantly degraded, especially if dissociated from the DNA. The constancy of genetic restriction in differentiated cells would then require an immediate replacement of histones from an existing intranuclear pool which, in turn, is constantly fed by histones made in the cytoplasm. This would result in a variable rate of histone synthesis which, unfortunately, simulates the situation expected if histones are to function in variable DNA restriction. Either situation can, then, result in marked differences in the biosynthetic rates of the individual histone fractions in various tissues.

Studies on histone biosynthesis can be divided into two general groups. One relatively small group presents data indicating that all common histones are synthesized at approximately equal rates. The other larger group describes the biosynthesis of individual histones to occur at different rates even after corrections have been made for the chemical composition of the fractions and for the amino acids used for labeling. The differences between the fractions are frequently considerable. In differentiated tissues, which do not undergo frequent mitoses, the nuclear nonhistone proteins are much more metabolically active than histones. As the mitotic rates increase, histones are biosynthesized more actively until, in rapidly dividing tissues, the histones are made at rates approximately equal to those of the nonhistone proteins.[109,159,168,522-525] The more rapid turnover of the nonhistone proteins raises a question of their possible contribution to the specific activities of the histone fractions, which they may easily contaminate.[141,144,164,526] However, since any nonhistone protein contamination present in large quantities would change the amino acid composition, electrophoretic behavior, and solubility of histones, the specific activity of the contaminating protein would have to be extremely high. An interesting example where the contamination of histones led to erroneous claims of their tissue specificity in neoplastic tissues appears in the reported discovery of a cationic protein which incorporated labeled lysine (RP2L). This RP2L protein has been described in great detail as tumor specific.[108,109,152,308] However, according to MacGillivray and Monjardino,[527] the RP2L is a nonhistone protein constituent of cell nuclei and is not specific for the neoplastic growth. Its labeling with radioactive lysine apparently reflects the metabolic state of the cells.

The possibility of histone contamination causing apparent differences in the biosynthetic rates of the histone fractions was first pointed out by Laurence and Butler[528] who investigated ^{14}C-lysine and ^{14}C-arginine incorporation into histone fractions Fl (I) F2a (IIbl and IV), F2b (IIb2), and F3 (III) from rat liver and several experimental tumors. The differences in the biosynthetic rates between the fractions, noted initially, decreased or disappeared altogether after each fraction was extensively purified. They concluded that the histones were most likely biosynthesized at closely similar rates. Similarly, Umana et al.,[529] Bustos-Valdes and Dounce,[525] Dick and Johns,[530] and Bondy[531] found the main histone fractions to be at metabolic equilibrium in rat liver, rat thymus, and developing chick brain, respectively. In tissue cultured Chinese hamster ovary (CHO) cells, the incorporation of ^{14}C-protein hydrolysate was nearly identical for all the histone fractions with perhaps the exception of the F2a2 (IIb1) histone, which exhibited a significantly lower uptake in thymidine blocked cells as compared with the rest of the fractions.[532] A later report from the same laboratory concerning more detailed studies on F1 (I) histone biosynthesis and turnover concluded that this fraction does not follow the temporal schedule found for the remaining histones. The incorporation data suggested the existence of an extrachromatin pool of F1 (I) histones.[533]

In contrast to the relatively few reports which described approximately equal temporal rates of biosynthesis for all the main histone fractions, experimental evidence indicating differences in such rates was much more abundant. Although the full extent of histone heterogeneity was not yet realized, several investigators noticed that not all the histones incorporated labeled amino acids to a similar extent.[109,534-536] Due to the different procedures utilized, it has been found difficult to discuss all the work concerning the biosynthesis of the histone fractions in various tissues and animals without confusing the issue at least to some extent. Therefore, the next several paragraphs point out which sources of contamination were most likely for the employed procedure. This was not meant to imply that the histone preparations actually were contaminated and that the differences in the labeling of the individual histones could not be real.

Using chemical fractionation supplemented by chromatography on carboxymethyl cellulose, Hnilica and Busch[50] found differences in the incorporation of ^{14}C-lysine into the histones of Walker tumor cells. The lowest specific activity, corrected for lysine content, was found in F1 (I) and increased in the order of F2a (IIb1 and IV), F2b (IIb2), and F3 (III). The specific activity of the latter fraction was about twice that of F1 (I). A similar finding (considerably higher incorporation of ^{3}H-lysine into the arginine-rich histones) for human adrenal carcinoma was more recently reported by Jungmann et al.,[537] who used chromatography on carboxymethyl Sephadex for the final purification of their histones. It has been shown that chromatography on carboxymethyl cellulose does not fractionate the F3 (III) histones satisfactorily.[59] The amino acid composition of the arginine-rich histones isolated by Jungmann et al.[537] suggested that chromatography on carboxymethyl Sephadex appeared to have similar disadvantages. Therefore, the higher specific activity of these histone fractions may be caused by their contamination with other proteins.

In normal and regenerating rat liver and Novikoff hepatoma, histone fractions F1 (I), F2a (IIb1 and IV), F2b (IIb2), and F3 (III) were reported to take up ^{14}C-lysine at different rates.[168] In liver, the F1 (I) and F3 (III) incorporated the label more than twice as efficiently as the fractions F2a (IIb1 and IV) and F2b (IIb2). The differences in labeling decreased in regenerating liver (36 hr after hepatectomy) and were nearly eliminated in Novikoff hepatoma. Dissimilar specific activities between histone fractions F1 (I), F2b (IIb2), F3 (III), and F2a (IIb1 and IV) from Ehrlich ascites carcinoma were seen by Ohly et al.[538] when the cells were incubated in vitro with ^{14}C-lysine. As in Walker carcinosarcoma,[50] the specific activity of the very lysine-rich histones was lowest and that of the arginine-rich histones highest. Both Hnilica et al.[168] and Ohly et al.[538] used essentially the same isolation procedure, which consisted of the selective extraction of the arginine-rich and lysine-rich histone groups[34] followed by fractionation on Sephadex G-75.[162-164] The disadvantage of gel filtration is that the peaks emerging first (i.e., F1 or I and F3 or III) were most likely to be contaminated by nonhistone proteins because of the relatively large size of their protein complexes. Although there was hardly any difference in the specific activities of histones and nonhistone proteins in Novikoff hepatoma, the latter were about 3 to 5 times more active in liver.[168] Hence, contamination by nonhistone proteins could have caused the higher specific activities of Fl (I) and F3 (III) histones in rat liver. However, the low specific activities of F1 (I) histones seen by Ohly et al.[538] in Ehrlich ascites carcinoma cannot be explained by such a contamination.

Another technique frequently employed for the fractionation of histones has been the selective extraction of the arginine-rich histones F2a + F3 (IIb1, IV, and III) with acidified ethanol.[34] The lysine-rich group F1 + F2b (I and IIb2) could then be obtained from the residue with 0.25 *N* HC1. Each group was fractionated chemically (acetone, ethanol, or trichloroacetic acid precipitation) into the individual main histone fractions. Histones fractionated by this technique or its modifications exhibited differences in their labeling behavior. In experiments with isolated calf thymus nuclei, the specific activity of F1 (I) histones was found by Reid et al.[213,485] to be only about half the activity of other histones. The nonhistone nuclear proteins were nearly ten times as active as the histones. Considerably lower specific activities of the very lysine-rich histones were also found in goat liver.[539] On the other hand, in rat liver undergoing neonatal carcinogenesis induced by dimethyl nitrosamine (three-and five-day-old rats), as well as in the control animals, the F1 (I) histones were labeled at considerably higher levels than the fractions F2a (IIb1, IV), F2b (IIb2), and F3 (III). The differences in specific activities were nearly eliminated in post-weaning rats (four to five weeks old).[540] No significant differences in the uptake of ^{3}H-leucine by histone fractions F1 (I), F2b (IIb2), and F2a (IIb1 and IV) – F3 (III) were noticeable in embryonic chicken brains.[531] The pattern of biosynthesis of histones in chicken reticulocytes has been even more controversial. While Henderson et al.[541] found only small differences in the uptake of ^{14}C-lysine by the fractions F1 (I), F2a (IIb1, IV), F2b (IIb2), and F3 (III), Freedman et al.[542] reported a maximal labeling (^{14}C-lysine) of the F3 (III) histones which decreased in the order of F2b (IIb2) $>$ F1 (I) $>$ F2a (IIb1 and IV). The results of both groups were corrected for the lysine contents of the histone fractions. This biosynthetic pattern was further complicated by the reported complete absence of label in histones I (F1) and II (F2a2 and F2b) in chicken reticulocytes incubated with ^{14}C-leucine.[543] Since all the histones discussed in this paragraph, except for those of Sadgopal and Kabat,[543] were obtained by modifications of the same method, the differences in labeling are difficult to explain.

The chromatographic separation of histones on Amberlite IRC-50® (or CG-50®), first introduced by Luck et al.,[58] yielded yet more controversial results. Using this method, Chalkley and Maurer[544] found considerable differences in the incorporation of ^{14}C-leucine into the histone fractions of rat liver, calf endometrium, pea cotyledons, and tissue cultured tobacco pith cells. Tissues with low rates of DNA replication biosynthesized only the arginine-rich histones III (F3) and IV (F2a1). The lysine-rich fractions I (F1) and II (F2a2 and F2b) became labeled only in tissues undergoing DNA synthesis. A quite similar situation, also detected by the chromatographic procedure on Amberlite IRC-50, was described by Sadgopal and Kabat[543] for chicken reticulocytes. The absence of lysine-rich histone synthesis in resting tissues was grossly discrepant with all other reports in the literature. As was pointed out by Stellwagen and Cole,[144,545] the much higher labeling of the arginine-rich histones III (F3) and IV (F2a1) separated by chromatography on Amberlite IRC-50 might be due to the presence of contaminating acid soluble nonhistone proteins. In resting tissues, these proteins are known to turn over much faster than histones. The relatively rapid increase in guanidine hydrochloride concentration used to elute the arginine-rich histones might precipitate a simultaneous elution of the contaminating proteins together with the arginine-rich fractions and, thus, increase their apparent specific activities. The incorporation of label into the histones in nondividing tissues might be below the detection level of this technique. Since part of the nonhistone protein contaminant eluted with the void volume, any significant amount of radioactivity eluted in this position should be cause for interpreting the specific activities of arginine-rich histones only with great caution. Since the biosynthetic rates of the histones and nonhistone proteins in rapidly dividing tissues were known to be almost equal, the nonhistone protein contamination of their arginine-rich histones prepared by chromatography on Amberlite IRC-50 could not be detected without amino acid analysis of the fractions.

The electrophoretic fractionation of histones, especially if performed in the presence of concentrated urea or sodium dodecyl sulfate, has been found to offer a suitable tool with which to obtain reliable information on the biosynthesis of histone proteins. Unless linked by chemical bonds, protein aggregates are known to dissociate in the presence of both reagents. This property, therefore, tends to eliminate erroneous radioactivity from the nonhistone proteins contaminating histones. Using polyacrylamide gels containing 6 *M* urea, Marzluff et al.[546] studied the biosynthesis of histone fractions during the hormone dependent proliferation of epithelial mammary cells. Both histone and DNA synthesis were stimulated by insulin and all the fractions increased their specific activities coordinately. There were differences in the specific activities of the main histones, suggesting individual rates of biosynthesis for each histone fraction. McClure and Hnilica,[547] who were interested in the temporal relationship of DNA and histone synthesis during the cell cycle of Chinese hamster DON-C cells, studied the incorporation of newly synthesized histone fractions into chromatin. The specific activities of the histones were determined for fractions separated by gel electrophoresis in the presence of 6 *M* urea. These conditions decreased the possibility of nonhistone protein contamination of histones and, furthermore, employed in vitro labeling experiments which yielded histones with high specific activities, thus improving the statistical significance of the experiments. The fractions F1 (I) and F3 (III) incorporated the label more than twice the rates seen for the fractions F2b (IIb2), F2a1 (IV), and F2a2 (IIb1). The ratios of the specific activities did not change during the cell cycle. These results resembled the findings of Marzluff et al.[546] discussed earlier. Higher specific activities for the F1 (I) and F3 (III) histones were also found in regenerating rat liver, especially during the peak of DNA synthesis.[267] Here, however, because of the low incorporation of ^{14}C-lysine into the histones in vivo, starch gel electrophoretic separation of the chemically fractionated histones was employed. Since no urea was present in the starch gels or in the samples, contamination of the histones, especially of the F3 (III) fraction (electrophoretically slowest), cannot be excluded. Such a contamination may have accounted for some of the observed differences in biosynthesis between the histone fractions of Ehrlich ascites cells reported by Holoubek,[548] who also used the starch gel electrophoretic technique for separating the histone fractions.

The F1 (I) very lysine-rich histones have been shown to exhibit a number of special properties relative to the other fractions (tissue specificity, heterogeneity, metabolic variability, binding to chromatin DNA, etc.) and were, hence, considered worthy of separate discussion. After the discovery of their heterogeneity, Cole and associates initiated detailed studies on the chemical and biological properties of this fraction. When isolated thymocyte nuclei were incubated with a mixture of ^{14}C-amino acids, an active biosynthesis of F1 (I) histones could be detected. However, the specific activities of the four F1 (I) subfractions in this tissue, were practically the same.[485] An entirely different situation was found when rat mammary gland cells were labeled in vivo or in vitro. During late pregnancy and early lactation, histone classes I (F1), II (F2b or IIb2 and F2a2 or IIb1) and III (F3 or III and F2al or IV) exhibited similar biosynthetic rates.[549] However, the lysine-rich F1 (I) and arginine-rich F3 (III) and F2al (IV) histones had consistently higher labeling rates than the moderately lysine-rich fractions F2a2 (IIb1) and F2b (IIb2), i.e., a situation also reported by several other investigators in various tissues.[168,267,546,547] One of the very lysine-rich histone F1 (I) subfractions had a higher rate of synthesis than the others. Moreover, in mammary gland organ cultures, Hohmann and Cole[550] demonstrated that the burst of DNA synthesis resulting from the incubation of the culture with insulin, hydrocortisone, and prolactin was accompanied by a similar burst of very lysine-rich histone biosynthesis. There were dramatic changes in the incorporation of ^{14}C-lysine into certain of the five F1 (I) histone subfractions as a result of the hormone treatment. In a more detailed report, these changes in the biosynthetic rates of the F1 (I) subfractions were correlated to the in vivo situation in lactating mice.[551] The finding of differences in the labeling of the individual F1 (I) histone subfractions in neoplastic tissues[552] pointed to the possibility that the heterogeneity of the very lysine-rich histones may be associated with the regulation of cellular metabolism. Since subfractionation of the F1 (I) histones was achieved by only a shallow increase in the guanidine hydrochloride concentration used for the elution of these proteins from Amberlite IRC-50 columns, a coelution of nonhistone protein contaminants was most unlikely.

The considerable discrepancies between reports concerning the biosynthesis of the histone fractions are difficult to explain entirely on the basis of contamination with nonhistone proteins of enhanced metabolic activity. The contamination cannot be excessive, since most of the fractions were extensively purified. Hence, the contaminating proteins would have to show extremely high turnover rates. Although the existence of such proteins cannot be excluded, no experimental support for their presence in chromatin has been reported. Perhaps the most plausible explanation of the differences in the biosynthetic rates of the histones in the same tissue is the existence of extrachromatin pools of at least some of the fractions. From the irregularities of the biosynthesis and turnover of F1 (I) histones in tissue cultures and from data obtained on X-irradiated cells, Gurley et al.[532,553,554] suggested that there is an extrachromatin pool of histones F1 (I) and F3 (III) and, perhaps also of the other fractions. It appears that under normal conditions, the biosynthesis and turnover of histones are well coordinated with those of the DNA. This coordination is not absolute however, and individual histones seem to be able to escape it, at least temporarily.

Histone Turnover

The remarkable metabolic stability of chromosomal DNA stimulated speculations as to whether the chromatin complex, mainly DNA and histones, turns over in unity or whether histones and other chromosomal proteins are metabolically dissociated from the DNA. Among the first attempts, Prescott and Bender[555,556] labeled Chinese hamster fibroblasts with a mixture of 6 amino acids for 8 hr. After transfer into nonradioactive medium, metaphase cells were collected during 6 to 8-hr intervals up to 57 hr. The metaphase cells were fixed in 5% formalin and air-dried from suspension in 3:1 ethyl alcohol-acetic acid. During evaporation, the cells burst and the cytoplasm dispersed. The remaining metaphase chromosomes were analyzed for labeled protein by autoradiography. The labeled protein segregated dispersively between chromatids at the second metaphase. The labeling intensity became more diluted as the cells entered subsequent cell division with only some chromosomes labeled in the third metaphase. By the fourth metaphase, the label was no longer detectable. Although the precursor amino acids were by no means selectively incorporated into the histones, the results could be interpreted to indicate that dissociation of chromosomal proteins and DNA occurred during the cell cycle. Random distribution of chromosomal proteins after cell division was also described in other tissues.[557-559]

Among the first biochemical attempts to determine the turnover of histones in vivo were the studies of Piha et al.[560,561] who found that in mouse brains and livers, histones had the slowest turnover rates of all the protein fractions studied. The average half-lives of liver and brain histones were 18 and 54 days, respectively. The high values found for brain histones appear consistent with the greater stability of the cell population in this tissue. The liver histones, on the other hand, turned over much faster than indicated from cellular turnover rates of liver. In both tissues histone fractions of considerably longer half-lives could be detected; the longest observed were 117 days for brain and 105 days for liver. Over a period of 2 to 8½ months in the brain and 5 to 8½ months in the liver, the decrease in histone specific activities was small, corresponding to a replacement rate of about 0.6% daily. Similar and small turnover rates of histones and DNA were also found in developing mouse embryos.[561]

Quite similar conclusions concerning the turnover rates of histones relative to the DNA were reached by Byvoet[562] who injected rats with 5-^{125}I-iodo-2′-deoxyuridine to label the DNA and ^{14}C-lysine to label the histones. The histones were fractionated into two fractions, i.e., the very lysine-rich fraction F1 (I) and other histones (the rest of the main fractions). The turnover rates were calculated from the first order decay curves. The ratios of the half times of F1 (I) histones and the DNA as well as those of "the other histones" to the DNA were close to unity for rat spleen, kidney, thymus, small intestine, and several transplantable tumors. Long-term experiments with liver and kidney showed the histone to DNA half time ratios to be slightly less than unity (approximately 50 to 80% of the ratios found for fast growing tissues.) The half-lives of histones in neoplasms were consistently 40 to 60% larger than those of DNA, indicating a possible reutilization of the labeled lysine in malignant cells. However, the author disregarded these differences as insignificant and concluded that histones and DNA turn over en bloc at approximately equal rates. The halflife for rat liver histones and DNA of approximately 20 days is in excellent agreement with the 18-day values reported by Piha et al.[560,561] in mouse liver. The metabolic integrity of the histone and DNA complex in chromatin was not altered in liver parenchymal cells during alternate enzyme induction and repression resulting from dietary changes or injection of phenobarbital.[563]

Much shorter half-lives of histones in rat liver were observed by Laurence and Butler,[528] who studied histone and DNA metabolism in liver and several ascites tumors of these animals. Once labeled, the histones turned over slowly. In liver, the whole histone lost about 50% of its activity in about one week. Although the rates of biosynthesis of all the histone fractions were essentially the same, the specific activities of the ethanol-HCl soluble histones, i.e., F2a1 (IV), F2a2 (IIb1), and F3 (III) decreased only by 24% in seven days, indicating a more rapid turnover of the lysine-rich histones F1 (I) and F2b (IIb2). The authors did not explain how, if all the histones are biosynthesized at equal rates, the lysine-rich fractions can disappear faster than the arginine-rich ones. Half-lives of seven days for labeled rat liver DNA and histones were also reported by Murthy et al.[564] According to these authors, the short half-lives of both DNA and histone increased after 2 weeks to about 30 days. The biphasic decay curves for rat liver histones may have been caused by metabolic changes or other stimuli. In developing chicken brain,[531] the histones decayed with half-lives of about 4 to 5 days during the prehatching period of embryogenesis. After hatching, the rates increased to about 19 days. In the same tissues, the DNA half-lives were approximately 23 to 50 days, respectively, indicating much higher turnover rates for histones than DNA, both before and after hatching. In chicken brain, all the histones were biosynthesized and turned over at approximately equal rates.[531]

In general, disregarding some discrepancies in the half-lives of histones in liver, these proteins, once synthesized, are quite stable and all the fractions appear to be turning over as a group. Slow turnover rates of histones in animal tissues have also been reported by other investigators.[534,546,565] In contrast to histones, the nonhistone proteins appear to be turning over quite rapidly[531] with no correlation to the much slower turnover of the DNA.[566,567]

As can be expected from a system where the cellular growth rates and amino acid pools can be better regulated, the in vitro experiments with cells grown in tissue cultures were less controversial. In HeLa cells synchronized by a thymidine block, Spalding et al.[568] found active synthesis of basic proteins concomitant to the DNA replication. The basic nuclear proteins which were separated into three bands by polyacrylamide electrophoresis (presumably histones) displayed similar metabolic patterns. Small differences were interpreted by the authors to indicate independent biosynthesis of these three fractions. All three proteins turned over, as was evidenced by a decline of their specific activities over a period of 16 hr. One of the three bands appeared metabolically unstable, whereas the other two were conserved.

In onion root meristem nuclei, histones were shown to be conserved from one cell cycle to another.[569] Working with P815 cells (mouse mastocytes) and HeLa cells, Hancock[570] demonstrated the conservative turnover of histones and DNA during considerable time periods. The cells, labeled for three to four generations and then transferred into unlabeled medium, maintained their content of ^{14}C-histone relative to ^{3}H-DNA essentially unchanged during several further generations. No loss or irreversible dissociation of the major histone fractions from chromatin occurred during growth and mitosis. No differences could be found in the quantitative ratio of histones to DNA, or in the relative amounts of the histone fractions separated by polyacrylamide gel electrophoresis between the chromatin of cells in interphase or in mitosis.

Detailed information concerning the biosynthesis and turnover of histones in CHO (Chinese hamster ovary) cells grown in tissue cultures was contributed by Gurley and Hardin.[532,533,571] As discussed earlier in this chapter, the biosynthetic rates of histones in these cells were quite similar and each fraction was biosynthesized concomitantly with DNA replication. During exponential growth, the F1 (I) histone halflife was calculated at 74 hr, while histone fractions F2a1 (IV), F2a2 (IIb1), F2b (IIb2), and F3 (III) did not show turnover for three cell generations. When DNA synthesis was inhibited by a thymidine block, histones F2a1 (IV), F2a2 (IIb1), F2b (IIb2), and F3 (III) continued to be biosynthesized and turned over with half-lives of 60, 58, 55, and 68 hr, respectively. The turnover pattern of F1 (I) histones was more complex, suggesting the existence of a nonchromatin pool of these proteins.[532] Later, Gurley et al. presented evidence for extrachromatin pools of both the F1 (I) and F3 (III) fractions.[553;554]

Although not in complete concert, the data on histone turnover are less controversial than those concerning the biosynthesis of individual fractions. In general, histone synthesis coincides with the time of DNA replication, although it may not entirely depend on it. All the histone fractions appear to be conserved over long time periods with turnover rates similar to those of the DNA. The differences in the turnover of some histone fractions reported by some authors can be readily explained by the existence of pools for histones F1 (I) and F3 (III) and perhaps also for the rest of the fractions. It can be reasonably expected that some of the histones in chromatin may become damaged by proteases during cellular life processes, requiring subsequent replacement of the damaged protein. Except for a sensitive feedback control of individual histone biosynthesis, the pool of histones not bound to the DNA offers a rapid means for replacing the damaged fractions.

Regulation of Histone Synthesis

In view of the uncertainties concerning the biosynthetic rates of histone fractions, it may be precocious to speak about the regulation mechanisms governing histone biosynthesis and turnover. Nevertheless, experiments have been carried out in several laboratories studying the influence of various drugs and chemicals on DNA, RNA, and protein synthesis in living cells. Studies in systems where DNA replication was inhibited to a large extent by 5-fluoro-deoxyuridine,[569,572] X-irradiation,[573,574] hydroxyurea,[575-577] 1-β-D-arabinofuranosyl cytosine,[578] or hydrocortisone,[579] showed that amino acid incorporation into histones was decreased only partially or not at all. This was in disagreement with a system using isolated small polysomes for in vitro histone synthesis. If the DNA synthesis of the cells was inhibited prior to the isolation of small polysomes, they rapidly lost their capacity to make histones in vitro,[502-505,509] indicating the prerequisite of DNA replication for histone biosynthesis. On the other hand, experiments with synchronized populations of *Tetrahymena pyriformis*[244,580] can be interpreted to indicate the need of histone synthesis prior to DNA replication. Hnilica and Trojan[581] inhibited DNA synthesis in Novikoff hepatoma cells by injecting colchicine into the tumor-bearing rats. The incorporation of labeled lysine into histones was substantially inhibited before the arrest of DNA synthesis (3 to 6 hr). No incorporation of ^{3}H-thymidine could be detected 12 hr after colchicine injection. At this time however, lysine incorporation into the histone fractions began to recover and reached essentially normal values [with the exception of the F2a (IV and IIb1) histones, which remained depressed] 24 hr after the administration of colchicine. At this time, there was no noticeable incorporation of labeled thymidine into the tumor DNA. This again could be interpreted to indicate that, although in normal conditions the DNA and histone synthesis is firmly coupled, the histones can be biosynthesized independently. Unfortunately, it has not been possible to specifically inhibit histone synthesis to determine whether DNA replication would also cease.

When Ehrlich ascites cells were infected with Maus-Eberfeld virus[582,583] the biosynthesis of their histones, as measured by the incorporation of labeled arginine, increased significantly, while thymidine incorporation into the DNA remained unchanged. However, since the specific activities of all the histones increased simultaneously, it was possible that the observed increase might have been caused by changes in the arginine pool sizes following infection. This was also suggested for the infected cells by a similar, but smaller, increase in the specific activities of the nuclear nonhistone proteins which contain considerably less arginine than most of the histone fractions. HeLa cell infection with herpes simplex virus resulted in a general inhibition of protein synthesis, including the histones.[584]

In polyoma virus-infected mouse embryo cells,[585] primary mouse kidney cells, and hamster BHK 21 cells,[586] as well as in SV 40 infected mouse 3T3 cells,[586] the increased rate of DNA synthesis caused by virus infection was accompanied by an increase in the incorporation of amino acids into the histone proteins. Similar experiments performed on SV 40 infected mouse 3T3 cells, which were X-irradiated (5,000 r), showed that the virus infection not only promoted DNA and histone synthesis, but also induced considerable mitotic activity following the wave of DNA and histone synthesis.[586] According to the authors, the contact inhibited 3T3 cells were arrested in the G_1 period of the cell cycle[587] prior to irradiation and the virus infection somehow released the block and promoted the progression of the irradiated cells into mitosis.

The inhibition of RNA synthesis and its effects on the incorporation of labeled amino acids into the histones have also been investigated. Based on autoradiographic studies, Suskind[588] concluded that actinomycin-D inhibited the synthesis of lysine-containing proteins in the nucleoli of HeLa cells. The inhibition of RNA synthesis in chicken reticulocytes by actinomycin D (or by 5,6-dichloro-1-β-D-ribofuranosyl benzimidazole) was followed by a rapid decrease in histone biosynthesis, although hemoglobin synthesis remained unchanged.[542] The biosynthesis of lysine-rich histones was inhibited more than the arginine-rich histones. The onset of inhibition was rapid, i.e., the incorporation of label into the lysine-rich fractions was reduced more than 50% in 20 min, indicating the relative short halflife of m-RNA coding for these histones. Short halflives of histone m-RNA (60 to 90 min) were also found in HeLa cells by Borun et al.[503] and Zampetti-Bosseler et al.[577] In vivo, actinomycin D inhibits histone synthesis in a similar fashion.[581;589] Curiously the metabolism of F1 (I) histones was not affected by actinomycin D in Chinese hamster cells.[532] Puromycin also inhibited histone biosynthesis[483-485,577] with a simultaneous inhibition of the incorporation of labeled thymidine into DNA,[569,577] indicating that histone synthesis may be necessary for DNA replication. On the other hand, the inhibition of arginine-rich histone synthesis by canavanine in HeLa cells increased the rate of DNA synthesis.[590] If the cells were first infected with poliovirus, canavanine nullified the decrease of DNA synthesis normally observed after the infection.

In general, the experiments with specific inhibitors of protein, DNA or RNA synthesis suggest that histones are biosynthesized by mechanisms similar to those for other proteins. How closely histone synthesis is coupled to DNA replication and vice versa is still controversial with experiments reported in the literature supporting both the tight coupling of these two events as well as their complete independence. Only more experiments can decide whether these controversies are peculiar to certain cellular types or whether they are the results of nonspecific effects of the chemicals used to modify the biosynthetic processes in living cells.

Temporal Relationships Between DNA and Histone Synthesis

Although the constancy of the DNA content in the nuclei of a given vertebrate species was established more than two decades ago,[591,592] it was only later that the close relationship of DNA and histone biosynthesis was temporally related to the cell cycle activities. The parallel increase of DNA and basic proteins in the nuclei of growing animal or plant cells was first demonstrated by Bloch and Godman[120,399] and by Alfert.[593] The close relationship between histone and DNA biosynthesis was further confirmed by elegant experiments with the regeneration band in the macronucleus of the *Euplotes amoeba.*[520,555,594-596] Bloch et al.[569] analyzed the temporal pattern of ^{3}H-lysine and ^{3}H-arginine incorporation into the chromosomes of onion root meristem cells and found not only a close coincidence of chromosomal protein and DNA biosynthesis, but also an additional period of chromosomal protein synthesis late in the interphase portion of the mitotic cycle. Similar evidence indicating that the biosynthesis of basic nuclear proteins commenced with the onset of DNA synthesis and ended with its termination, based mainly on cytochemical and autoradiographic data, was also presented by other investigators.[597-604]

Since it was impossible to correlate the cytochemical and autoradiographic measurements to the biosynthesis and turnover of the individual histone fractions, attempts were made to establish cellular systems capable of synchronous mitotic activity. One of the first sytems used in studies on the temporal relationship between histone and DNA synthesis was regenerating liver, mostly in rats. Although the mitotic synchrony of postoperative liver is poor, it was possible to establish that the maximal biosynthesis of all the histones coincides with, or slightly precedes, the maximal incorporation of labeled thymidine into DNA.[267,529,534-536,605-611]

In a protozoan, *Tetrahymena pyriformis,* considerable synchrony of both cell multiplication and DNA replication could be achieved by exposing the cells to a heat shock.[612,613] Interestingly, the stationary phase population of this protozoan maintained a histone to DNA ratio of 2:1, while the logarithmic growth phase has a ratio of 1:1.[244,580] This unusual phenomenon indicating the independence of histone and DNA synthetic processes was probably the result of an incomplete inhibition of DNA synthesis in the arrested cells.[614,615] During DNA replication, there was an active synthesis of all the histones and some differences in the rates of individual fractions.[244] Another primitive system, the slime mold *Physarum polycephalum,* could be synchronized by simple manipulations and a change in nutrient medium.[616] The histone to DNA ratio remained constant (about unity) during the cell cycle with slightly more histone present during mitosis. The biosynthesis of all the histones continued even after DNA replication was blocked by 5-fluorodeoxyuridine. Histone synthesis in the absence of DNA replication continued for an extended period of time and resulted in a final threefold excess of histones over the DNA. The ratios of the individual histones changed with the physiological and reproductive states of the cultures. There was a slight decrease (0.8 ratio) of histone relative to the DNA in starving and sporulating plasmodia.[617]

Cultured mammalian cells can be synchronized by blocking biochemical pathways leading to DNA synthesis[618-620] or by selective dislodgement of cells arrested in metaphase.[621] In HeLa cells synchronized by the latter method, Robbins and Borun[502] found very low histone synthesis in the G_1 period. Most of the histones were biosynthesized during the S phase of the cell cycle. Studies on the biosynthesis of histones by isolated polysomes using inhibitors of DNA synthesis demonstrated a strict dependence of histone biosynthesis on DNA replication.[502-506] A conclusion that histones and DNA are biosynthesized and turn over simultaneously was also reached in several other laboratories.[547,560-563,570,619] On the other hand, Gurley and Hardin[533] found that although in Chinese hamster ovary cells (CHO) the major period of histone biosynthesis and turnover can be identified with the S phase of the cell cycle, there was significant histone biosynthesis in the G_2, M, and G_1 periods. In HeLa cells, Sadgopal and Bonner[622] also reported considerable biosynthesis of histones outside the S period of the cell cycle, confirming the similar observations of Spalding et al.[568] and Littlefield and Jacobs.[623] Reports on the temporal dissociation of histone and DNA synthesis in cells exposed to various metabolic inhibitors were discussed earlier in this chapter. It appeared that although these two biosynthetic events are closely correlated in cells during normal growth and reproduction, their coupling is not absolute.

Since many cell synchrony systems require biochemical intervention into the normal physiology of the cultured cells (thymidine, 5-fluorouracil, colcemid, isoleucine starvation, etc.), the dissociation of histone synthesis from DNA replication during the cell cycle may be caused by interference with biosynthetic pathways. It has been reported that the use of thymidine to synchronize cells in the culture does not absolutely block the cells at the G_1-S boundary, but allows a very slow rate of

DNA synthesis.[624-626] Also, the synchrony obtained by this method is relatively low.[627] Such conditions of biochemical imbalance together with the presence of asynchronous cells in the population may account for the biosynthesis and accumulation of histones during the cell cycle other than the S period. It is noteworthy that most of the experiments reporting histone biosynthesis outside the S phase of cell cycle were performed on thymidine blocked cells.

In addition to incomplete synchrony and the physiological effects of the chemicals used to synchronize cellular populations, the existence of extrachromatin histone pools may contribute to the disparities in DNA and histone biosynthesis. Such pools[532,553,554] which can be expected to be carefully balanced during normal living conditions, may escape their control mechanisms as the result of biochemical or other interferences used for synchronization. If this is true, the controls for histone synthesis would appear more labile than those for DNA replication. Such lability could also explain some of the differences in biosynthetic rates of individual histone fractions reported by many investigators. If histones are used for permanent DNA repression in differentiated cells, they would be expected to be biosynthesized during DNA replication. This would permit instant repression of the DNA segments designated by cytodifferentiation to remain inactive. While DNA replication without histone synthesis would be detrimental to the internally ordered state of the cells in a multicellular organism, premature histone synthesis would most likely result only in their degradation or in a minor metabolic imbalance.

Basic Protein Synthesis During Gametogenesis and Embryogenesis

It is not known what compels germ cells in sexually reproducing organisms to undergo meiosis or reduction division. The biochemical processes governing chromosome pairing, crossing over and other events of meiosis are yet to be determined. While histone and DNA synthesis are temporally related (in most cells) during the S period of the cell cycle, in meiosis, DNA synthesis terminates before or during the leptotene stage, i.e., DNA replicates during the premeiotic S phase and during early meiotic prophase.[628-633] In addition, intracellular variations in the quantitative relationship between histone and DNA contents are known to occur during meiosis. In the cricket *Acheta domesticus L.*, the ratio of histone to DNA varies during the timetable of spermatogenesis. The average histone content at the leptotene-zygotene stage corresponds to 3C (4C for the DNA content) and increases with time to 4C in zygotene-pachytene (4C DNA content).[630] While DNA synthesis is practically completed during the premeiotic S period, histones continue to be synthesized during leptotene-zygotene. In pachytene-diplotene, histone synthesis terminates and the DNA to histone ratio is near unity, where it remains for the rest of meiosis. Similarly, in the hemipteran *Pyrrhocoris apterus* the DNA amount per nucleus doubles in the premeiotic interphase, while the termination of histone doubling is delayed to pachynema.[635] During meiotic prophase, the inhibition of protein synthesis under conditions not affecting RNA synthesis results in the rapid inhibition of DNA synthesis. Since the formation of the synaptinemal complex does not occur in the absence of such DNA synthesis, it apparently can not proceed when deprived of specifically required proteins. Biosynthesis of the nuclear proteins is essential for the synthesis of DNA and a combination of these events must occur during zygonema and pachynema to achieve successful chromosome pairing and chiasma formation.[634]

Since the chromosomes are deficient in histone protein during their pairing, Bogdanov et al.[630] suggested that this may be the signal for the onset of this event. Sheridan and Stern,[420] on the other hand, consider the presence and biosynthesis of a special meiotic histone, which appears to be synthesized only during the premeiotic period, to be involved in the progression of meiotic prophase events. It should be mentioned at this point that most of the information concerning DNA and nuclear protein synthesis during meiotic division has been obtained by cytochemical and cytophotometric procedures and almost exclusively on male germ cells.

Spermatogenesis

As discussed earlier, the terminal stages of spermatogenesis are accompanied by an interesting replacement of histones by arginine-rich proteins in many living species. Although there is an ample amount of cytochemical evidence about this transition, the best biochemically documented system is the appearance of protamines during the terminal stages of sperm maturation in many species of fish.

The appearance of protamines during spermatogenesis in fish was first studied by Miescher, who discovered this peculiar group of proteins almost a century ago. In the yet unpolluted river Rhine, the salmon fish *(Salmo salar)* were born in fresh water, matured in the ocean and, sexually mature, started their upstream journey back to the breeding grounds along the Rhine river. During this period, the fish accepted no food and their originally small testes enlarged many fold, filling with ripe spermatozoa. During the six to nine months of their upstream travel, the final stages of sperm maturation took place. The histones originally present in the spermatocytes were completely replaced by more primitive protamine. This replacement which occurs in many fish species was observed by Miescher and confirmed by others. Felix et al.[636] analyzed the basic protein contents in testes of the brook char *Salmo fontinalis* at 10-day intervals during the sperm maturation, which begins approximately 90 days before spawning. The nucleoprotamine appeared suddenly, about 40 days before spawning. The late appearance of protamines replacing histone protein during maturation of spermatozoa was confirmed in Chinook salmon *(Oncorhynchus tshawytscha)* by Alfert.[637] The protamines are biosynthesized after the cessation of DNA synthesis. It was assumed by early investigators that protamines represent the degradation products of more complex somatic proteins and histones. More recent investigations demonstrated however, that their de novo synthesis is regulated by mechanisms common to other cellular proteins.

Our present knowledge of protamine biosynthesis derives mostly from the work of Dixon and his associates. Ingles et al.[638] studied the biosynthesis of protamines in maturing testes of steelhead or rainbow trout *(Salmo gairdnerii)* and other salmon species. At an early stage of sperm maturation, polyacrylamide gel electrophoretic patterns of the basic nuclear proteins became more complex with the addition of a broad band of slower mobility than somatic histones. Later samples, taken 52 days after the induction of spermatogenesis by injections of the Chinook salmon pituitary extract, showed an additional electrophoretically fast protein band corresponding to the protamines in the protein pattern. The protamines could be labeled by ^{3}H-arginine and their biosynthesis was inhibited by cycloheximide and puromycin and to a lesser extent by chloramphenicol. Actinomycin D did not inhibit protamine synthesis at all over a four-hour period, indicating a long life of the m-RNA coding for the biosynthesis of protamines. At the concentrations used by the authors, the actinomycin D extensively inhibited the incorporation of ^{3}H-uridine into the RNA. Subsequently, Ling et al.[500,501,639] demonstrated that protamines are synthesized in the cytoplasm and rapidly transferred to the nucleus. Working with the rainbow trout *(Salmo gairdnerii)*, Ling and Dixon[639] followed the distribution of the slowly sedimenting disomes (120 s) on which the protamines were assembled during the maturation process of this species. The appearance and concentration of these disomes coincided with the period of rapid protamine synthesis. In the livers of this species, the polysome profiles did not change during sexual maturation. The disomes from spermatid cells did not dissociate in low Mg^{++} media during isolation and were RNase resistant. From the incorporation ratios of ^{14}C-arginine and ^{14}C-lysine by the disomes, the authors concluded that during the maturation process, the disomes are largely engaged in protamine synthesis. Experiments with actinomycin D indicated that the biosynthesis of m-RNA for protamines takes place before the start of protamine synthesis and is conserved (metabolically stable) through the final stages of sperm maturation. The synthesis of protamines was initiated by *N-formyl-methionyl*-t-RNA, i.e., a mechanism similar to numerous other protein synthesizing systems.[640] The protamines in the rainbow trout *(Salmo gairdnerii)* can be fractionated into three related components differing in their amino acid composition.[641] All of these three components were found in spermatid nuclei during maturation and the amount of each component depended on the stage of spermatogenesis. As the maturation progressed, component I decreased with a simultaneous increase of component III. This indicated that the biosynthesis of each of the protamine components was controlled separately and that the individual protamines might play specific roles during spermatogenesis. During the replacement, the histones were most likely removed from the DNA by proteolysis.[642] In addition to the mechanism of protamine biosynthesis, Dixon and co-workers also studied protamine phosphorylation and its bearing on the transfer of protamines from cytoplasm to the DNA in the nucleus. The phosphorylation of histones and protamines will be discussed in the next chapter.

Although the reason for the replacement of histones by arginine-rich protamines or similar proteins is unknown, it has been suggested that the more basic arginine-rich protein may be more efficient than histones in the genetic inactivation and packing of the DNA in mature spermatozoa. This suggestion received support from the studies of Marushige and Dixon,[360] who observed a decrease in the templating activity of spermatid chromatin paralleling the replacement of histones by protamines during sperm maturation. Chromatin isolated from testes at the beginning of maturation had approximately 10% of its DNA available for transcription. This changed to only about 2 to 3% of open template at stages 3 and 4. At this time the nonhistone protein content of chromatin decreased with a simultaneous increase of histones. In the next stages (5 and 6), histones began to be replaced by protamines and the templating activity of chromatin further decreased. Native nucleoprotamine did not contain any histone and was completely inactive as a template for the RNA synthesis in vitro. The mechanism of histone displacement has not been elucidated. Although it was reported[198, 641a,642a] that protamines can displace histones from chromatin in vitro, it is more likely that the actual mechanism is more complicated than simple electrostatic interaction and involves a selective proteolysis of the histones.[642]

Embryogenesis

Since not all the DNA contained in the eukaryote genome is transcribed in differentiated cells, a mechanism for the selection of genes to be transcribed at specific time periods must exist. The control mechanism, most likely invoked during ontogeny, maintains the differentiated state of the cells in an organism. It was suggested earlier in this chapter that histones may function as repressors of such permanently inactivated genes. If this suggestion is correct, the early embryonic state of eukaryotic organisms should evidence considerable variations in both the number of histones present in the nucleus and their biosynthetic rates.

First indications that histones may be absent from early embryos or replaced by less basic proteins came from cytochemical studies. According to Bloch and Hew,[122,499] four different kinds of basic proteins can be found in the snail *Helix aspersa* during its life cycle. The most prevalent "adult type histones" are present in embryos after the gastrula stage, adults, and spermatogonia. These proteins are characterized by positive staining with fast green, bromphenol blue, and eosin. Acetylation of the free amino groups (mainly on the ϵ-nitrogen of lysine) decreases their stainability considerably. Spermatid nuclei contain arginine-rich histones which also stain positively with all three stains, but are not affected by deamination. Spermatozoa contain protamine which stains with bromphenol blue and eosin, but not with fast green. Cleaving eggs (male and female pronuclei) and embryos before gastrulation contain "cleavage histones" which stain positively only with bromphenol blue. Deamination or acetylation eliminates the staining completely, suggesting that these proteins mostly contain lysine. Similar findings were reported for frog embryos[429] and during *Drosophila melanogaster* development.[397]

In mouse embryos, both the nuclei and cytoplasm stain equally well with alkaline fast green during the 1- and 2-cell stage.[643] Beginning at the 4-cell stage and continuing to the blastocyst, nuclear staining increases as the intensity of cytoplasmic staining decreases. Acetylation has no effect on staining prior to the 4-cell stage after which it reduces the fast green staining considerably. This suggests that the lysine-rich nuclear proteins increase during early development. In the frog (*Rana pipiens*), the early cleavage embryo interphase nuclei do not stain with alkaline fast green, while mitotic chromosomes are stainable.[396] Since Feulgen staining for DNA is also negative in interphase nuclei, the lack of histone staining is most likely due to their dispersal over a large area in the large interphase nuclei.[430]

Immers et al.[644] used ferric acetate staining at pH 2 (Hale staining) to detect protein groups with free sulfate, phosphate, and certain carboxyl groups with low pK values. The reaction was visualized by the formation of Prussian blue after the addition of ferrocyanide. Positive staining depends on the availability of free phosphate groups in chromatin or chromosomes. The removal of histones increased the staining, as can be expected from the exposure of the additional DNA phosphate groups to the stain. In experiments with the sea urchin *Paracentrotus lividus*, interphase nuclei remained unstained in early developmental stages even after pretreatment with trypsin, which efficiently removes histones from chromatin.[645] In early blastula, interphase nuclei became Hale positive following trypsin treatment. In swimming blastula, pretreatment with acetate buffer alone showed a light staining of the interphase nuclei which increased in intensity and became less pH dependent with further development. While prophase chromosome stainability varied according to the pretreatment used and depended on the developmental stage, metaphase and anaphase chromosomes were stainable with pretreatment at any developmental stage. Treatment with DNase or RNase did not affect the Hale staining in early blastula cells. If the sections were pretreated with trypsin, the staining was completely abolished. Interestingly, in late blastula, the presumptive ectoderm nuclei which later form nervous tissue, epidermis, etc., stained more intensively than the presumptive endoderm (gut) or mesoderm (connective tissue, muscle, etc.) nuclei. Although Hale staining procedure lacked specificity, the results indicated that the nuclei of the early embryonic stages contained proteins tightly bound to the DNA. This binding was weakened in metaphase chromosomes and during development. In an extension of this work, the Hale staining results were correlated to the ultrastructure of sea urchin embryo nuclei during embryogenesis.[646] During early cleavage, when Hale staining could not be induced in interphase nuclei, the chromatin was highly coiled and dense. During blastula, when Hale staining can be induced by either buffer or trypsin treatment, the chromatin was much more dispersed. That polysaccharides may participate in the condensation and unstainability of early blastula or unfertilized eggs nuclei was suggested by Bäckström[428] who found that digestion with hyaluronidase increased the staining for the nuclear basic proteins.

Although the cytochemical methods lack specificity necessary to distinguish the chemical nature of proteins involved in DNA restriction during early embryogenesis, they indicate that these proteins may be different from histones found in later developmental stages. It can be concluded that in most of the studied embryos, the mitotic chromosomes contain basic proteins stainable with fast green or picric acid-bromphenol blue. Nuclei from blastula and higher stages stain well with fast green even during interphase, while those nuclei from pronuclear and early cleavage stages are negative. It is impossible to decide presently whether the lack of staining occurs because the chromatin is too diffuse for adequate staining or whether the basic proteins are modified either chemically or by interactions with electronegative macromolecules. Either modification should, however, decrease the interaction of histones with DNA, which is in contrast to the highly condensed and transcriptionally inactive chromatin DNA observed in the unfertilized eggs or early embryos of the sea urchin.[27,646]

Better conclusions about the chemical nature and composition of nuclear basic proteins can be reached if these proteins are isolated and characterized. Unfortunately, the biochemical evidence about histone biosynthesis during early embryogenesis is controversial. Most investigators agree that typical histones are present in the nuclei of blastula or gastrula embryos depending on the particular animal species. Analytical studies (polyacrylamide gel electrophoresis, chromatography, isolation, and analysis of histone fractions, etc.) show that in advanced embryos, the histones are similar to those in adult tissues.[159] Quantitative changes in the concentration of three histone fractions in chick embryos have been reported during the first five days of incubation.[647] However, the qualitative distribution of histones and their chemical composition did not change during the embryonic development in chickens[328,329] or during the development of tadpoles.[320] Fambrough et al.[100] found considerable variations in the content of very lysine-rich histones F1 (I) during the development of pea embryos. Since this fraction is the most susceptible to procedural artifacts caused by shearing and proteolysis during the isolation of chromatin[41-43,101,102] the quantitative studies on F1 (I) histones must be interpreted with caution. There is no information, at the present time, about the composition and properties of histones in early stages of vertebrate embryogenesis.

Of all the animals, sea urchins have been the most extensively used in embryology laboratories all over the world. Their eggs can be fertilized in sea water, where further development also takes place, and it is possible to work with large quantities of temporally well synchronized sea urchin embryos. Since the period of embryonic development from fertilization to gastrulation is characterized by rapid DNA replication and frequent mitoses, an active synthesis of histones can be expected during this time to meet the need for these proteins in chromatin.

Stimulated by the reports about the absence of lysine-rich histones in sea urchin embryos during the early developmental stages,[236] Orengo and Hnilica[239] analyzed the 5% perchloric acid extracts of the nuclei from the sea urchin *Strongylocentrotus purpuratus*. At this concentration, perchloric acid solubilizes only the very lysine-rich histones F1 (I).[36] The amino acid composition of such proteins, which is shown in Table 12, revealed the absence of the very lysine-rich histones in these embryos before the blastula stage. Instead, a peculiar acid soluble, serine-rich protein was obtained from the nuclei of developing embryos. In blastula and later, a typical very lysine-rich histone appeared and could be extracted according to the protocol used for other organisms.[239,648]

Silver and Comb[649] have shown that a group of acid soluble proteins was selectively synthesized

TABLE 12

Amino Acid Composition of Proteins Soluble in 5% Perchloric Acid (Very lysine-rich histones Fl or I) From Developing Embryos of *Strongylocentrotus purpuratus*[239]

	Developmental Stage			
Amino acid	Egg	64 cells	Hatching blastula	Prism
Lysine	**13.5**	11.7	**25.5**	**25.3**
Histidine	1.7	2.3	1.3	1.6
Arginine	2.6	2.8	3.4	2.8
Aspartic acid	7.9	7.9	**4.6**	**4.4**
Threonine	4.4	4.3	5.0	4.5
Serine	**15.5**	**17.4**	**9.0**	**9.5**
Glutamic acid	**11.7**	**13.7**	**10.3**	**10.2**
Proline	4.3	3.7	6.1	6.7
Glycine	**13.9**	**13.2**	6.3	6.6
Alanine	**11.1**	8.6	**15.6**	**15.5**
Valine	4.0	3.6	3.9	4.5
Methionine	0.5	0.7	0.2	0.2
Isoleucine	1.9	2.1	2.1	2.2
Leucine	3.3	5.0	3.7	3.3
Tyrosine	1.3	1.0	0.7	1.0
Phenylalanine	2.2	1.8	1.2	1.8

All values are expressed as percent of total moles of amino acids recovered. The serine values were corrected (10%) for hydrolytic losses. The main features distinguishing each fraction are printed in boldface type. Tryptophan was not determined.

at the beginning of blastulation in *Lytechinus variegatus*. This protein had an amino acid composition unlike that of histones, but could be cochromatographed with the calf thymus histone group F2 (mixture of F2a1 or IV, F2a2 or IIb1 and F2b or IIb2). In their more recent contribution, Benttinen and Comb[434] reported the presence of a basic protein in the 32-cell stage embryos of *Lytechinus variegatus* with an electrophoretic mobility corresponding to the moderately lysine-rich fraction F2b (IIb2) from calf thymus tissues. During early embryogenesis, this band predominated in the electrophoretic patterns. While it appeared in gastrulae, it was eventually replaced at later stages by a pattern attributable to normal somatic histones. These findings were confirmed by Johnson et al.[650] Unfortunately, the techniques employed for the isolation of nuclei or chromatin do not exclude the possibility of histone contamination by basic cytoplasmic proteins known to abound in early embryonic life. Since Benttinen and Comb[434] did not report an amino acid composition to substantiate their claim that the protein they described was a lysine-rich histone, the identity of this protein band remains to be determined. The basic protein was not seen in unfertilized eggs, which contained instead a protein migrating in the electrophoretic position usually assigned to the F2a1 (IV) histone. It was suggested that the unfertilized eggs might contain an extranuclear pool of F2a1 (IV) histone.[434]

Labeling experiments on developing *Strongylocentrotus purpuratus* embryos[239,648] revealed that after a 30 min labeling period, no nuclear acid soluble proteins with the chromatographic properties of histones (Amberlite IRC-50) could be detected in the embryos during the first three cleavages. Proteins eluting in the position of the arginine-rich histones F3 (III) and F2a1 (IV) first became labeled between the four-to-eight cell stage and early blastula. The lysine-rich histones started to accumulate in the nucleus only at blastulation and increased during further development (Figure 29). These findings were further substantiated by an analysis of labeled histones from developing embryos by polyacrylamide gel electrophoresis.[511] Significant labeling of the nuclear histones could be detected in the 32 to 64 cell stage, while zygote histones showed only slight labeling. Again, proteins corresponding to the arginine-rich histones F3 (III) were labeled first, followed by the labeling of all the other

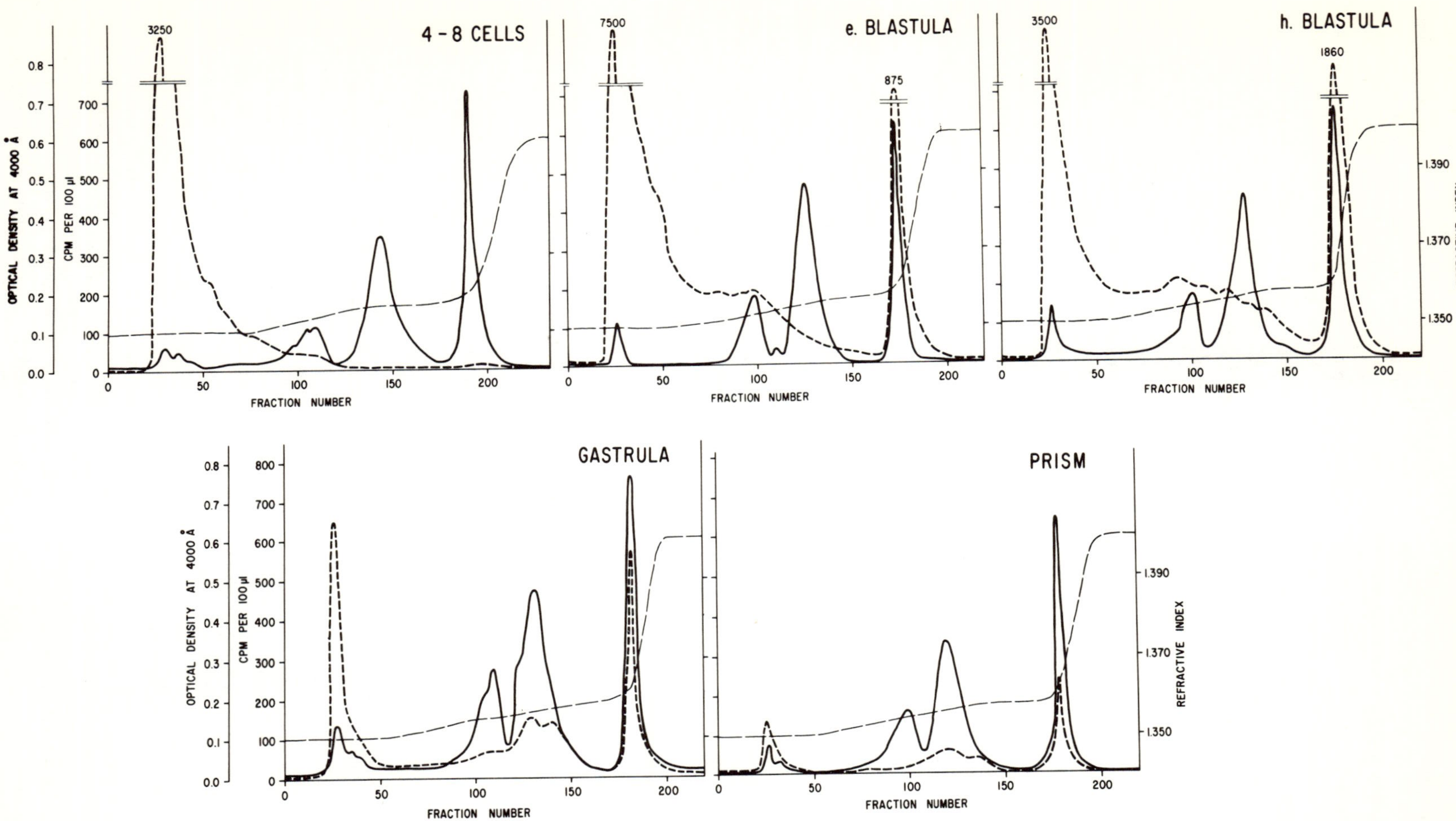

FIGURE 29. Chromatographic comparison (Amberlite CG 50 with guanidine-HCl gradient) of acid soluble proteins from isolated nuclei of *Strongylocentrotus purpuratus* embryos at several developmental stages. The embryos were pulsed for 30 min with ^{14}C-labeled amino acids; isolated nuclei were first washed with 5% perchloric acid and then extracted with 0.4 *N* H_2SO_4 (proteins in this Figure). To all protein extracts, 10 mg of calf thymus histones were added as a marker (solid line). Abscissae: fraction number; ordinates: (left) absorbency at 4,000 A; (right) refractive index of guanidine-HCl gradient. – absorbency at 4,000 A; – – – cpm/0.1 ml of effluent.[239]

fractions in the hatched blastula. Interestingly, during early development, the active synthesis of proteins electrophoretically similar to histones was detected in the cytoplasm, suggesting that histones may be biosynthesized very early in development and not released to the nucleus until the hatching blastula stage.

A decreasing complexity of the histone electrophoretic patterns from the sea urchin *Strongylocentrotus dröbachiensis* and the loach *Misgurnus fossilis L.*, accompanied by an increase of lysine-rich histones relative to the arginine-rich fractions, was described by Gineitis et al.[651,652] Using these two species, Vorobyev et al.[653] further confirmed that the electrophoretic patterns of the histones became less complex during the progress of development. At gastrulation, the patterns resembled those of calf thymus histones. There was an active biosynthesis of histones in blastula cells as well as all the higher embryonal stages. The arginine content of the histones decreased and the lysine content increased with the progress of differentiation. These data supported the findings from the author's laboratory,[239,511,648,650] since the increase in the synthesis of lysine-rich fractions after hatching would result in an apparent decrease of the arginine-rich histone content. In amino acid composition of the embryonal sea urchin histones (*Strongylocentrotus purpuratus*) were quite similar to the corresponding fractions from calf thymus.[204,239,311]

The schedule for the nuclear appearance of individual histones in the newt *Triturus pyrrhogaster* was, though delayed, very similar to that seen in the sea urchin embryos. Asao[237] analyzed cytoplasmic and nuclear acid soluble proteins labeled with ^{14}C-amino acids at several developmental stages. Using chromatography on Amberlite IRC-50 for histone fractionation, no labeled histones could be detected in the blastula or early gastrula nuclei. The arginine-rich histones F3 (III) first became labeled in gastrula, reaching a constant level during late gastrulation. The lysine-rich histone fractions F1 (I) and F2 (mixture of F2a1 or IV, F2a2 or IIb1, and F2b or IIb2) first appeared in late gastrula and their content increased during development from the late gastrula to the tail bud stages. The amino acid composition of the newt embryonal histones was different from the calf thymus stereotype. Whether this indicates true species specificity or a contamination was not investigated.

A more detailed study of regional changes in histones in post-blastula embryos proved quite interesting.[238] Histones were extracted from isolated newt embryo chromatin and analyzed on Amberlite IRC-50 microcolumns. It was found that all the histone fractions were present in the animal pole presumptive ectoderm in early gastrula. However, the presumptive endoderm and mesoderm at the same developmental stage contained only the arginine-rich fractions. With the progress of development, the mesodermal cells attained a complete set of histones, but not until the neurula stage did the lysine-rich fractions appear in the presumptive gut. Of special interest was the finding that the dorsal lip area, which first invaginates during gastrulation, had the lowest histone to DNA ratio of any part of the embryo. This area is the primary embryonic organizer, and, when transplanted to any other part of a normal embryo, it will induce the development of a second embryo. Since Asao[237,238] used chromatography on Amberlite IRC-50 to fractionate embryonal histones from newt, his results were directly comparable with the experiments on sea urchins reported by Orengo and Hnilica.[239,648] There was good agreement in that the arginine-rich histones appeared first, followed by the other fractions.

Active synthesis of both the arginine-rich and lysine-rich fractions in the gastrula of *Arbacia punctulata* was described by Thaler et al.[433] Both groups of histones incorporated ^{14}C-valine at approximately similar rates. Crane and Villee[654] used two different concentrations of actinomycin D to investigate late blastula stage histone biosynthesis in sea urchins. At a concentration of 2 μg/ml, the actinomycin D inhibited the labeling of F1 (I) histones by 60 to 70%, while the other fractions remained unaffected. An actinomycin D concentration of 20 μg/ml caused the greatest reduction of F1 (I) histone synthesis with only a minimal inhibition of the other fractions. It was concluded that F1 (I) histone biosynthesis depends on rapid m-RNA synthesis, while the other fractions utilize stored maternal messages. Since actinomycin D did not interfere with the embryonic development of sea urchins until the gastrula stage, the results could suggest that the F1 (I) histones are either not made before gastrulation or are not essential for early embryonic life.

Nuclear protein synthesis during the early embryonal stages of *Arbacia punctulata* was also studied by Kedes et al.[509] Autoradiography and a crude fractionation of nuclear proteins (into 0.25 *N* HCl soluble and insoluble) led to a conclusion that 40 to 60% of all the proteins synthesized during cleavage accumulate in the nuclei. Most of these proteins were found associated with chromosomes and did not turn over during further development. More than 40% of the nuclear proteins synthesized during early embryonic stages were soluble in acid. Using the small or slowly sedimenting polysomes which accumulate in the cytoplasm during the cleavage period, Kedes et al.[509] also studied the incorporation of ^{14}C-lysine and ^{3}H-tryptophan into nascent peptides as an indication of lysine-rich histone biosynthesis. The small polysomes assembled basic proteins quite efficiently, but only during active DNA synthesis. If DNA replication was inhibited with hydroxyurea, the biosynthesis of basic proteins on small polysomes decreased rapidly. This was similar to the experiments on HeLa cell polysomes[502-505] which led to the conclusion that histones are assembled in the cytoplasm of HeLa cells on small polysomes and that their biosynthesis depends on active DNA replication.

Blocking RNA transcription with actinomycin D inhibited the accumulation of small polysomes in the cytoplasm of cleaving embryos, suggesting the dependence of histone biosynthesis on new, rather than maternal, m-RNA. If the proteins assembled on small polysomes were indeed the lysine-rich histones, the inhibition of their biosynthesis by actinomycin D agreed well with the in vivo experiments of Crane and Villee.[654] The accumulation of newly synthesized proteins in the nuclei of actinomycin D-treated embryos was diminished by 20 to 40% and the labeling of acid soluble proteins was inhibited with a particular severity. However, since cleaving embryos can undergo many cell divisions in the presence of actinomycin D,[655] the basic proteins synthesized during early developmental stages are not necessary for the embryos to reach the early gastrula stage.

Moav and Nemer[510] further characterized the proteins synthesized on small polysomes in sea urchin embryos. In early blastula (i.e., the time when activation of histone synthesis appears to take place) the small polysomes (three to seven ribosomal aggregates) represented more than 70% of all polysomal ribosomes. Proteins synthesized by these polysomes behaved like typical histones when analyzed by gel filtration on Sephadex G 100, chromatography on Amberlite IRC-50, and polyacrylamide gel electrophoresis. Similarly, the incorporation of labeled lysine, arginine, and tryptophan indicated that these proteins are most likely histones. Their biosynthesis closely followed the rate of cell formation as can be expected from the relationships of histone and DNA synthesis established in many experimental models. Large polysomes from developing embryos also assembled proteins, but their composition and behavior was different from those assembled on small polysomes.

The active biosynthesis of basic proteins, presumably histones, during the cleavage stages poses a question as to why these basic nuclear proteins, which represent more than 40% of total protein synthesis,[509] cannot be detected by the fractionation of acid soluble proteins obtained from isolated sea urchin embryo nuclei. Although, according to Kedes et al.,[509] these basic proteins are conserved, they do not accumulate as distinguishable histones during the first several cleavages. It is possible that the early embryogenesis of sea urchins may be accompanied by the biosynthesis of special "cleavage histones," as first reported by Bloch and Hew[122] in the snail. Because of their lesser basicity, these proteins may escape detection as typical histones. Orengo and Hnilica[239,648] have reported a considerable labeling of the acid soluble nuclear proteins during the early embryonal development of *Strongylocentrotus purpuratus*. The labeling of these proteins decreases greatly during the late blastula and gastrula stages. The acid soluble nuclear proteins are not typical histones, however, since during chromatography on Amberlite IRC-50 they elute with the void volume (run-off peak). In their more recent studies, Johnson and Hnilica[511] have shown that histone-like proteins are actively biosynthesized in the cytoplasm during early development, but do not accumulate in the nucleus until later. The absence of typical histones in nuclei from early embryonal stages is also evident from the amino acid composition of nuclear acid soluble proteins, which is shown in Table 13.

If typical histones are absent in the nuclei of cleaving embryos before early blastula, this should be reflected in the extent and nature of the genetic restriction of the DNA in early embryos. Experiments with inhibitors of RNA synthesis have

TABLE 13

Amino Acid Composition of Acid-Soluble Proteins from Developing Embryos of *Strongylocentrotus purpuratus*[311]

Amino acid	Proteins obtained by a method used for extraction of							
	Arginine-rich histones				Lysine-rich histones			
	Egg	Blas	Gast	Plut	Egg	Blas	Gast	Plut
Lysine	**7.6**	**9.1**	**9.9**	**9.6**	**11.2**	**16.3**	**17.3**	**17.4**
Histidine	1.7	2.1	1.9	3.0	0.7	2.1	1.9	1.0
Arginine	**6.0**	7.7	**9.6**	**9.5**	**9.8**	**6.5**	7.3	**5.4**
Aspartic acid	**9.7**	8.2	**6.3**	**6.9**	**9.5**	**6.9**	**5.4**	**5.5**
Threonine	5.0	5.5	5.5	5.2	5.0	4.7	5.4	5.2
Serine	6.7	6.7	5.9	5.4	6.6	7.4	7.0	6.6
Glutamic acid	**11.4**	**10.8**	**9.4**	**9.2**	**10.0**	**9.2**	8.9	**8.7**
Proline	4.2	4.6	3.7	3.5	5.9	5.8	4.9	5.2
Glycine	7.8	8.7	9.4	9.2	10.5	9.0	7.7	7.8
Alanine	8.7	8.4	10.3	9.8	7.7	8.9	12.7	16.4
Half-cystine	0.4	0.3	1.7	0.8	0.5	1.5	–	0.7
Valine	6.3	6.5	6.3	6.6	5.5	5.5	5.9	6.2
Methionine	2.5	1.2	1.4	1.6	2.3	1.5	1.3	1.1
Isoleucine	5.5	5.0	5.1	4.9	3.9	4.3	4.4	3.6
Leucine	8.8	9.0	8.7	9.0	6.0	5.4	5.8	5.1
Tyrosine	3.1	2.8	2.7	2.9	2.5	2.1	2.1	1.9
Phenylalanine	4.2	3.0	2.5	3.9	2.3	2.0	1.7	1.4

All values are expressed in mol %. Tryptophan was not determined and all serine values were corrected (10%) for hydrolytic losses. The arginine-rich histone-like proteins were obtained by extraction with absolute ethanol – 1.25 *N* HCl (4:1 v/v). The lysine-rich histone-like proteins were obtained from the residue by extraction with acid (0.2 *N* HCl).

demonstrated that prior to the blastual stage, protein synthesis in sea urchin embryos is mainly regulated at the polysomal level.[655-657] Little, if any, RNA synthesis is necessary for the embryos to reach early gastrula stage. During gastrulation, transcriptional controls become important in regulating specific protein synthesis.[657] Although in normal embryos, limited RNA synthesis begins early in development,[655,658,659] this RNA does not differ significantly from stored maternal RNA species.[660-663] After blastulation, new species of RNA appear.[662,663] The appearance of new RNA species and the activation of transcriptional controls takes place approximately at the time when all the histone fractions are first detected in the isolated nuclei.

Using external RNA polymerase enzyme from *Micrococcus luteus*, Johnson and Hnilica[27] studied the ability of nuclei or chromatin isolated from various developmental stages of *Strongylocentrotus purpuratus* to serve as templates for in vitro RNA synthesis. Unfertilized eggs and early zygotes showed about 2 to 3% of their nuclear DNA available for transcription. This increased during late blastulation and stabilized at 5 to 6% in pluteus and later stages. These results are very similar to the studies of Marushige and Ozaki[388] who found approximate doubling of the templating activity between blastula and pluteus chromatins. If the DNA restriction seen in unfertilized eggs and early zygotes nuclei is due to histone-DNA interactions in their chromatin, the removal of histones by trypsin was expected to increase the templating efficiency of such treated nuclei. It was shown by Allfrey et al.[645] and confirmed in our laboratory[664] that trypsin digestion of nuclei will increase their DNA-dependent RNA synthesis severalfold. When sperm nuclei or chromatin, as well as nuclei or chromatin from blastula or gastrula embryos, were treated with trypsin, their efficiency to template for RNA in vitro increased dramatically.[27] However, no change in the templating efficiency of unfertilized egg or early cleavage stages nuclei resulted from their exposure to trypsin. Disruption of isolated nuclei with ultrasound did not increase their sensitivity to trypsin or their ability to support in vitro RNA synthesis. Hence, it appeared that either the nuclear histones in unfertilized eggs or early embryos were not accessible to trypsin or that they were not the typical histones found in the later developmental stages and adults. As noted earlier,[509] the nuclear proteins synthesized in early cleavage stages represented more than 40% of the total embryonal protein synthesis activity. It has not been possible from present evidence to concretely identify such nuclear proteins with the lysine-containing proteins synthesized on the small cytoplasmic polysomes. One could easily speculate that these latter proteins might represent a new class of repressor proteins used by the early embryos and later replaced by the common histones.

As can be seen from the discussed research, the literature on histone biosynthesis during animal development is currently in a phase of rapid growth. Although some issues are still controversial, it appears that in unfertilized eggs and in early embryonic stages, DNA may be restricted by proteins different from typical histones. The complete complement of adult type histones first appears during blastulation in sea urchins and even later in other animal species. The arginine-rich fractions make their appearance first, before the lysine-rich histones. Once established, the qualitative histone pattern in differentiated tissues remains unchanged, with the exception of some specialized cells such as nucleated erythrocytes, spermatozoa, etc. These specialized cells, containing tissue specific histone fractions, exhibit very limited, if any, RNA synthesis in vivo. The exact biochemical nature of the macromolecules restricting the DNA during early embryonal life remains to be determined. The relatively late appearance of typical histones during embryogenesis is compatible with their possible role as permanent DNA repressors.

Chapter 5

METABOLIC MODIFICATIONS OF HISTONES

It has been made evident from the large collection of experiments, which will be discussed later in this book, that chromosomal DNA must be free of histones in order to function as a genetic template for RNA synthesis. Several authors suggested that if histones participate in the variable restriction of DNA, which is regulated by the physiological necessities of the cellular life cycle, there must exist a mechanism which would allow their selective removal from the DNA. Allfrey et al.[645] found that if histones were removed from isolated nuclei by digestion with trypsin, the RNA synthesis supported by de-histonized nuclei increased considerably. Restoration of the histones abolished the increase in transcription. Arginine-rich histones inhibited in vitro RNA synthesis even more than the lysine-rich fractions. Later, Allfrey and Mirsky[665] found that by making the lysine-rich histone "arginine-rich" by partial guanidylation of the ϵ-*N*-amino groups of lysine, the inhibitory activity of these modified proteins increased. It was also discovered that another modification of histones, i.e., the partial acetylation of ϵ-*N*-amino groups of lysine, decreased the inhibitory effect of these proteins on in vitro RNA synthesis. The decrease in inhibitory activity of histones was proportional to the extent of their acetylation.[666] These initial studies on the biologically active chemical modification of histones opened a completely new chapter in the biochemistry of these proteins by stimulating an extensive search for chemical modifications of histones which would occur in vivo. It was anticipated that such modifications were of significance in the control of the transcriptional activity of chromatin. Presently, several modifications of histones have been demonstrated to occur in vivo; the acetylation, phosphorylation, methylation, and thiolation of certain amino acids being the main ones.

Acetylation

More than a decade ago, it was discovered that proteins can have their NH_2-terminal amino acid acetylated. Acetyl NH_2-terminal groups were found in tobacco mosaic virus protein, ovalbumin, cytochrome C, etc. In 1963, Phillips[222] reported that the NH_2-terminal amino acids in histone fractions F1 (I) and F2a (F2a1 or IV and F2a2 or IIb1) were acetylated. Later, he substantiated his findings by the isolation of the NH_2-terminal acetyl peptides from the calf thymus histone fractions F2a1 (IV) and F2a2 (IIb2).[272] It became obvious, however, that if histone acetylation was to play any significant role in genetic restriction, it could not be limited to the acetylation of only the terminal amino acid residues. Even before the determination of the complete amino acid sequence of the F2a1 (IV) histones from calf thymus and pea seedlings, which established unequivocally the presence of labile acetyl groups on certain lysine residues, Gershey et al.[284] and Vidali et al.[667] reported that calf thymus histone fractions F2a1 (IV) and F3 (III) contained appreciable amounts of metabolically labile acetate. The fraction F2a2 (IIb1) had only a small amount of labile acetate, while histones F1 (I) and F2b (IIb2) did not contain any. The labile acetate sites were labeled with acetate-1-C^{14} by incubation of the isolated nuclei in vitro. All of the labeled acetate was found in the form of ϵ-*N*-acetyllysine. Pogo et al.[668,669] used 2 *M* neutral hydroxylamine to differentiate between NH_2- and O-acetyl groups in rat liver histones. Only amino-bound acetyl was found in fractions F2a1 (IV) and F2a2 (IIb2), but both NH_2- and O-acetyls were present in F3 (III) histone. ϵ-*N*-acetyllysine was also found in the F3 (III) fraction and, in only minute amounts, in F2a2 (IIb1) histone. The existence of labile O-acetyl groups in the F3 (III) histone was also reported by Nohara et al.,[670,671] who used partially purified pigeon liver enzyme for the in vitro acetylation of histones, and by Gallwitz and Sekeris,[672] who used intact isolated rat liver nuclei. In contrast to the above, Vidali et al.[667] did not find any O-acetyl-C^{14} incorporated into F3 (III) histones during the incubation of isolated thymocyte nuclei with labeled acetate.

The acetylation of histone is an enzymatic reaction. Since puromycin fails to inhibit the incorporation of metabolically labile acetyl groups into histone, the acetylation event occurs after the completion of the molecules biosynthesis. Furthermore, histones can be acetylated in vitro in the presence of acetyl transferases and appropriate conditions. Histone acetylation is reversible and the incorporated labeled acetyl groups are lost in a short time.[650,666,673] Allfrey et al.[666,674,675] reported that acetyl coenzyme A is the donor of acetyl groups for histone acetylation. This has been confirmed by other investigators.[670,672,676-680] Interestingly, it has been found that histones can be nonenzymatically acetylated in the presence of acetyl-CoA.[672,678,681] Although all the histone fractions can be acetylated non-enzymatically, the F2 group of histones obtained by carboxymethyl cellulose chromatography (most likely corresponding to F2a1 or IV, F2a2 or IIb1, and F2b or IIb2 histones) evidence the highest degree of acetylation. This is different from the enzymatic acetylation system, where the arginine-rich histones F3 (III) and F2a1 (IV) are acetylated almost exclusively. The nonenzymatic acetylation of histones is optimal at relatively high pH values (9 to 10) and the acetate is incorporated entirely into the ϵ-*N*-acetyllysine. In addition to the histones, polylysine, polyarginine, albumin, and lysozyme also become acetylated to a variable extent.

The results of numerous experiments concerning the enzymatic acetylation of histones both in vivo or in vitro agree that the arginine-rich histones F3 (III) and F2a1 (IV) are the best acceptors of labile acetyl groups.[293,294,512,537,650,666-668,670-672,674,675,677,679,682-695] The acetylation occurs at the amion group on the fifth carbon (ϵ) of lysine. The F3 (III) histone molecule contains at least three ϵ-*N*-acetyllysine residues. According to Miller et al.,[695] the sites of acetylation are identical in F3 (III) histones from calf thymus and mouse mammary gland. The second principal acceptor of labile acetyl, the F2a1 (IV) histone, contains at least two such residues. During spermatogenesis in the rainbow trout (*Salmo gairdnerii*), the F2a1 (IV) histone from the testis is acetylated variably at four lysine residues (positions 5, 8, 12, and 16).[696] As was already mentioned, in addition to ϵ-*N*-acetyllysine, histones F1 (I), F2a1 (IV), and F2a2 (IIb1) have an NH_2-terminal serine which is also acetylated. These acetyl residues are, however, metabolically stable and, as indicated by the sensitivity of NH_2-terminal acetylation to puromycin,[512,673,694] the three fractions are acetylated during their biosynthesis. It has been suggested[673,694] that the NH_2-terminal acetyl serine present in the F2a1 (IV) and F2a2 (IIb1) fractions may serve in the initiation of biosynthesis of these histones. Interestingly, during sperm maturation in trout, the NH_2-terminal serine in the F2a1 (IV) and F2a2 (IIb1) histones is both acetylated and phosphorylated.[201]

Although there has been much agreement about the acetylation of arginine-rich histones, metabolically unstable acetylation of the lysine-rich histones F1 (I) and F2b (IIb2) was also reported. In synchronized Chinese hamster ovary cells (CHO) all of the histone fractions became acetylated during the cell cycle when the cells were grown in the presence of labeled acetate.[697] The acetate contents of the fractions F2a (IIb1 and IV), F2b (IIb2), and F3 (III) increased rapidly, reaching maxima late during the S period of the cell cycle. The F1 (1) histones attained maximal acetylation earlier, approximately in mid-S. All fractions evidenced high turnover of their acetyl groups, the turnover rates of lysine-rich histones F1 (I) and F2b (IIb2) were much higher than those of the arginine-rich histones F3 (III) and F2a1 (IV). In vitro acetylation of all of the histone fractions was also observed by Bondy et al.[680] who isolated the acetylating enzyme from the nuclei of rat liver and brain. The enzyme was quite specific for histones and acetylated histones F3 (III), F2a (IIb1 and IV) and F2b (IIb2) more efficiently than the F1 (I) fractions. The apparent acetylation of lysine-rich histones was quite peculiar, since no acetylated amino acids were reproted in the sequence of the F2b (IIb2) histone from calf thymus.

In addition to the partial purification of histone acetylating enzyme by Bondy et al.,[680] other investigators attempted the purification and characterization of histone acetylases. Pigeon[670] and rat liver[678] tissues yielded crude enzymatic fractions catalyzing the transfer of acetate from acetyl CoA to histones. More recently, Gallwitz[676,693,698,699] purified acetyltransferases from isolated nuclei of various rat tissues. There were three acetyltransferases in the rat thymus, two of which (B1 and B2) preferentially acetylated histone F2a1 (IV), whereas the third enzyme (A) was more specific for the F3 (III) fraction. Furthermore, acetyltransferases isolated from various rat organs were tissue specific, as was evidenced by their different elution patterns from DEAE cellulose columns.[693,699] All the enzymes transferred acetyl groups from the acetyl CoA to the nitrogen on the ε-carbon of lysine. The acetyltransferase enzymes appeared to be tightly associated with chromatin,[679,692] since they resist most of the usual extraction procedures. According to Racey and Byvoet,[692] the specific acetylation of arginine-rich histones occurred only if the histone proteins were bound to the chromatin DNA. When free, the lysine-rich fractions could also be acetylated, although to a lesser extent than the arginine-rich histones. Although the acetylation rates of histones change during the cell cycle,[697] these changes were not regulated by the availability of transacetylases, since their activity was found to remain relatively constant in synchronized Chinese hamster ovary (CHO) cells.[700]

As could be anticipated from the rapid turnover of acetyl groups in arginine-rich histones, enzymes deacetylating histones must be present in living cells. Such an enzyme was reported by Inoue and Fujimoto[701,702] to be present in calf thymus extracts. In contrast to the transacetylases, the deacetylase enzyme seemed to be present in the cytoplasm and not in the nuclear fraction. Mouse liver had little deacetylase activity and exhibited an inhibitory effect on the deacetylation by the thymus extracts. The calf thymus deacetylase appeared to be heterogeneous by DEAE cellulose chromatography. Biologically acetylated histones were more efficiently deacetylated by this enzyme than acetylated histones prepared by chemical acetylation. In rat liver, the histone deacetylase activity was found to be three to six times greater than in Novikoff hepatoma.[703] This, together with the fact that the chromatin-bound histone transacetylase in Novikoff hepatoma was 70% less active than that of rat liver,[704] might explain the very long (25 hr) half-lives of ^{14}C-acetyl groups in Novikoff hepatoma histones as compared to only 2 to 3 hr in normal liver.[684,690] However, the very long half-lives of acetyl groups in Novikoff hepatoma were not specific for neoplastic growth, since the acetyl groups turned over at very similar rates in slow, moderately or fast growing Morris hepatomas.[690] The hepatoma turnover rates were, at most, twice those found for normal livers from several rat strains.

Because the acetylation of histones lowered their capacity to inhibit RNA synthesis in vitro[666] and decreased the stability of the DNA-histone complexes, as evidenced by thermal denaturation measurements, Allfrey suggested that histone acetylation might be a necessary prerequisite for the activation of previously repressed genes. To support this proposal, Allfrey et al.[674,689] studied the acetylation of histones in several systems both in vivo and in vitro. It was known that the chromatin from thymocyte nuclei could be fractionated into a condensed fraction which is relatively inactive in RNA transcription and into an extended, or diffuse fraction, where most of the RNA synthesis takes place.[705] The condensed form of chromatin showed a much lower incorporation of labeled acetate than did the extended or diffuse part.[673,675] When acetyl-l-C^{14}-CoA was used, both forms of chromatin became labeled to the same extent. A similar situation occurred when isolated nuclei were incubated with acetyl-l-C^{14}-CoA directly. This observation was interpreted to indicate that the two forms of chromatin differed not in their transacetylase contents, but in their ability to convert acetate to acetyl CoA. An increased acetylation of the transcriptionally active chromatin in puffing areas of the polytene chromosomes from *Chironomus tentans* was also reported.[674] It was later shown, however, that the uptake of ^{3}H-acetate was not localized at either existing or newly induced puffs. Hence, it appeared that puff induction and a subsequent activation of RNA transcription did not require histone acetylation.[706-708]

During the maturation of nucleated erythrocytes in birds and other species, the transcriptional activity (RNA synthesis) of their DNA decreases progressively. Simultaneously, the chromatin becomes more condensed and considerably decreases its ability to acetylate histones.[709] Similarly, the inhibition of RNA synthesis in polymorphonuclear leukocytes exposed to phytohemagglutinin or certain antigens in the culture medium produces a rapid proliferative growth.[710] This induction is accompanied by a considerable increase in RNA synthesis.[689] Before the cells begin to increase their RNA synthesis, profound structural changes occur in their chromatin which are reflected in a changed stainability of histones[711-713] and an increased DNA binding of actinomycin D and acridine orange.[714-718] The phytohemagglutin stimulated lymphocyte system has been used by Allfrey et al.[674,689] to study histone acetylation relative to RNA synthesis. An increased rate of acetate incorporation into histones preceded the anticipated increase in the rate of RNA synthesis during the initial hours of growth. The turnover of previously incorporated ^{14}C-acetate was slower in the stimulated lymphocytes than in the control cells. In contrast to the above, however, the turnover of acetyl groups was increased in horse leukocytes in which phytohemagglutinin inhibited RNA synthesis.[685] The acetylation of histones was not inhibited by puromycin and was stimulated without a concomitant rise in histone biosynthesis.[685,689] Moreover, the acetate incorporated into histone turned over, whereas labeled amino acids did not. Histone acetylation in phytohemagglutinin stimulated human lymphocytes was also confirmed by the autoradiographic analyses of Mukherjee and Cohen.[719] The incorporation of labeled acetate into nuclei increased immediately after induced stimulation, but the maximum effect was only observed after about 24 hr. The level of RNA synthesis, as measured by the incorporation of labeled uridine, began to increase at 1 to 2 hr and evidenced a maximal rate at about 48 hr after the exposure to phytohemagglutinin. Desai and Foley[294] compared the extent of histone acetylation in several different lines of human lymphocytic cells. Although there were some variations, the distribution of radioactive acetate among the individual histones was very similar in all the cell lines. An increase in histone acetylation preceding the rise in RNA synthesis was also seen in spleen cells stimulated with erythropoietin[687] and, together with a rapid turnover of the incorporated acetate, in HeLa cells and other tissues.[682,720] Contrary to the reports on the increased acetylation of histones in phytohemagglutinin stimulated lymphocytes, Monjardino and MacGillivray[721] recently observed a decrease in the acetylation rates of histones during the first 4 hr, i.e., prior to the increase of RNA synthesis. According to these authors, the stimulation of histone acetylation observed by Allfrey and others may have been due to the effect of leucoagglutinating impurities present in their phytohemagglutinin preparations which might have caused an increased acetate transport across the lymphocytic membranes.

Another system known to undergo a relatively abrupt period of genetic activation is the compensatory hypertrophy or regeneration of rat or mouse liver. Pogo et al.[668,669] studied histone acetylation in regenerating rat liver and found the rate of histone acetylation to increase during the initial 5 hr of regeneration, while its turnover (deacetylation) was decreased compared to sham operated controls. The time of the incorporation of labile acetate preceded the reported appearance of new RNA species[722] in regenerating mouse liver and the increase of its transcriptional activity.[668,723,724] During a subsequent portion of the regeneration schedule, when DNA was actively replicated, NH_2-terminal acetylation could be detected. Only the arginine-rich histones, F3 (III) and F2a1 (IV), became acetylated several hours after the hepatectomy. Wilhelm and Hnilica[691] recently reinvestigated the acetylation of histones at 2½ and 20 hr after partial hepatectomy in rats and compared the histones from nucleolus-associated and extranucleolar chromatin. In normal rats, histone acetylation was more active in the nucleolar chromatin, whereas in the hepatectomized animals both types of chromatin were nearly equal. There was no difference in the incorporation rates of acetate into sham operated or regenerating livers 2 to 5 hr after hepatectomy, except for the F2a1 (IV) histone whose specific activity increased considerably at this time. Later, during active DNA synthesis (20 hr after the surgery) histone acetylation increased significantly due to the NH_2-terminal acetylation of newly synthesized histones. The F2a2 (IIb1) histone was the most actively acetylated at this time.

The administration of cortisol to rats has been shown to increase both the templating activity of liver chromatin[725,726] and the rate of RNA synthesis in liver.[727,728] The newly synthesized RNA contained new species,[729,730] previously not present in liver, which suggested that the hormone derepressed specific parts of the DNA which were transcriptionally inactive in the normal liver. The activation of the genetic templates in liver by cortisol also resulted in an increase of histone acetylation within a short time.[731,732] Analogous to the reports on acetylation patterns in regenerating rat liver, the acetate bound to the arginine-rich histones turned over rapidly in the livers of cortisol treated rats after the administration of this hormone. Libby[677] observed the stimulation of histone acetylation in vitro in cell-free extracts of rat uterine tissues by estradiol 17-β. A similar stimulation was recently reported to occur in vivo.[733,734] Histone acetylation was also observed to be increased in hormonally activated mammary gland explants.[694] Here, the arginine-rich F2a1 (IV) histone was the most affected by the hormone treatment, while the arginine-rich fraction F3 (III) was not influenced.[695] Gallwitz and Sekeris,[735] on the other hand, were unable to confirm any correlation between histone acetylation and increased RNA polymerase activity in rat liver nuclei treated with cortisol.

The nucleoli found in the nuclei of living cells are well known to preferentially transcribe ribosomal precursor RNAs and few, if any, other RNA species.[736-738] Nucleoli fractionally isolated from nuclei continue to synthesize RNA in vitro with a ribosomal RNA-like base composition.[664] Despite the fact that isolated nucleoli contain a significant amount of associated or perinucleolar chromatin, ribosomal RNA hybridizes with less than 0.5% of the perinucleolar chromatin DNA. It can be concluded from the above that the major portion of the perinucleolar chromatin is transcriptionally inactive. While hybridization experiments indicate that ribosomal RNA is transcribed only from 0.005 to 0.05% of the total nuclear DNA,[738] the extranucleolar, or nuclear chromatin is much more active, transcribing between 5 to 20% of the nuclear DNA content.[182,739-741] Thus, as a minimal estimate, there is at least a tenfold difference between the transcriptional activity of nucleoli and the remainder of the nuclear DNA. If histone acetylation is correlated to RNA transcription, then the acetylation of histones should be higher in nuclear chromatin than in perinucleolar chromatin. Wilhelm et al.[683] have analyzed the acetylation of histones in these two chromatin fractions in Novikoff hepatoma cells and found that active acetylation occurs on already biosynthesized histone molecules. The labeled acetate is incorporated primarily into the arginine-rich histones F3 (III) and F2a1 (IV). However, there are no significant differences in the rates of acetylation between the histones isolated from nucleolar or nuclear (extranucleolar) chromatin. Similar experiments have been carried out recently on regenerating and normal rat liver, with essentially the same results.[691] It appears that histone acetylation is not directly related to the amount of active templates in these two chromatin fractions.

The experimental evidence accumulated on the relationships between gene activation and the acetylation of histones does not suggest unequivocally that these two processes are directly connected. Although an increase of histone acetylation can be seen to occur in systems undergoing activated RNA synthesis, the reason for the increase in acetylation may rest with changes in the acetate pools resulting from the treatment. It is known that hepatectomy considerably alters the pool sizes of many precursor molecules.[742] This, together with the increased permeability of the cellular membranes and changes in the metabolism of many macromolecular species may lead to increased acetylation of histones concomitant with, but not related to, the increase of RNA synthesis. Changes in membrane permeability do occur[743] and have been suggested to be the reason for an increased histone acetylation in phytohemagglutinin stimulated lymphocytes.[721] Thus, the relationship between the increased RNA synthesis and histone acetylation may be only circumstantial. Similarly, the exposure of target cells to steroid hormones causes many changes in their metabolism, including modifications in the permeability of cellular membranes. There are reports in the literature which failed to detect significant differences in the acetylation of histones in naturally regulated systems. The formation of puffs in polytene chromosomes and its associated activation of RNA and protein synthesis does not appear to be accompanied by an increase in histone acetylation.[706,708] Nucleolar-associated (perinucleolar) and nuclear (extranucleolar) chromatin acetylate their histones at approximately equal rates in Novikoff hepatoma and regenerating rat liver.[683,691] Interestingly, in normal liver, the histone acetylation of the nucleolar-associated chromatin is found to be slightly higher than in the nuclear chromatin, which represents most of the nuclear DNA. It is noteworthy that most of the nucleolar-associated chromatin is transcriptionally inactive heterochromatin[744-747] and, therefore, should not be acetylated. In developing embryos of *Strongylocentrotus purpuratus*, histones have been found to be actively acetylated during all the embryonal stages investigated (early blastula and later), but only small variations between the specific activities of the various histones occur during the developmental schedule.[650,748] Similar observations have been reported for the metamorphosis schedule of *Rana catesbeiana.*[749] Finally, histones are not the only nuclear proteins which are acetylated. Several nonhistone proteins in chromatin incorporate labeled acetate very actively.[683,691] Acetyl groups incorporated into these nonhistone proteins are metabolically unstable and the acetylation process responds considerably to changes in the physiological conditions of the animals. Presently, it is not possible to conclude unequivocally that histone acetylation leads to structural or other changes in chromatin which would result in the activation of DNA transcription.

Phosphorylation

Another mechanism frequently suggested to function in the modification of chromatin transcriptional activity is the enzymatic phosphorylation of histones. Ord and Stocken[227] and Kleinsmith et al.[228] have shown that ^{32}P-phosphate incorporation into rat histones occurs both in vivo and in isolated rat thymus nuclei in vitro. The main acceptor of the phosphate is reported to be the very lysine-rich F1 (I) histone. The phosphate groups are attached through an ester bond to the hydroxyl residue of serine[224,227,228] and, to a lesser extent, of threonine.[227,228] Histone phosphorylation has been reported to occur in Chinese hamster ovary cells (CHO);[268,554,750] various rat tissues including tumors,[200,224,226,236,266,267,564,751-760] lymphocytes,[688,761,762] and erythrocytes;[763,764] and the tissues of various animal species,[230,537,765-767] including trout testis[201,366,768] and sea urchin embryos.[236] The phosphorylation occurs after histone biosynthesis. For this reason, puromycin does not inhibit the incorporation of ^{32}P-phosphate into these proteins.[228,768] The most extensive histone phosphorylation occurs in the F1 (I) fraction,[224,226,228,231,236,752,753,755,757] although in regenerating rat liver[200] and goose erythrocytes[764] the F2a2 (IIb1) has been reported to be the most active of all the fractions. As discussed in the chapter on histone structure, the phosphorylation site of the F1 (I) histones exhibits considerable tissue and species specificity.[219,766,767] In addition to the very lysine-rich F1 (I) histones, the occurrence of phosphorylation in the other histone fractions has also been reported for various tissues,[200-202,227,229,266-268,537,554,564,755,763,765] including the erythrocyte-specific histone F2c (V).[763,764]

As was pointed out by Shepherd et al.[750] and Balhorn et al.,[760] reports on the phosphorylation of histones must be evaluated cautiously. Histone fractions frequently contain significant amounts of phosphorus-containing contaminants, such as oligonucleotides,[164,278,750] phosphoproteins, and other contaminating nonhistone proteins,[164,206,224,266,298,753,764] inorganic phosphate,[751,760] and various phospholipids.[750] Therefore, for the most unambiguous results, all the known criteria of purity should be determined in conjunction with analytical determinations of the acid precipitable, base labile phosphate, or phosphoserine for the histone fractions under study. Although prohibitive in time and frequently impossible because of the minute quantities of the histone fractions available for analysis, the most decisive criterion of histone phosphorylation is the determination of the actual peptide sequence involved in the phosphorylation. Since the criteria of purity for the phosphorylated histones have only rarely been met, the following discussion of the potential biological significance of histone phosphorylation is complicated by the possible phosphorylation of histone contaminants. This is particularly so since contaminant phosphorylation may be greater than the phosphorylation of histones.

Stevely and Stocken[769] observed a turnover of histone phosphate in incubated rat thymus nuclei and also showed that the phosphorylation of histones may modify the repressed sites on chromatin DNA. The very lysine-rich F1 (I) histone phosphorylated in vitro (34.2 mμ mol P/mg protein) inhibited in vitro DNA-dependent RNA biosynthesis less than a sample obtained directly from the animals (22.4 mμ mol P/mg protein). They also found that, on the average, there were two phosphates present per molecule of rat liver F1 (I) histone (phosphate-rich fraction obtained by chromatography on DEAE cellulose).[752,769] Histone phosphorylation also decreased the degree of template inhibition in the in vitro DNA polymerase reaction.[752] However, the opposite results were obtained by Chae et al.,[770] who studied the templating activity of rat liver chromatins containing histones phosphorylated to different extents. Maximally phosphorylated chromatin (in vitro) and chromatin reconstituted with maximally phosphorylated histones (in vitro) had the same templating activity for *E. coli* RNA polymerase as the corresponding native chromatin containing only marginally phosphorylated histones. As will be discussed later, the interaction of DNA with histones changes the DNA conformation (determined by circular dichroism experiments). Phosphorylated F1 (I) histones affected the DNA conformation considerably less than de-phosphorylated histones.[771]

X-irradiation of experimental animals markedly decreased the phosphorylation of the F1 (I) fraction.[236,752] The dose-response curve showed a remarkable likeness to a similar curve obtained by Ord and Stocken for the incorporation of ^{32}P-phosphate into DNA after irradiation.[772] According to Pawse et al.,[773] in regenerating rat liver, where X-irradiation decreases the peak of DNA and histone synthesis after hepatectomy, both the phosphorylation of F1 (I) histones and the amount of F1 (I) histone kinase were substantially decreased 22 hr after the surgery if the animals were X-irradiated during the early G_1 phase. This could be interpreted to indicate that histone kinase induction occurred before DNA synthesis and was sensitive to x-ray irradiation during the early hours after hepatectomy (G_1 period). Recently, Gurley and Walters[554] studied the effects of X-irradiation on the turnover and phosphorylation of histones in Chinese hamster ovary (CHO) cells. Significant and rapid histone phosphorylation was detected only in fractions F1 (I) and F2a2 (IIb1) after the histones had been purified by chromatography on Amberlite IRC-50 or preparative gel electrophoresis. While the phosphorylation of F2a2 (IIb1) histone was insensitive to X-irradiation, the phosphorylation of F1 (I) histones decreased by 50% one hour after exposure of cells to 800 rad. The turnover of F1 (I) histones was completely arrested for 3 hr after the irradiation and was then resumed at an accelerated rate. A continued DNA mass increase indicated that DNA replication was not affected by the irradiation treatment. On the other hand, the incorporation of ^{3}H-thymidine into DNA followed an inhibition and recovery pattern similar to that observed for F1 (I) histone turnover and phosphorylation. This indicated, that F1 (I) histone phosphorylation does not correlate with DNA replication per se, but may instead be associated with the biosynthesis of some radiosensitive DNA precursor which does not reflect the actual DNA replication. The authors suggested that the association of F1 (I) histone turnover with its phosphorylation may be connected to some kind of nucleocytoplasmic transport system, since F1 (I) histone was found in polysomes[774] together with a specific protease degrading this fraction.[775] Increased phosphorylation of the F2a2 (IIb1) histone 15 hr posthepatectomy was reported by Sung et al.[200] Considerable changes in the phosphorylation of two F1 (I) subfractions were also seen during the compensatory hypertrophy (regeneration) of rat liver.[760] In the F2a2 (IIb1) rat liver histone, the radioactive phosphate was incorporated at a single site which was identified as the NH_2-terminal serine of this fraction. This residue site was also acetylated.[200] A similar situation was seen in F2a1 (IV) and F2a2 (IIb1) histones in maturing trout testis.[201]

When the biosynthesis of histones, measured by the incorporation of ^{14}C-lysine, was compared with phosphorylation of the individual histone fractions, a considerable increase in the uptake of ^{32}P-phosphate by the lysine-rich histones F1 (I) and F2b (IIb2) was observed in regenerating rat liver during the first 6 hr after hepatectomy.[267] This increase approximately coincided with the appearance of new RNA species in regenerating mouse liver which were similar in their hybridization characteristics to the RNA in mouse embryos.[722,776] An increase of the transcriptional activity of liver nuclei[723] or chromatin[724] can also be seen at this time in hepatectomized rats. It was reported that both the net phosphorus content and the phosphorylation rate of F1 (I) histones increased at the time of accelerated DNA synthesis in regenerating rat liver.[236,752,760,769] This was most likely due to the increase in the number of F1 (I) molecules phosphorylated and to the induction of histone kinase biosynthesis.[753,757,773,777]

The incorporation of inorganic phosphate into histone fractions appeared to be tissue specific and greater in growing than in resting cells.[200,230,267,752,753,756-768] The total phosphate content and the rate of F1 (I) histone phosphorylation increased during both the early preDNA synthesis phase and the subsequent period of accelerated DNA synthesis in lymphocyte cultures stimulated to grow and divide by treatment with phytohemagglutinin or 6-*N*,2′-o-dibutyryl adenosine 3′,5′-cyclic monophosphate.[688,778] The initial increase in RNA synthesis by stimulated lymphocytes was reported to be accompanied by an increased rate of phosphate turnover in F1 (I) histones, but not by a net increase of their phosphate content.[688,761,778] According to Cross and Ord,[778] at least part of the mechanism may reside in the cyclic AMP activation of histone phosphokinase during the early phase of lymphocyte stimulation. The rate of phosphate incorporation (turnover) might correlate with the increased transcriptional activity of chromatin during early stimulation (activation), while the quantitative increase in F1 (I) phosphate content might occur only at the time of DNA replication. The administration of hydrocorticosterone and ACTH to rats did not increase histone phosphorylation in liver,[226] although the opposite was also reported.[564] Glucagon, on the other hand, caused a marked increase of F1 (I) histone phosphorylation within 1 hr after treatment. Insulin had a similar effect.[226] The phosphorus was incorporated at only one serine residue in the F1 (I) histone sequence. The residue was identified by the isolation of the phosphorylated peptide resulting from trypsin digestion of this fraction. Shepherd et al.[268] studied the incorporation of ^{32}P-phosphate into histones during the cell cycle of Chinese hamster ovary (CHO) cells. Only the F1 (I) and F2b (IIb2) histones were phosphorylated. The phosphate content of the very lysine-rich F1 (I) histone rose steadily during the entire cell cycle, while the F2b (IIb2) was phosphorylated only during the G_1 and S periods, reaching a maximum at the mid-S. After this time, the specific activity of this fraction declined rapidly, which indicated a time-specific correlation of F2b (IIb2) phosphorylation to some events in the first part of the cell cycle, possibly early DNA replication. The variations in the phosphorylation of histones could not be correlated to the activity of histone phosphokinase which remained relatively constant through all periods of the cell cycle.[779]

Langan[224] established that the phosphorylation of histones was enzymatically performed by a histone kinase and a less specific phosphoprotein kinase, both present in the cytoplasm. Using the histone kinase to saturate total histone extracts (a mixture of the five fractions) with ^{32}P-phosphate, he found that the F1 (I) and F2b (IIb2) histones accepted up to 1 mol of phosphate per mol of each fraction. The other fractions were phosphorylated to a much lesser extent.

A peptide containing 2 lysine residues and 11 other amino acids including phosphoserine was isolated from calf thymus histone F1 (I) by Langan.[224-226] Chae et al.[770] observed that the F2a1 (IV) fraction, in addition to the F1 (I) and F2b (IIb2) histones, was also considerably phosphorylated in vitro. Substantial phosphorylation of all the histone fractions was also reported by other investigators.[759,780,781] Similar to the reports of Chae et al.,[770] the F2b (IIb2) histone accepted more phosphate than the F1 (I) fraction. When the histones were associated with DNA in chromatin, their phosphorylation in vitro decreased to the point where practically only the F1 (I) histones were being phosphorylated. On the other hand, the presence of histones caused a five- to tenfold increase in both the rate and extent of the phosphorylation of nuclear nonhistone proteins (phosphoproteins).[782] The F1 (I) fraction was most active in this stimulation. According to Schiltz and Sekeris,[783] the enzymatic phosphorylation of chromatin in vitro required both Mg^{++} and Mn^{++} and was most efficient at high pH values (9.6). The specific activity of the nonhistone proteins was about 20 to 30-fold higher than that of the histones.

It has also been found that cyclic AMP considerably enhances the phosphorylase b kinase activity from rabbit skeletal muscle.[784] Adenosine cyclic 3′,5′-monophosphate at low concentration (10^{-7} *M*) caused a four- to sixfold increase in the rate of F1 (I) histone phosphorylation[225] in vitro. The net extent of phosphorylation (33 to 40 nmol/mg protein) remained unchanged, however, with or without the addition of this hormone. Cyclic AMP also stimulated the phosphorylation of F1 (I) histones in vivo.[236] Since the same labeled peptide (containing phosphoserine) was found in the F1 (I) histones phosphorylated in vitro and in vivo,[225,226,231] the mechanism of stimulation by cyclic AMP appeared to be the same in both situations and probably concerned the activation of the histone kinase.[778] This would also explain the observed stimulation of the rates of histone phosphorylation in rat liver by glucagon and insulin,[226] since both hormones produced increased cyclic AMP levels in the livers of the experimental animals.

Phosphoprotein kinases associated with chromatin are heterogeneous.[780,781,785] Yamamura et al.[780] recently fractionated a histone kinase from rat liver into two components (B1 and B2). The B1 enzyme required cyclic AMP for its activity, while the B2 enzyme was not activated by this compound. Both kinases incorporated labeled phosphate predominantly into serine residues, but some phosphate was also incorporated into the threonine residues of the histone molecules. The cyclic AMP-independent enzyme B2 was rapidly inactivated by storage or heat. Most likely, this histone kinase was the one responsible for the phosphorylation of the additional site(s) on the F1 (I) histones recently observed by Langan.[786] In contrast to the primary phosphorylation site in position 40 of the F1 (I) amino acid sequence (Figure 8), the additional site of phosphorylation was located in the COOH-terminal fragment of the F1 (I) molecule, in the sequence (THR, SER, GLY_2 ALA)-GLY-SER(PO_4)-PHE-LYS, i.e., serine residue 106 in Figure 9. According to Siebert et al.,[777] the cyclic AMP-independent enzyme B2 accumulated in the nucleus, while the cyclic AMP-sensitive enzyme B1 appeared in the cytoplasm. It was the nuclear enzyme B2 which increased its activity 6-fold 22 hr after partial hepatectomy in rats. The cyclic AMP-dependent enzyme could be dissociated into its subunits in the presence of low concentrations of histone. Such a dissociation resulted in the conversion of the dependent enzyme into a cyclic AMP-independent kinase.[787] An enzyme involved in the metabolism of histone-bound phosphorus (a phosphatase specific for histones and protamines) has also been characterized in several tissues.[224,777,788] The activity of histone F1 (I) phosphatase did not change during the first 22 hr of liver regeneration in rats.[773]

Recently, Jergil et al.[767] phosphorylated in vitro F1 (I) histones from trout testis, liver, and erythrocytes, rabbit liver, and calf thymus. The phosphorus content of the individual F1 (I) histones exhibited considerable tissue and species specificity both in the extent of phosphorylation and in the number of phosphorylation sites identified by the isolation of ^{32}P-labeled tryptic peptides. Disregarding a number of minor radioactive bands, which may have originated from an incomplete cleavage by trypsin, their data indicated compositional differences in the phosphopeptides derived from the very lysine-rich histones F1 (I) in a variety of tissues and species. The phosphopeptide patterns of the F1 (I) histones from rabbit liver and calf thymus were almost identical in that both showed three major radioactive peptides. These peptides most likely represented the sites of primary phosphorylation in three of the F1 (I) histone subfractions present in these tissues. Specific labeling patterns were shown for F1 (I) histones in the remaining tissues. The results obtained for the F1 (I) histones from the nucleated erythrocytes of trout, frog, and chicken must be interpreted with some caution, however, since it was reported that bird erythrocytes may not contain these proteins[352] and that the erythrocyte specific F2c (V) histones in the frog are soluble in 5% perchloric acid.[358] Although Jergil et al.[767] used 5% trichloroacetic acid, there remained the possibility that the phosphopeptides in F1 (I) histones from nucleated erythrocytes might represent the in vitro phosphorylation of the F2c (V) fraction or the F2c (V) and F1 (I) mixture.

All four serine residues in protamine are phosphorylated during its biosynthesis by a specific enzyme, protamine kinase, and appear in the newly synthesized protein as phosphoserine.[366,789] The study of Marushige et al.[768] demonstrates that the phosphorylated protamines enter the nucleus and are incorporated into the chromatin of maturing spermatozoa. There is no appreciable loss of phosphate during this time and the high phosphoserine content of protamine does not interfere with its binding to the DNA, although its affinity for the DNA is decreased. Later, in mature sperm, the protamine molecules are slowly dephosphorylated. The authors suggest that the phosphorylation of protamines is a mechanism necessary to facilitate their transport into the nucleus.

During the replacement of somatic histones by sperm protamines in spermatocyte chromatin, the first histone to be displaced is the F2a1 (IV) fraction. Four of its lysine residues are acetylated and oné serine residue is phosphorylated during this process.[201] The NH_2-terminal serine in the F2a1 (IV) histone bears both acetyl and phosphate residues. In addition, the ϵ-*N*-amino groups of lysines 5, 8, 12, and 16 are also acetylated. The helical wheel projection of the first 18 residues of the F2a1 (IV) histone from calf thymus brings all the modified amino acids together and results in the elimination of electrostatic binding between this part of the F2a1 (IV) histone molecule and DNA (Figure 30). According to Sung and Dixon,[201] this may initiate local dissociation or "unzipping" of the F2a1 (IV) histones which, perhaps, results in the removal of this histone and its replacement by protamine. Alternatively, the local dissociation of the NH_2-terminal portion of F2a1 (IV) molecule may facilitate its degradation by a specific protease. Since the NH_2-terminal serine in F2a2 (IIb1) histone is also acetylated and

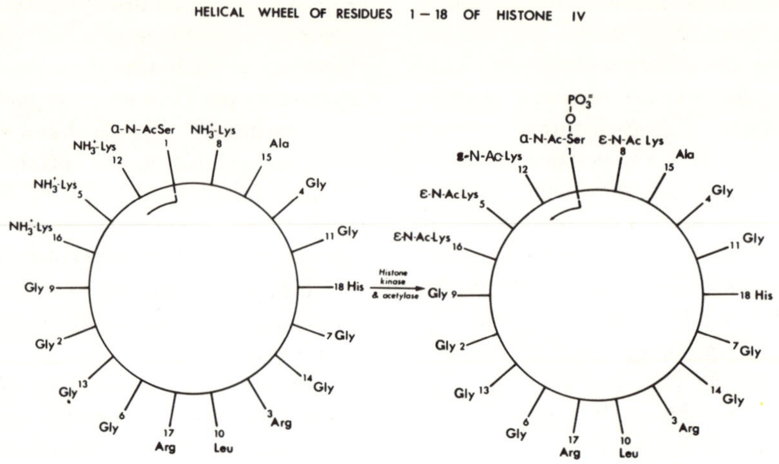

FIGURE 30. Helical wheel of the first 18 amino acid residues NH_2'-terminal) of arginine-rich histone F2al (IV). On the outer periphery of the wheel the positions of known enzymatic modifications of this portion of the sequence of histone F2al (IV) are shown. This accumulation of modified lysine residues can initiate the dissociation of this histone from the DNA during its replacement by protamines.[201,201a,696]

phosphorylated and the sequence of the first six amino acid residues is identical in both the F2a1 (IV) and F2a2 (IIb1) fractions, such a mechanism may serve in the removal of the F2a2 (IIb1) histones. During sperm maturation, the F1 (I) histone becomes phosphorylated in its COOH-terminal region. The sequence of this phosphopeptide in the rainbow trout testis is LYS-SER(PO_4)-PRO-LYS.

In circumstances similar to those of acetylation, it is difficult to exactly assess the biological significance of histone phosphorylation. Although the mechanism of phosphorylation appears to be more sophisticated in that there exist at least two kinase enzymes, each of which is specific for a different site on the F1 (I) histone, the biological reason for such a change remains to be discovered. In contrast to the acetylation of histones, only about 1% of the histone molecules present in the F1 (I) population are phosphorylated. This creates considerable difficulties in the isolation and identification of the phosphopeptides obtained by the enzymatic digestion of histones. The relatively numerous reports suggesting possible relationships between the phosphorylation of histones and an increase in the transcriptional activity of chromatin must be interpreted with reservation, since histones are easily contaminated with non-histone proteins. Many of these latter proteins very actively incorporate phosphate into their serine and threonine residues. Moreover, these nuclear phosphoproteins are tissue specific and their phosphorylation changes with cellular activities.[761,790-792]

Although not strictly within the scope of this book, it is noteworthy that in another functional complex of nucleic acids with proteins, the ribosomes, the proteins are modified by phosphorylation.[793-795] This modification appears to regulate the ability of ribosomes to function in protein synthesis.[794] In addition to chromatin phosphoproteins and histones, RNA polymerase, which also participates in the process of genetic transcription in chromatin, can be phosphorylated with a simultaneous increase of its activity.[796] The metabolism of protein-bound phosphorus is, therefore, quite complicated and it will need some effort and time before the many pieces of the chromatin protein phosphorylation puzzle can be assembled to reveal a meaningful pattern.

Methylation

The presence of ε-*N*-methyllysine in histones was first reported by Murray,[285] who detected small quantities of this amino acid in calf thymus histone fractions IIaa, IIa (F2a2), III (F3), and IV (F2al). It was found that histone methylation occurred after their biosynthesis and that methionine acted as the donor of the methyl groups to lysine.[285,666] In isolated calf thymus nuclei, the methylation proceeded through an S-adenosyl-methionine intermediate.[797] In addition to ε-*N*-methyllysine, ε-*N*-dimethyllysine and ε-*N*-trimethyllysine were also found in histones from calf thymus and chicken erythrocytes.[286,798,799] According to Paik and Kim,[286] about 5% of the lysine residues of the calf thymus arginine-rich histone F3 (III) were methylated. Later, these authors found ω-*N*-methylarginine in histones and other proteins.[800,801] This new amino acid accounted for about 1.3% of the arginine residues in the slightly lysine-rich histone from rat liver as compared to only 0.3% of the arginine residues in various cytoplasmic protein fractions. In calf thymus histones, only trace amounts of this amino acid could be detected.[800] As for the methylation of lysine residues, S-adenosyl-methionine was the methyl donor in this reaction. The methylation and possibly ethylation of arginine residues in rat liver nuclear proteins was also reported by Friedman et al.[802] Another methylated amino acid, 3-methylhistidine, was found in the avian erythrocyte histones F1 (I) and F2c (V) by Gershey et al.[803] Histones were methylated more rapidly in immature erythroid cells than in the mature erythrocytes. However, the amounts of this amino acid were rather small. The presence of all three methylated basic amino acids in Novikoff hepatoma arginine-rich histones was shown by Byvoet.[804] A comprehensive review of the enzymatic methylation of various proteins, including histones, was recently published by Paik and Kim.[805]

The amino acid sequence analysis of the calf thymus F2al (IV) histone molecule indicates that the lysine residue in position 20 is partially methylated.[281,282] The F2al (IV) histone from pea seedlings is not methylated, which indicates a considerable specificity in histone methylation that is possibly due to the specificity of methylating enzymes.[806] On the other hand, the F3 (III) histone molecule has methylated residues in both the calf thymus and pea seedlings. There is more ε-*N*-dimethyllysine than ε-*N*-monomethyllysine in this fraction.[274,298,301] As with other chemical modifications of the histones, methylation occurs at specific sites on the "already assembled" proteins. The amount of methylated histone molecules varies in individual tissues and, perhaps, with the functional activity of the cells.[805]

Both isolated nuclei and chromatin incorporated methyl groups into their proteins in vitro without any exogenous energy source, providing S-adenosyl-methionine was available in the incubation mixture.[807] The in vitro methylation process required Mg^{++} and was optimal at pH 9.0. Under these conditions, all the histones became methylated with specific activities decreasing in the order F3 (III) > F2a (mixture of IV and IIb1) > F2b (IIb2) > F1 (I). Only lysine was labeled under these conditions. On the other hand, mono- and di-ε-*N*-methyllysine was found in incubated calf thymus nuclei by Paik and Kim.[808] Although methylation of the lysine-rich histones has been reported,[807,809,810] most reports on the in vivo occurrence of methylated lysine derivatives indicated that methylation occurred primarily in the arginine-rich fractions of the histones.[214,281,282,285,806,810,811a]

Several protein methylases capable of methylating histones have been described. Some of them were present in cytoplasmic extracts, while others appeared in chromatin or nuclear preparations.[797,805,807,811a,811-816] Although these enzymes methylated certain of the amino acids in various proteins, about 70% of the incorporated methyl groups occurred indiscriminately among the various histone fractions. Guanidyl-methylated and *a*-*N*-methyl-guanidyl-methylated arginine were found in histones enzymatically methylated in vitro. A soluble cytoplasmic fraction from various rat tissues exhibited methylase activity (methylase II) that was capable of transferring methyl groups to unfractionated histones.[815] The arginine-rich histones and polyarginine showed the greatest capacity for methylation. Insulin, ribonuclease and lysine-rich histones were less active acceptors and nucleohistone (DNA-histone complex) did not accept any methyl residues. According to Kim and Paik,[816] protein methylase II required freely reactive carboxyl groups in the substrate proteins for their methylation. In vivo experiments on rat liver nuclear proteins demonstrated methylation and ethylation of all histone fractions.[817] The arginine-rich histones were more actively labeled than the lysine-rich fractions and the residual chromatin proteins exhibited considerable methylation. Interestingly, histones were found to be significantly methylated in vivo after the administration to the animals of the alkylating (methylating) carcinogen, dimethyl nitrosamine or methionine.[818]

Tidwell et al.[611] studied histone methylation

during the regeneration of rat liver. The peak of methylation followed that of the DNA synthesis indicating that histones were probably methylated in proportion to their biosynthetic rates and methylation occurred sometime after the biosynthesis of histones. Only the arginine-rich histones were methylated, with ε-*N*-dimethyllysine being the predominant form of the modified lysine residues. The increased methylation of arginine-rich histones after hepatectomy was recently confirmed by Benjamin.[810] Burdon and Garven[809,819] purified enzyme preparations from Krebs 2 ascites tumor cell chromatin which catalyzed the transfer of methyl residues to proteins, but only in the presence of added histones. Both the histones and nonhistone proteins became methylated with ε-*N*-trimethyllysine being the chief product of the methylation. There was an inverse relationship between DNA and protein methylation in the same system. Digestion of the DNA with DNase resulted in an accumulation of the trimethyl derivative at levels not usually encountered. The time course of DNA and protein methylation suggested that the protein methylation may be an intermediate step in methyl transfer to the DNA. A turnover of methyl groups in histones was also observed by Paik and Kim.[808]

Possible relationship of histone methylation to DNA replication and cell division is further strengthened by reports of higher histone methylase activities in embryonic and neoplastic liver as compared with normal adult liver.[820] There appears to be a considerable specificity in the methylation of lysine, histidine and arginine in rat liver, thymus, and Novikoff hepatoma, all labeled in vivo. While liver contains most of the ω-*N*-arginine, the content of this amino acid in thymus is about ten fold lower. The distribution pattern in Novikoff hepatoma is more akin to that in liver, but the tumor contains only approximately one third of the ω-*N*-methyl arginine content of liver. According to Burdon and Pearce,[821] cyclic AMP has no effect on histone methylation in vitro, while AMP is stimulatory. When bound to the DNA in chromatin, histone methylation is refractory to the AMP.

Methylation of histones during the cell cycle was studied by Shepherd et al.[822] in Chinese hamster ovary cells (CHO). The F1 (I) histone was not methylated. Other fractions were methylated differentially with respect to the three methylated lysine derivatives. The methyl content of the F2b (IIb2) histone was minimal during G_1, rose rapidly during the early S period, and then declined during mitosis. In synchronized HeLa cells, a maximum transfer of methyl residues was observed during the time of maximal condensation of chromatin, i.e., during the G_2, M, and very early G_1 periods of the cell cycle.[823] Although the F1 (I) histones were not methylated, a slight methylation was seen in fractions F2b (IIb2) and F2a2 (IIb1). In contrast, the arginine-rich histones F3 (III) and F2a1 (IV) were extensively methylated. The mono- and dimethyllysine derivatives increased in amounts, while the ω-*N*-methylarginine remained unchanged. An interesting observation was recently reported by Dobozy et al.[824] In human lymphocytes stimulated by phytohemagglutinin, chloramphenicol inhibited thymidine incorporation into DNA but had no effect on histone methylation. In patients hypersensitive to chloramphenicol, the methylation of histones in phytohemagglutinin-stimulated lymphocytes was considerably enhanced with the thymidine incorporation remaining suppressed.

Although there appears to be no direct correlation between the methylation of histones and the transcriptional activity of chromatin, the methylated basic amino acids seem to promote structural changes in chromatin architecture, perhaps by increasing the basicity of the histone molecules. The main transfer of methyl residues to chromatin histones occurs during the profound morphological changes of chromatin condensation.[611,822,823] However, any proposed biological function for histone methylation is, at this time, only speculative.

Thiolation

Early studies on isolated chromatin showed compositional contents in percent of dry weight of nitrogen, 15.57; phosphorus, 3.72; carbon, 45.60; and sulfur, 1.67. The sulfur was largely present as protein-bound. Although the presence of thiol groups in histones was also suggested some time ago,[825] Phillips[277] first showed that the cysteine content observed in arginine-rich histone preparations was not entirely due to their contamination with thiol-containing acidic proteins. About 70% of the total thiol groups in the cell nucleus were subsequently found to be associated with the F3 (III) histone fraction.[829] With the exception of a recent report by Subirana,[287] who identified cysteine in the F2a1 (IV) histone of echinoderms, all the reports describing the occurrence of sulfhydryl groups in histones were limited to the arginine-rich F3 (III) histone. In addition to the F3 (III) histone, the fractionation of acidic nuclear extracts by gel filtration on Sephadex G-75 yielded a preparation which contained significant amounts of thiol groups and showed a fast electrophoretic mobility in polyacrylamide gel electrophoresis. The exact chemical composition of this material was not determined, but it was suggested that it may contain peptides associated with nucleotides which can interact with certain histone fractions.[278,827,828]

A number of investigators have found thiol groups in the F3 (III) histone isolated from various species.[276,278,826] Hnilica and Bess[164] and Ord and Stocken[227] investigated the several forms of calf thymus histone F3 (III) usually detected in zone electrophoresis and concluded that they were aggregates of the same protein. Fambrough and Bonner[298] demonstrated the existence of one monomeric and one dimeric form of the F3 (III) histone in the pea seedling and oligomeric forms in calf thymus F3 (III) histone, which suggested the presence of 1 and 2 mol of cysteine in pea seedling and calf thymus F3 (III) monomers, respectively. Recent work on the amino acid sequence of calf thymus F3 (III) histone resulted in the isolation of a peptide fragment containing two cysteine residues associated with other neutral and hydrophobic amino acid residues and, hence, indicated that the cysteine residues were situated in the nonbasic portion of the F3 (III) histone molecule.[274] Lower animals, some plants, and the vertebrates up to and including rodents appeared to have only one cysteine residue per F3 (III) histone,[46,240,298,303] while higher vertebrate species had two.[46,240,298,303,830,831]

The sulfhydryl titration studies reported by Greenstein in 1944 first established that essentially all of the sulfhydryl groups in interphase chromatin occurred as free thiols.[830] No disulfide form of the F3 (III) histone has been observed in more recent studies on the interphase chromatin of various animal species.[303] Interestingly, during the phytohemagglutinin stimulation of lymphocyte proliferation, the thiol to disulfide proportion of F3 (III) histone, which remained constant during the early phase (20 to 30 min) at about 25% thiol, increased to 80% thiol during the stimulation of DNA synthesis (40 to 48 hr)[688] when the chromatin was most dispersed. In synchronized HeLa cells, Sadgopal and Bonner[830a,831] found a four-fold higher thiol content in the more extended interphase form of chromatin as compared to the condensed metaphase chromatin. These results indicated that the transition between the reduced and oxidized forms of the F3 (III) histone might be associated with, and perhaps the reason for, the condensation of chromatin. Experimental oxidation of the F3 (III) histone increased its ability to inhibit in vitro RNA biosynthesis,[832,833] but had no detectable effect with regard to in vitro DNA synthesis.[236] The higher repressor activity of the oxidized F3 (III) histone was in agreement with the higher disulfide content (51 to 53% of thiol) of condensed chromatin as compared with the low disulfide content (76 to 100% of thiol) found in extended chromatin, which was known to be responsible for most of the RNA transcription in vivo.[705,843] Even though oxidized F3 (III) histone was more inhibitory to the in vitro DNA-dependent RNA synthesis reaction, this mechanism did not seem to operate in vivo, since phytohemagglutinin stimulated lymphocytes showed no increase in the thiol content of their F3 (III) histones during the early period of newly stimulated RNA synthesis (20 to 30 min).[688]

In developing sea urchin embryos, the ratio of thiol to total thiol + disulfide content increased from less than 20% in the unfertilized eggs to about 60% just before cleavage.[236,756] More detailed studies showed that this ratio increased in acid extracts from washed nuclear fractions of the embryos during pronuclear fusion following the fertilization, during the DNA synthesis at early mitotic stages and in telophase.[835] This interesting fluctuation in the thiol content during the first cleavage of sea urchin *Echinus eculentus* could not be directly compared to the cell cycle in higher animals, since nuclear polypeptide material other than somatic histones was found in the sea urchin embryos during development. According to the authors, at least five thiol-containing basic proteins, several low molecular weight peptides containing thiol groups, 3-methyl histidine and methyllysine, and thiol-containing nonhistone proteins were present in nuclear acid extracts from the cleaving embryos. The importance of thiol groups for the early developmental schedule in sea urchins and other animals was discussed in detail by Brachet.[836]

Accurate measurements of the thiol content of F3 (III) histones in chromatin isolated from various sources meet with some difficulties. Since isolation of nuclei or cell disruption by shearing or ultrasound is frequently necessary to obtain chromatin preparations of sufficient purity, the F3 (III) histones may become partially oxidized during this procedure. To determine the amount of free sulfhydryl groups, the isolated chromatin is usually exposed to labeled p-mercuribenzoate, and extracted for histones which are then fractionated. The sulfhydryl content of F3 (III) histone then corresponds to its radioactivity. Obviously, the handling of chromatin between the homogenization of the tissue and the exposure of isolated chromatin or histones to p-mercuribenzoate may influence the sulfhydryl/disulfide ratio in individual samples.[46,240,298] There is presently insufficient evidence to evaluate critically the biological reasons for the changing thiol/disulfide ratio in F3 (III) histone during the cell cycle. In general, it appears that the increased oxidation (disulfide content) of the F3 (III) histone corresponds to the condensation of chromosomes during mitosis.

Ribosylation of Nuclear Proteins

As opposed to the major chemical modifications affecting histones, an in vivo modification of histones which is less well known, was found to be the binding of ADP-ribose polymer to these proteins. An enzyme present in the chromatin of mammalian tissues which can transfer the adenosine diphosphoribose moiety of NAD to the nuclear proteins with the simultaneous release of nicotinamide was described by several investigators. The ADP-ribose portion of NAD which became covalently attached to the nuclear proteins, presumably histones, then initiated the formation of ADP-ribose polymer (poly ADP-ribose). The DNA in chromatin did not participate in the ADP-ribosylation.[837-846] The binding of the ADP-ribose polymer to the histones was stable to 1.0 *M* guanidine hydrochloride and was not hydrolyzed during their extraction with mineral acids. Snake venom phosphodiesterase removed, at least partially, the histone-bound ADP-ribose polymer.

Burzio and Koide[847,848] have shown recently, that the incubation of chromatin with NAD, which results in the formation of an ADP-ribose polymer complex with histones, inhibited DNA synthesis. The inhibition of DNA synthesis occurred only in chromatin. Isolated DNA was not affected, indicating that chromatin proteins, most likely histones, mediated this inhibition. This assumption was further strengthened by experiments in which the exposure of chromatin previously labeled with ^{3}H-NAD to 0.6, 1.2 and 2.0 *M* NaCl gradually dissociated the ^{3}H-ADP-ribose polymer label. A parallel increase in the ability of chromatin to replicate its DNA in vitro was shown by the significantly increased incorporation of ^{3}H-TTP supported by exogenous DNA polymerase enzyme. It was concluded that the ADP-ribosylation of the nuclear proteins (histone fraction) induced changes in their conformation and affected their ability to interact with DNA, thereby inhibiting the templating capacity for its own replication. The transcription of RNA from DNA was not affected by this modification.[848]

Other Modifications of Histones

Srivastava[849] described the presence of hydroxyproline in lysine-rich histones from tobacco callus cells and suggested that proline hydroxylation might be another chemical modification of histones which could change their interactions with DNA. The very lysine-rich histone fraction they reported contained small, but significant amounts of hydroxyproline. Unfortunately, its amino acid composition indicated partial degradation and contamination by nonhistone proteins. Since the nonhistone protein fraction from this tissue contained considerably more hydroxyproline than the very lysine-rich histones, special efforts should have been made to eliminate the possibility that the hydroxyproline content of the histones represented contamination by this protein. It has been shown that histones, and especially the very lysine-rich fraction F1 (I) contain small amounts of acid-soluble nonhistone proteins which are quite difficult to remove.[141]

Although chemical modifications of histones offered the most logical mechanisms for altering their interactions with DNA and thereby changing its transcriptional activity, other mechanisms were suggested in the literature. These involve the interactions of both histones and DNA with small molecules such as steroid hormones or carcinogens, with various RNA species, nonhistone proteins, etc. The biochemical and regulatory aspects of such interactions will be discussed later.

Chapter 6

HISTONE INTERACTIONS WITH DNA

Almost every scheme attempting to explain the genetic restriction of DNA in differentiated cells employs variations of the same theme, i.e., the interactions between DNA and other macromolecules such as proteins, RNA or polysaccharides. Since histones are known to quantitatively restrict DNA transcription in vitro and, most likely, in vivo, many experiments have been reported in the literature concerning the specificity of interactions between DNA and basic proteins. Essentially, two criteria have been frequently used in the past. These are the modification of the ability of DNA to template either for its own replication or for RNA transcription, both in vitro and in vivo, and the changes in the thermal stability of double stranded DNA when associated with stabilizing molecules. Only more recently have studies on the conformation of DNA-protein complexes and their electrometric behavior appeared in the literature. Currently, the interaction studies can be divided into those performed on DNA-protein complexes formed in vitro and those using isolated chromatin. The latter approximate the in vivo conditions more closely.

In vitro Complexes of DNA with Other Molecules

Increase in the temperature of a solution containing DNA leads to a disruption of the bonds stabilizing the double stranded form of the DNA helix. Disruptive effects similar to the above can also be observed if a DNA solution is exposed to extreme pH values or if the interactions between the individual bases are modified by replacing the water with organic solvents. The denaturation of DNA normally proceeds in three overlapping steps. These are (1) the collapse of hydrogen bonded structures (mostly between AT and GC base pairs), (2) the disruption of base stacking, and (3) complete strand separation. The stability of the DNA double stranded structure depends partly on the base composition of the DNA (the stability increases with its G + C content) and partly on the molecules associated with the DNA. The pH and ionic strength of the solvent also contribute substantially to the stability of double stranded DNA. It exceeds the scope of this monograph to discuss in detail the multiple aspects of DNA denaturation. In addition, these have been reviewed by other investigators.[850,851]

The acidic nature of DNA indicates that multivalent cations will bind in near stoichiometric proportions to the phosphate back bone of this macromolecule. Small cations such as Na^+, Li^+, and Cs^+ (as well as Ca^{++}, Mg^{++}, and many others) stabilize the double-stranded structure of DNA and raise its thermal denaturation temperature.[850,851] At a neutral pH range, individual cations compete for the binding sites on DNA and the competition is almost entirely determined by the magnitude of their positive charge.[852,853] The basic amino acids lysine, arginine, and tetralysine, as well as polyamines, all bind to the DNA and, hence, contribute to molecular stabilization.[850,854-857] Relative to small cations, polyamines are more active on a molar basis in neutralizing the electrostatic repulsion of the phosphate groups. Their activity depends on the distance between the reactive amino groups.[858] While the complexes of small cations with the DNA are soluble, equimolar ratios of protonated nitrogen groups and negative phosphates on the DNA result in the precipitation of DNA-polyamine complexes[856,857] due to the formation of macromolecular aggregates.[855] While there is no preference in the binding of small cations to the DNA of any base composition, tetramethyl-, tetraethyl- and some small tetraalkyl-ammonium ions bind preferentially to A + T-rich bacterial DNA (62% A + T) as compared with DNA low in A + T (26% A +T) content.[853] Tetrabutyl ammonium ions are too large for close interaction with DNA and therefore exhibit no specificity. A similar preference for A + T-rich polynucleotides has also been reported for the binding of polyamines to DNA.[859] In chromatin, all cations compete for the binding sites on the DNA with histones or protamines and can, in sufficiently large concentrations, displace these proteins.

Optical absorption and proton magnetic resonance studies, together with other studies, show that in addition to arginine and lysine, tryptophan and other indole derivatives can bind stoichiometrically to nucleotides.[860,861] The interactions result in stabilization of DNA.[862] Similarly, certain other amino acids, such as methionine and alanine, have a stabilizing effect on the thermal denaturation of DNA.[862-865] According to Gabbay et al.,[865] electrostatic binding (nonrigid), hydrophobic and electrostatic binding (rigid), and intercalation are all involved in the attachment of the amino acids to the DNA. Equilibrium studies with diamine-containing alanine showed preferential binding to rA + rU as compared with rI + rC.[863] The extinction of histone phosphorescence (due to its tyrosine residues) after its association with DNA suggests a binding (perhaps intercalative) of tyrosine with DNA.[866] Of special interest is the interaction of DNA with certain acridine dyes and polypeptide antibiotics such as actinomycin D, mitomycin, anthramycin, etc. All these molecules stabilize DNA against thermal denaturation and have a profound effect on its templating activity. Since these compounds are frequently used as chemical probes into the structure of chromatin, their individual applications will be described in the discussion of DNA transcription.

Early investigations into the specificity of interaction between DNA and basic polypeptides such as histones, protamines, polylysine, or polyarginine indicated the preference of lysine-rich or arginine-rich proteins for regions on the DNA possessing certain base sequences.[381,867-870] Leng and Felsenfeld[870] and Shapiro et al.[871] employed conditions (1.0 *M* NaCl) most likely to enhance the interaction selectivity, i.e., conditions under which the electrostatic binding forces are relatively weak, and found that polylysine exhibited an almost perfect selectivity for (A + T)-rich DNA (precipitated complexes). This selectivity of polylysine for (A + T)-rich DNA was not limited to mammalian species, since the DNA from bacteria exhibited similar selectivity in interacting with polylysine. The selective interaction was also detected by thermal denaturation studies.[872] However, the specificity of polylysine binding to (A + T)-rich sequences of DNA could be changed. In the presence of the tetramethyl ammonium ion, polylysine interacted selectively with the G + C-rich DNA.[871] Polyarginine had a slight preference for G + C-rich DNA and a marked preference for the poly dG + dC homopolymer. As was already mentioned, the interaction of polylysine with DNA was found to be highly cooperative in 1.0 *M* NaCl.[870-874] Although the interaction was irreversible in solutions of low ionic strength at temperatures between 5° and 25°C and quantitative for polylysine: DNA ratios of less than the stoichiometric ratio of unity,[873] there was a rapid exchange of DNA in the soluble phase in 1.0 *M* NaCl.[870,871] Differences in the binding specificity of polylysine and polyarginine in the presence of NaCl were also found by other investigators.[875] According to Yuki and Fox[876] artificial arginine-rich or lysine-rich proteinoids interacted specifically with nucleotide homopolymers. In the absence of arginine, the lysine-rich polypeptide interacted preferentially with pyrimidine polynucleotides. When lysine was absent, the polypeptides reacted more readily with purine polynucleotides.

A molecular model (Figure 31) similar to that suggested by Wilkins[380] and Feughelman et al.[877] was proposed for the polylysine-DNA complex.[873] In this model, an extended polypeptide chain was wound around the DNA double helix. The pitch of both the polynucleotide and polypeptide helices showed the same spacing of one amino acid residue per one phosphate residue.[380,873,878,879] The complexes aggregated into larger units in solution. Using cationic homopolymers, Olins et al.[872] studied their interaction with DNA. As had been reported for histones and protamines, the basic polypeptides markedly stabilized the DNA against thermal denaturation and the magnitude of the stabilization and the specificity of the interactions varied with the chemical nature of the synthetic polypeptide. In further experiments concerning the interactions of DNA and salmine, or DNA and

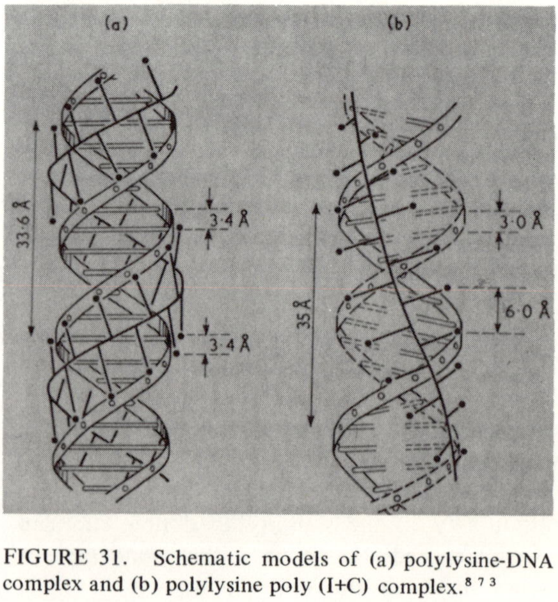

FIGURE 31. Schematic models of (a) polylysine-DNA complex and (b) polylysine poly (I+C) complex.[873]

polyarginine, these authors[880] concluded that the DNA-protamine (salmine) complexes were similar to those between the DNA and basic homopolypeptides, resembling most closely the complexes between DNA and polyarginine. Differences in the melting behavior of DNA-salmine and DNA-polyarginine complexes were attributed to the existence of arginine clusters in protamine molecules as opposed to the uniform distribution of arginine residues in polyarginine. This conclusion suggested the necessity of basic amino acids clustering in natural polypeptides (proteins) for their more specific interaction with DNA. The stabilization of DNA by polylysine was also described by other investigators.[873,874,881,882]

Wagner and Arav[883] used equilibrium dialysis to study the binding of 5′-ribo- and 5′-deoxyribonucleoside monophosphates to polylysine and polyarginine. They found that the binding of the nucleotides depended on the chemical nature of the polybase. There were few or no differences in the binding of individual pyrimidine nucleotides. The purine nucleotides, however, exhibited large differences among themselves and differed substantially from the pyrimidine nucleotides. The binding was cooperative.[884] As was already mentioned, in a reverse situation, small cations, arginine, lysine, and tetralysine bind equally to DNA of any base composition. Although the mechanism of the selective interactions between nucleic acids and basic polypeptides, particular amino acids, and certain other molecules has not been made clear (besides electrostatic interactions and intercalations, other forces are at play), experiments based on these models have demonstrated the feasibility of selective base sequence recognition on the DNA by proteins and other molecules.

In addition to selective binding, the interaction of basic polypeptides with DNA was known to significantly change the conformation of both macromolecules, resulting in many instances in a more stable and more ordered configuration.[447,456,460,871,885-894] In their latest work, Rifkind and Eichhorn[888] showed that the chemical nature of the nucleotide, as well as that of the polypeptide, governed the specificity of interaction. They found that the nucleotides not only stabilized (to various extents) the helical conformation of interacting polypeptides, but that the stabilization increased with the base stacking ability of the nucleotides. The different abilities of the nucleotides to promote conformational changes resulted from their different orientation around the polypeptides. Since the geometry for base stacking around a helical polypeptide was considered to be poor, the authors concluded that GMP appeared to be the only nucleotide that underwent appreciable base stacking. The interaction specificity of the basic polypeptides depended on the chemical nature of their side chains, an observation also made by Olins et al.,[872,880] and in studies on DNA-histone interactions which will be discussed subsequently. The helical conformation of polyarginine was more readily stabilized by interaction with polynucleotides than that of polylysine, while the stacking of GMP occurred much less readily around polyarginine than around polylysine.[888]

Since experiments on synthetic polypeptides and polynucleotides demonstrate both structural and chemical specificity of the complexes between these macromolecules, it was of considerable interest to examine similar studies on DNA and histones or protamines, i.e., substances germaine to the nucleus of eukaryotic cells. The isolation and analysis of chromatin from eukaryotic cells which retained most of their biological properties[23,895] removed any doubt that histones occurred in natural complexes with DNA and that such complexes were not dissociated during their isolation.[896,897]

The attachment of histones to DNA has been found to occur principally through the formation of electrostatic linkages between the amino groups of histone and the phosphate groups of DNA. Huang et al.[898] prepared artificial nucleohistones from DNA and isolated histone fractions by slow reconstitution in a decreasing NaCl gradient. The very lysine-rich histones F1 (I) stabilized the DNA against thermal denaturation better than the arginine-rich histones. The increased denaturation temperature (Tm) was histone specific, as could be expected from the various proportions of lysine and arginine in each of the fractions. Essentially identical results were reported by Hnilica and Billen,[899] who also found the F1 (I) histones to stabilize the DNA better than the other histone fractions. The presence of Mg^{++} almost eliminated the differences in DNA stabilization by various histone fractions. Preferential binding and stabilization of (A + T)-rich regions of the DNA by the F1 (I) histones was observed by Ohba[881] and Minyat et al.[900] However, no such preferential stabilization could be confirmed by Olins.[901] In a detailed study, Akinrimisi et al.[381] found that protamines and histones had a greater affinity for native than denatured DNA. The affinity of these proteins increased with their arginine content, being higher for protamines and lowest for the F1 (I) histone fraction. The binding of histones to DNA was considerably decreased at higher salt concentrations (0.3 to 0.7 *M* NaCl), but was not completely abolished in 1.0 *M* NaCl, which indicated that forces other than the electrostatic salt-like binding contributed to the interaction between DNA and the arginine-rich histones. Cooperativity in binding was also found for individual histone species, as well as for mixtures of histone fractions, by Shih and Bonner.[902] These authors prepared their complexes in the presence of 5.0 *M* urea. Each complex had a characteristic thermal denaturation temperature which reflected the amino acid composition of the particular histone fraction. The complexes of DNA with protamines, polyarginine, and polylysine had denaturation temperatures of 92.5°C, 98°C, and 99.5°C, respectively, as compared with temperatures of 74.3°C and 83.7°C for complexes with the very lysine-rich histone Ia (F1) and with the arginine-rich histone IV (F2a1), respectively. A similar melting behavior for DNA-histone complexes was also shown by Ansevin and Brown.[903] Here, the complexes were formed by a very slow addition of the histones to a cacodylate buffered DNA solution containing 3.6 *M* urea. Nucleohistones prepared by reconstituting DNA with histone fractions F1 (I), F2a1 (IV), F2a2 (IIb1), and F3 (III) each exhibited thermal denaturation temperature (Tm) increases in three or more steps, presumably corresponding to free DNA and several nucleohistone complexes with different binding stabilities. On weight basis, the complexing efficiency increased F1 (I) > F2b (IIb2) > F3 (III) $\simeq$ F2a1 (IV) $\simeq$ F2a2 (IIb1). At appropriate DNA:histone ratios, the stoichiometry of complex formation with histones other than F1 (I) approximated that of the arginine + lysine content vs. DNA-phosphate. Complexes formed at histone:DNA weight ratios of 0.75 showed considerable free DNA in the plots of derivative hyperchromicity at 260 nm vs. temperature. The F1 (I) nucleohistone was an exception, however. The amounts of free DNA in these nucleohistones decreased with the lysine content of the histones, i.e., in the order: F2a1 (IV) $\geq$ F3 (III) > F2b (IIb2) > F1 (I) = 0. This order was similar to that seen by Huang et al.,[898] Hnilica and Billen,[899] and others, for the extent of the inhibition of DNA dependent RNA synthesis in vitro by histones.

The manner in which the DNA and histone fractions were brought together in vitro apparently affected the stability properties of the resulting complexes. In experiments on complex reconstitution by gradient dialysis in the absence of urea,[898,899] the F1 (I) histones stabilized the DNA better than the other fractions. The very lysine-rich histones tended to cross-link individual DNA strands[904-908] and, thereby, precipitated the DNA more efficiently than the other histone fractions.[181,899,909-911] The precipitation of DNA by histones, especially the F1 (I) fraction, was sensitive to ionic strength and decreased with an excess (by weight) of F1 (I) histone to DNA.[181,912] Complexes of F1 (I) with T7 phage DNA made by reconstitution dialysis from concentrated urea-NaCl or guanidine hydrochloride solutions behaved during sedimentation analysis like naked DNA at low ion concentrations. Electron micrographs of this material revealed a loose network of DNA-like fibers. At higher ionic strengths, the fibers associated into doughnut and stem-like structures of considerable size. Similar doughnut-like structures were observed at higher salt concentrations for DNA complexes with the F2a1 (IV) fraction.[912] If reconstitution or thermal denaturation was carried out in the absence of urea, it thus appeared possible that the excessive degree of DNA stabilization by F1 (I) histones represented the melting behavior of DNA cross-linked to various extents. In native chromatin, the F1 (I) histones seemed to contribute but little to the stabilization of the DNA structure, since their selective removal only slightly affected the flow birefrigence, flow dichroism, and sedimentation behavior of the resulting nucleohistone.[177,895] Interestingly, actinomycin D binding experiments demonstrated that F1 (I) histones most likely bind in the large groove of DNA.[901]

As suggested by x-ray diffraction[913-917] and infrared spectroscopy studies,[453] DNA assumes the B configuration when complexed with histones or basic polypeptides. More recent investigations indicate, however, that the interaction of basic proteins with DNA caused its partial unstacking and deformation (C conformation).[918,918a] In the C form the DNA is more densely packed along its helical axis. Circular dichroism measurements of DNA and F1 (I) histone complexes indicated that the conformation of undenatured DNA changed as a result of binding to the histones.[919] In 0.14 *M* NaF, the circular dichroism spectrum of DNA became progressively distorted during the addition of increasing amounts of F1 (I) histones. Such conformational changes, which may have resulted from changes in the hydration of DNA, did not occur when heat-denatured DNA was complexed to the F1 (I) histones. Deformation of the double stranded DNA structure was also demonstrated in complexes of DNA with polylysine.[871,890] More recently, Shih and Fasman[918] extended the circular dichroism studies to complexes formed with DNA and the F2a1 (IV) histone fraction. The circular dichroism spectra of the DNA depended on the conditions of reconstitution. In complexes reconstituted from concentrated urea and NaCl to a NaCl concentration of 0.14 *M*, the DNA spectrum was nonconservative and resembled that of the A form of DNA or double stranded RNA. These complexes were extremely stable. The addition of 0.1% sodium dodecyl sulfate restored the circular dichroism spectrum to that of native DNA. Complexes reconstituted by the removal of the urea from 0.015 *M* NaCl did not change their DNA spectra. Also, an excess of histones over DNA in the complexes caused the spectra to look more like that of native DNA. The conformational changes observed for DNA complexed with the F2a1 (IV) histones were different than changes produced by association of the DNA with the F1 (I) fractions.[919,920] It was possible that the differences observed were due to various extents of DNA hydration resulting from interaction specificities of the F1 (I) and F2a1 (IV) histones with individual DNA base pairs. The F2a1 (IV) histone in association with DNA was partially α-helical.

Reconstitution of the F1 (I) histones to DNA was also effected by the presence of cations. While guanidinium-HCl changed the circular dichroic spectra of the DNA in the reconstituted nucleohistone the most, $NaClO_4$, KCNS, and NaI were the least able to do so. Several anions, on the other hand, did not significantly differ from one another in their effect on the circular dichroism spectrum. Besides the specific salt effects, the conformation of the DNA was also affected by the method by which the F1 (I) histones and DNA were reconstituted,[920] which was an indication that specific associations between the interacting components might have taken place during reconstitution. Alteration of the DNA conformation during its association with histones was also reported by other investigators to be proportional to the amount of DNA-bound protein.[922] Further work from Fasman's laboratory revealed considerable specificity in the F1 (I) histone binding to DNA, since phosphorylation of this fraction (one site per molecule) decreased the deformation of circular dichroism spectra.[771] Furthermore, when the F1 (I) histone was cleaved with bromosuccinimide, the two fragments behaved very differently in their interactions with DNA.[923] The NH_2-terminal fragment, which was less basic, did not change the DNA spectra, whereas the basic COOH-terminal fragment deformed the DNA conformation even more efficiently than the intact F1 (I) protein. However, these changes could not be directly correlated to the lysine contents of both fragments. If both fragments (NH_2- and COOH-terminal) were interacted simultaneously with DNA, the inactive NH_2-terminal fragment considerably enhanced the deformation effect of the COOH-fragment. However, in intact F1 (I) histone, the inactive NH_2-terminal portion had a moderating effect on the interaction of the COOH-terminal part with DNA, as can be concluded from the larger effect of this fragment when compared with that of intact F1 (I) histone. In the absence of DNA, these two fragments did not interact with each other. Perhaps, in intact F1 (I) the NH_2-terminal part of the molecule folded back on the COOH-terminal part and modified its interaction with the DNA. The interaction of the histone half molecules with DNA was also studied by Li and Bonner.[924] The findings that the interaction with individual histone fractions and with parts of their molecules has a profound effect on the DNA structure in such complexes might have considerable impact on an understanding of the regulation of the structural and transcriptional properties of the DNA in chromatin and chromosomes.

Chromatin

It is known that the functional form of the chromosomal DNA in eukaryotes is associated with other macromolecules such as histones, nonhistone proteins, lipids or lipoproteins, RNA, etc.

This complex has been called nucleohistone or chromatin and represents the cytologically indistinct form of the individual chromosomes of the cell. The development of methods for the isolation of nuclear chromatin in a biochemically functional state resembling that of the native material in vivo has initiated a steady stream of publications concerned with the various properties and functional mechanisms of chromatin. Although a wealth of information is now available regarding the isolation, structure, and function of chromatin, the relationship between the physical and chemical structure of chromatin and the mechanism of selective gene activation is only beginning to be understood.

The Isolation and Composition of Chromatin

The nucleohistones, which in an oversimplified statement are salts of the histones and nucleic acids, were first observed by Kossel[3,830] nearly a century ago in material extracted from the nuclei of several animals. In 1892, Lilienfeld[8,9] isolated a similar material from the tissues of thymus glands and, although several reports of nucleohistone isolation appeared between 1893 and 1910,[830] it was not until 1940 that Claude and Potter[925] first reported the isolation of chromatin threads per se from the interphase nuclei of living cells. The latter method was adapted by Mirsky and his associates for their early biochemical studies on nucleoprotein complexes. Based on the cumulative efforts of early investigators, in 1959 Zubay and Doty[48] devised a method for the isolation of chromatin from calf thymus gland tissues and characterized this chromatin with regard to a number of physical and chemical parameters. Their method entailed a repeated homogenization of the thymus tissue in 75 mM NaCl containing 24 mM EDTA and a final solubilization of the chromatin material in distilled water. The isolated nucleohistone, when dissolved in 0.7 mM phosphate buffer at pH 6.8, consisted of highly asymmetrical molecules completely dispersed in solution and containing DNA at approximately 50% of its weight. This preparative procedure was used with some variations in many laboratories.[49,705,926-929] Since the dispersion of nucleohistone in water frequently resulted in a partial proteolytic degradation of the histones, dilute buffers containing EDTA were subsequently found to be preferable. Although it was possible to obtain good chromatin from whole calf thymus tissue by the method of Zubay and Doty,[48] in tissues containing cells with considerable amounts of cytoplasm, it was necessary to use isolated nuclei for the preparation of undegraded chromatin.[182]

The most notable and widely accepted procedure for the isolation of chromatin from the tissues of almost any eukaryotic animal and many plant species was developed by Marushige and Bonner.[930] It consisted essentially of a slow speed centrifugation and washing of blendor disrupted cells or isolated nuclei with dilute saline-EDTA solutions and, finally, with 50 mM Tris buffer (pH 8.0). For a final purification, the chromatin was suspended in 50 mM Tris buffer and centrifuged through a crude sucrose density gradient of Tris buffer to 1.7 *M* sucrose. This step could be repeated to further decrease the contamination of chromatin by cytoplasmic material. The gelatinous pellet resulting from the density centrifugation was resuspended in buffer, dialyzed against the same buffer overnight and, finally, sheared to increase the solubility of the preparation. In tissues known to contain proteolytic enzymes, dialysis should be omitted, since it increases the possibility of proteolytic degradation of chromatin. Moreover, the DNA might also become damaged during this step. The isolated chromatin preparations were not turbid and could be stored frozen (preferably at −40°C or lower) for several months. The DNA in such chromatin was found to be stabilized against thermal denaturation by the histone proteins. The chromatin was also electrophoretically homogeneous and carried a negative electric charge.[23,895] These electrophoretic properties of chromatin could be utilized for its further purification.[931]

The composition of typical chromatin consists of almost equal proportions of DNA and histones and smaller amounts of nonhistone proteins and RNA. The chemical composition of chromatin prepared from a variety of sources is shown in Table 14. Although the ratio of histone to DNA in most of the chromatin samples and in the isolated nuclei is near unity, the histone content can increase considerably during special conditions of growth or DNA replication.[244,580] As previously noted, there are five major histone fractions present in the cells of higher organisms and all five are present in chromatin. The absolute quantities of each fraction can vary between individual tissues. This is especially noticeable during early embryonic development. However, in most chromatin preparations, the five histones are present in almost equal weight ratios.[29,100,206]

In contrast to the circumstances pertaining to the histone content of chromatin, the nonhistone protein content was found to be much more variable. Indeed, the very existence of the nonhistone proteins as a class was recently challenged by Sonnenbichler et al.,[932-934] who analyzed the proteins which remained firmly associated with DNA after acid extraction of isolated chromatin. These proteins were said to represent a considerable part of the chromatin protein complex and could be partially solubilized in acid containing urea.[932] Later, these authors[933,934] concluded that the acid treatment of the chromatin artifactually produced the nonhistone proteins which were firmly associated with the DNA. However, their suggestion that nearly all of the chromatin nonhistones were artifacts resulting from the acid extraction of histones is hardly acceptable in light of the present day knowledge of the biochemistry of the nonhistone chromatin proteins.

The extent of the variability of the nonhistone content of chromatin was known some time ago to be subject to preparative influences. Nearly two decades ago, Butler and Davison[935] cautioned that the amount of nonhistone protein ascribed to a particular nuclear or chromatin preparation will vary inversely with the investigator's success in completely extracting the histones. The arginine-rich histones, in particular, are known to be difficult to extract quantitatively. Such contamination of the "residual" nonhistone proteins by histones can be easily detected by subjecting the sample to polyacrylamide gel electrophoresis in the presence of SDS. The histones, excepting the F1 (I) and dimer F3 (III) fractions, are of low molecular weight and are well separated from the significantly heavier nonhistone proteins. In an opposite sense, it is also apparent that the nonhistone content of chromatin will vary with the more or less successful removal of contaminating nonchromosomal nuclear nonhistone proteins as well as cytoplasmic contaminations. While the older literature must be evaluated with caution in the above respect, the more modern techniques of chromatin isolation and purification tend to minimize the presence of nonchromosomal proteins in such preparations. In general, the nonhistone protein content of chromatin appears to parallel its transcriptional activity, since it was found to be relatively high in tissues active in RNA synthesis and quite low in repressed chromatin (Table 14). The nonhistone proteins were also shown to influence the specificity of transcription. The varying aspects of the contribution of nonhistone proteins to the transcriptional specificity of DNA restriction will be discussed subsequently in this monograph.

Chromatin Structure

It was already known to Kossel and his contemporaries that the complex of histones and DNA, as isolated from nuclei, is quite insoluble in salt solutions of physiological strength. The solubilization of chromatin by exposure to salt solutions of high ionic strength (e.g., 2.0 *M* NaCl) was found to be due to the dissociation of the primary electrostatic binding between the histones and DNA resulting in a mixture of these molecules in the solution.[93,113] A dilution of the salt-dissociated chromatin solution to physiological salt concentration (0.14 *M* NaCl) causes a reassociation of histones and DNA in the form of long fibers.[113] The reconstituted chromatin differs, however, from the original chromatin sample in its physicochemical behavior[92,214,942] and transcriptional specificity.[739,740]

In water, chromatin forms a gel that progressively swells on dilution and ultimately dissolves into a viscous dispersion. The behavior of chromatin in water depends on the concentration of the chromatin, which indicates a strong interaction of the constituent macromolecules.[943] The interaction caused by the negative charges of the macromolecular complexes can be decreased by the addition of salts to reduce the polyelectrolyte effect. Unfortunately, the chromatin aggregates and ultimately becomes insoluble in dilute salt solutions[944] and more or less dissociates at higher concentrations. On the other hand, isolated chromatin dissolves well in concentrated urea solutions with a simultaneous increase of its intrinsic viscosity.[945] The increase of viscosity,[945] the electron microscopic images[946] of chromatin dissolved in urea, and its sedimentation behavior[946] all suggest that, in urea solutions, chromatin exists in an extended, rod-like conformation which is unlike the supercoiled structure assumed in water.

The ability of chromatin to form rigid gels in water is regarded by many investigators as a

TABLE 14

Composition of Chromatin Isolated from Various Tissues

Organism	Tissue	DNA	RNA	Histone (acid soluble)	Nonhistone (residual)	Reference
Pea	Embryonic axis	1.00	0.26	1.03	0.29	895
Pea	Vegetative bud	1.00	0.11	1.30	0.10	895
Pea	Growing cotyledon	1.00	0.13	0.76	0.36	895
Barley	Leaf 7 days	1.00	0.23	0.98	0.56	936
Barley	Leaf 25 days	1.00	0.63	1.03	0.39	936
Neurospora crassa	Mycelia	1.00	0.88	–	1.24	73
Strongylocentrotus purpuratus	Blastula	1.00	0.04	1.04	0.48	388
Strongylocentrotus purpuratus	Pluteus	1.00	0.08	0.86	1.04	388
Milkweed bug (*Oncopeltus fasc.*)	68-hr embryo	1.00	0.43	0.91	4.70	937
Milkweed bug (*Oncopeltus fasc.*)	92-hr embryo	1.00	0.29	1.02	3.20	937
Milkweed bug (*Oncopeltus fasc.*)	140-hr embryo	1.00	0.06	1.11	1.65	937
Trout testis	Stage 1	1.00	0.05	1.08	0.59	360
Trout testis	Stage 3	1.00	0.03	1.38	0.25	360
Trout testis	Stage 5	1.00	0.02	0.61	0.18	360
Trout testis	Nucleoprotamine	1.00	<0.01	0.54	0.04	360
Chicken embryo	5 day liver	1.00	0.17	0.80	1.70	928
Chicken adult	Liver	1.00	0.10	0.80	1.37	928
Chicken embryo	4 day brain	1.00	0.29	0.91	2.09	928
Chicken adult	Brain	1.00	0.18	0.78	2.05	928
Chicken embryo	1.5 day	1.00	0.36	0.81	2.82	928
Chicken embryo	11 day	1.00	0.16	1.20	0.36	928
Chicken embryo	3.5 day erythrocyte	1.00	0.10	0.84	1.16	928
Chicken adult	Erythrocyte	1.00	0.01	0.82	0.48	928
Carp (*Cyprinus carpio*)	Erythrocyte	1.00	–	1.06	0.42	938
Pike (*Esox Lucius*)	Erythrocyte	1.00	–	1.09	0.53	938
Trout (*Salmo irideus*)	Erythrocyte	1.00	–	1.00	0.37	938
Rat	Liver	1.00	0.04	1.00	0.67	895
Rat	Liver	1.00	0.10	1.20	1.00	182
Rat	Liver	1.00	0.13	1.20	1.60	931
Rat (Buffalo)	Liver	1.00	–	2.09	1.54	939
Rat	Liver	1.00	0.11	1.04	0.67	940
Rat	Hepatoma[a] 5123c	1.00	0.11	1.21	0.89	940
Rat	Hepatoma[a] host liver	1.00	0.11	1.20	0.86	940
Rat	Hepatoma[b] 5123c	1.00	0.11	1.38	0.75	940

TABLE 14 (continued)

Organism	Tissue	DNA	RNA	Histone (acid soluble)	Nonhistone (residual)	Reference
Rat	Hepatoma[b] host liver	1.00	0.10	1.50	0.59	940
Rat	Novikoff hepatoma	1.00	0.13	1.16	1.00	895
Rat	Novikoff hepatoma	1.00	0.18	1.10	0.57	940
Rat	Hepatoma HDEN	1.00	–	1.45	0.40	939
Rat	Morris hepatoma 5123	1.00	0.15	1.06	1.64	941
Mouse	Ehrlich ascites carcinoma	1.00	0.23	1.50	0.72	931
Calf	Thymus	1.00	0.06	1.40	0.24	931
Calf	Thymus	1.00	<0.01	1.14	0.33	895
Calf	Thymus	1.00	0.06	1.00	0.50	182
Man	HeLa carcinoma	1.00	0.09	1.02	0.71	895
Man	HeLa[c] carcinoma	1.00	0.72	1.05	2.37	830
Man	HeLa[d] carcinoma	1.00	0.12	1.29	1.56	830
Man	HeLa[e] carcinoma	1.00	0.10	1.03	0.72	830

[a]Chromatin prepared according to the method of Dingman and Sporn.[928]
[b]Chromatin prepared according to the method of Marushige and Bonner.[930]
[c]Chromatin prepared by precipitation with magnesium acetate.
[d]Unsheared
[e]Sheared

characteristic of enzymatically undegraded chromatin samples.[926,947–954] Since DNA at similar concentrations does not form cross-linked gels, the formation of gels by isolated chromatin in water must be due to its protein content. The formation of gels is most likely assisted by proteins cross-linking adjacent nucleohistone strands and interlocking the entire structure. Electron microscopy, dissociation, reconstitution, and templating studies on isolated chromatin all suggest that the very lysine-rich F1 (I) histones may participate in the cross-linking and interlocking of the individual nucleohistone fibers in chromatin gel structures.[93,904,907,908] However, other mechanisms have also been suggested. According to Dounce and Hilgartner,[947] the nonhistone proteins are linked to the DNA by covalent bonds and the cross-linking of the chromatin fibers is realized through disulfide bridges between the cysteine residues of the individual DNA-bound proteins. In this model, the histones associate with DNA in a parallel fashion with very little cross-linking of the DNA strands. The presence of disulfide linked nonhistone proteins appears to be essential to the gel-forming properties of nucleohistone, since the exposure of chromatin to appropriate reducing agents disrupts its structure.[947,955,956] In addition to gel formation, the possibility that the condensation of chromatin and chromosomes may depend on the oxidation of sulfhydryl groups in the chromosomal proteins has been supported by experimental evidence. Several investigators have reported that the nonhistone chromatin proteins, as well as the arginine-rich F3 (III) histone fraction, contain more disulfide groups in metaphase chromosomes than in interphase chromatin.[227,236,298,830a,831,833] Especially, the acid-soluble nonhistone proteins appear quite extensively polymerized through disulfide bridges in metaphase chromosomes, while the corresponding proteins from interphase chromatin do not display such polymerization.[830a,831] The molecular weight of the nonhistone proteins cross-linking chromatin fibers has been estimated to be approximately 400,000 daltons or higher.[956] According to several reports, there exists in addition to the usual nonhistone proteins a class of nonhistone proteins which tenaciously bind to DNA and are characterized by being relatively rich in dicarboxylic acids and serine.[957-964] The possibility is considered in these reports that the DNA-protein binding in this case occurs through the serine residues of the proteins.[964] It is not known at present whether these proteins serve the structural functions suggested by Hiltgartner and others.[947,955,956]

The molecular weight of chromatin in its native form is very large and is determined by the length of the individual DNA molecules as well as by the extent of interaction of the individual strands. During isolation, the chromatin undergoes breakage into smaller fragments with average molecular weights between 2.5 and 35 x 10^6 daltons. The molecular weight reflects both the gel forming property of chromatin and its solubility, since the soluble fractions average lower molecular weights than the gels. The passage of isolated chromatin through Sephadex gel bead columns disrupts the macromolecular chromatin aggregates and fragments the chromatin.[946,965] Limited fragmentation of the chromatin does not seem to affect its templating activity and melting behavior.[23,895,966] In electron micrographs, the 30 A wide fibrils of "soluble" nucleohistone, which result from shearing in the preparative method of Zubay and Doty,[48] look quite different from chromatin fibrils isolated by methods of minimal shear. This, according to Ris,[967] supports the view that the structure of this type of solubilized nucleohistone is artifactual and is not representative of the native structure of chromatin. The tendency of nonsolubilized chromatin fibers to associate with one another to form networks is in accord with the contention that gel formation is a natural property of deoxynucleoprotein.[967]

Electron microscopic analysis of the structure of various isolated nucleoproteins showed that the DNA core of the chromatin was surrounded with protein in the form of a more or less evenly distributed, continuous sheath.[48,946,968-970] A characteristic surface structure was evident in the 100 A diameter nucleoprotein fibers.[967] Such surface irregularities, though reported by several authors, could not be interpreted with regard to the exact arrangement of the DNA and histone in the fiber due to technical limitations. On the other hand, electrometric and spectrophotometric titration data demonstrated that, in calf thymus nucleohistone, about two thirds of all arginine and lysine residues are not accessible for titration at pH 11.8, which indicates that they are involved in ionic bonding to the phosphate groups on the DNA.[971] Similar estimates of free lysine and arginine residues in chromatin histones were obtained by acetylation experiments.[972] Since the carboxyl, imidazole, and tyrosyl groups were found to be freely accessible to hydrogen and hydroxyl ions during the titration of nucleohistone, it was concluded that they are on the surface of the nucleoprotein complex and in contact with the solvent.[971] In a recent study, electron micrographs of isolated calf thymus nucleoprotein (chromatin) molecules showed them to consist of long fibrils whose diameter varied from 80 to 120 A and averaged about 100 A in thickness.[970] There was a considerable content of X- and Y-formed junctions evident in most of the electron micrographs, which suggested that many of the strands were branched (Figure 32). In addition to the basic 80 to 120 A fibrils, electron microscopy of chromatin prepared by the Anderson critical point technique[973] showed knoby extensions of the fibrils which projected 80 to 200 A from the fiber axis and probably represented short side branches (Figure 33). Digestion of the nucleoprotein fibers with pronase extended their structure into a single double stranded DNA molecule which, in native chromatin, was irregularly folded because of its interaction with histones.[968] The mechanical stretching of nucleoprotein fibrils caused them to become progressively thinner until they reached a minimum of about 25 to 30 A in diameter. This value approached the minimum diameter of the basic fibers in rat thymus nucleoprotein reported by Itzhaki and Rowe[969] and was reminiscent of the 30 A fiber diameter reported by Zubay and Doty.[48] It is noteworthy that the basic fibers of 100 A average thickness were also shown to become partially uncoiled when exposed to urea. This was evidenced by the reduction of the average fiber diameter of approximately 100 to 200 A to about 40 A. This transition was very uniform and was accompanied by the disappearance of the knob-like structures (Figure 34).[946] Since a similar extension and narrowing of nucleoprotein

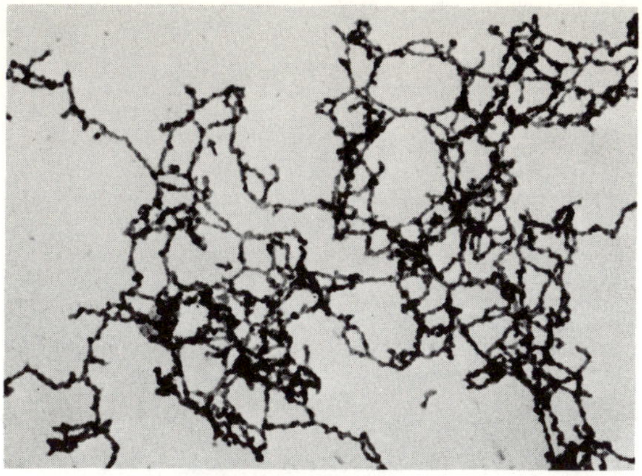

FIGURE 32A.

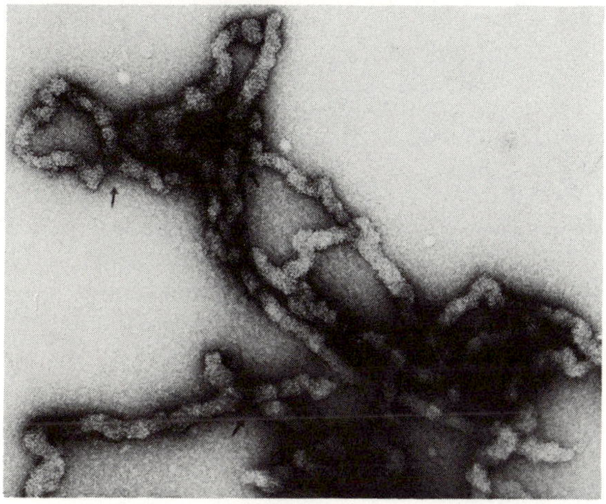

FIGURE 32B.

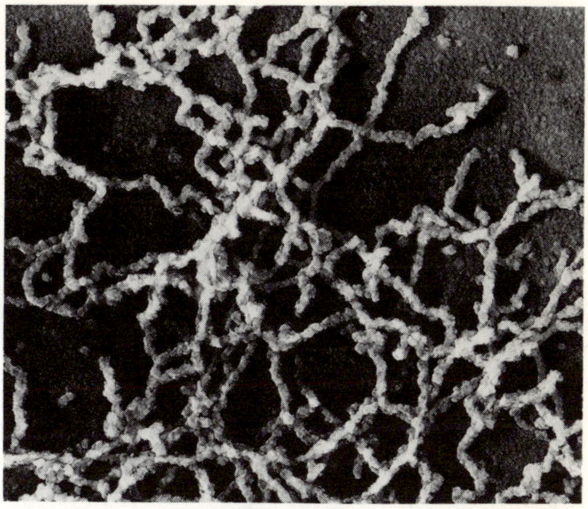

FIGURE 32C.

FIGURE 32. Chromatin from erythrocyte nuclei of *Triturus viridescens.* (A) Spread on water. These fibers which fill the nucleus represent condensed inactive chromatin and contain DNA and histones. Arrows point to places where side branches join other fibers. (B) Chromatin fixed in 10% neutral formalin and negatively stained with 1% uranyl acetate. Arrows indicate side branches and contact points in the network. (C) Chromatin fixed in ethanol and shadowed with carbon-platinum.[1023]

fibers was produced by its exposure to 0.6 *M* NaCl (Figure 35), which removes selectively the F1 (I) histones, it is possible that the lower order coiling of the nucleoprotein fiber in aqueous solutions (100 A average thickness) results from interactions between individual F1 (I) molecules within the nucleoprotein. However, as discussed earlier, these coiling (or folding) interactions could be overcome in concentrated urea solutions and, apparently, by certain mechanical stresses without the actual removal of the F1 (I) histones.

Based on a best fit analysis of their experimental data to a computer simulation, Bram and Ris[970] proposed a model of nucleoprotein which approximates a long cylinder of 80 to 120 A diameter having axial projections of 120 to 250 A in length. It was suggested from the small-angle x-ray scattering data that the nucleoprotein fiber structure consisted of a single DNA double-helix molecule that was irregularly supercoiled or folded with an average pitch of 45 A and cross section radius of gyration of 30 A to make a thicker fiber. The nonuniformity of the nucleoprotein coiling is in good agreement with the physicochemical observations concerning DNA and histone interactions discussed earlier.

During the cell cycle of eukaryotic cells, the chromatin nucleoprotein assumes various states of condensation. In the interphase portion of the mitotic cycle, a part of the chromatin remains persistently condensed, while the remainder undergoes dispersal into an intranuclear network of finely-distributed threads that do not again condense until the following mitosis. This phenomenon of heterologous condensation was first noted by the German cytologist Heitz, who referred to the condensed, noncycling chromatin as heterochromatin and to the diffuse alternate form as euchromatin. Heterochromatin itself has been recently divided into facultative and constitutive heterochromatin. The facultative form remains transcriptionally inactive most of the time. A typical example of this type of chromatin is the heterochromatinization of one of the two X chromosomes in female mammals. Constitutive heterochromatin, on the other hand, is found in

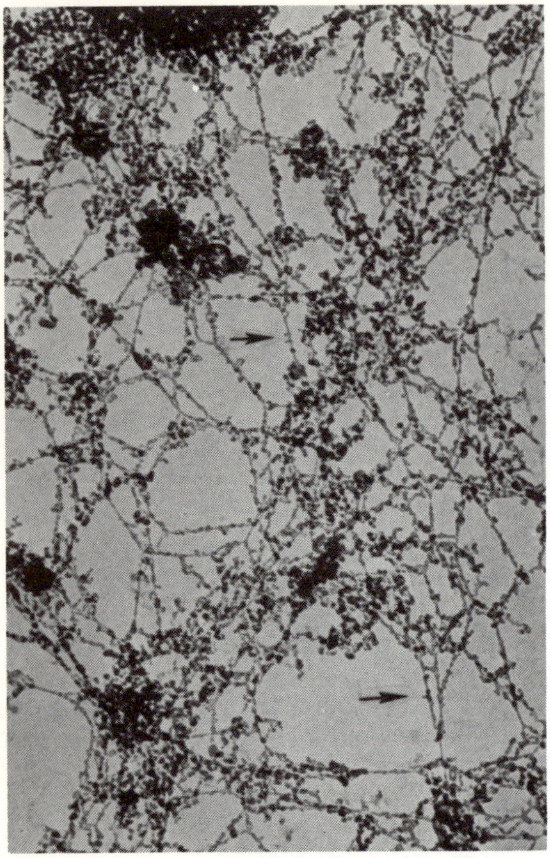

FIGURE 33A.

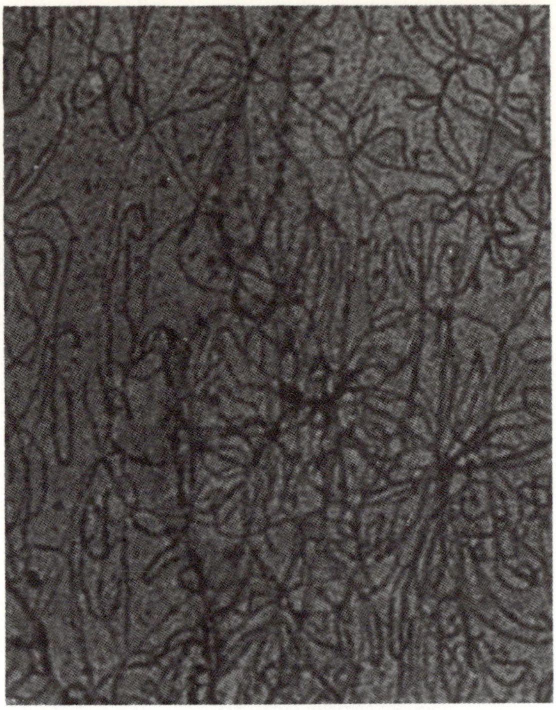

FIGURE 34. Calf thymus chromatin in 4 *M* urea. Compare with structures shown in Figures 32 and 33 and notice the loss of higher organization. The 100 to 200 A fibers became disaggregated to a uniform network of 40 A fibers (X 75,000).[946]

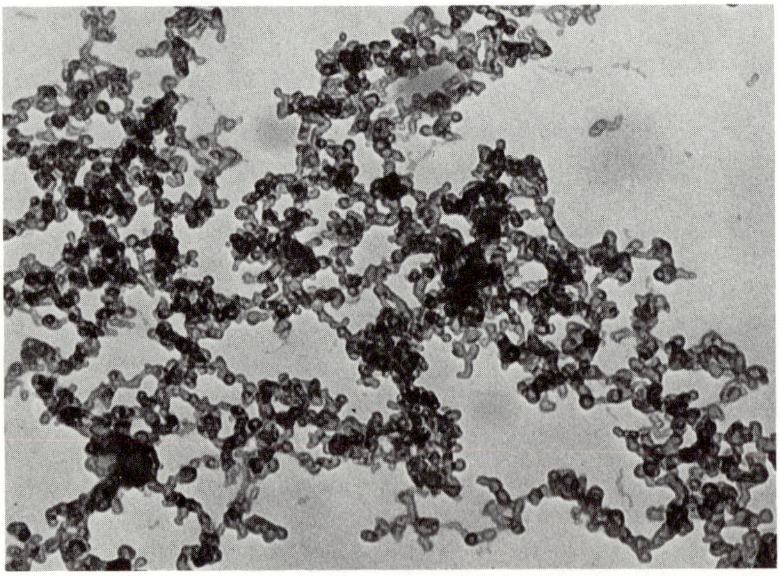

FIGURE 33B.

FIGURE 33. (A) Fibrils (100 A) from calf thymus chromosomes isolated in saline–EDTA at pH 6.5 and dissolved in water. Critical point dried. Note knobby protuberances marked by arrows (X 42,800).[970] (B) Isolated rat liver chromatin. Critical point dried. Basic fiber diameter 100 to 200 A (X 115,000). Courtesy of B. R. Brinkley and M. E. McClure.

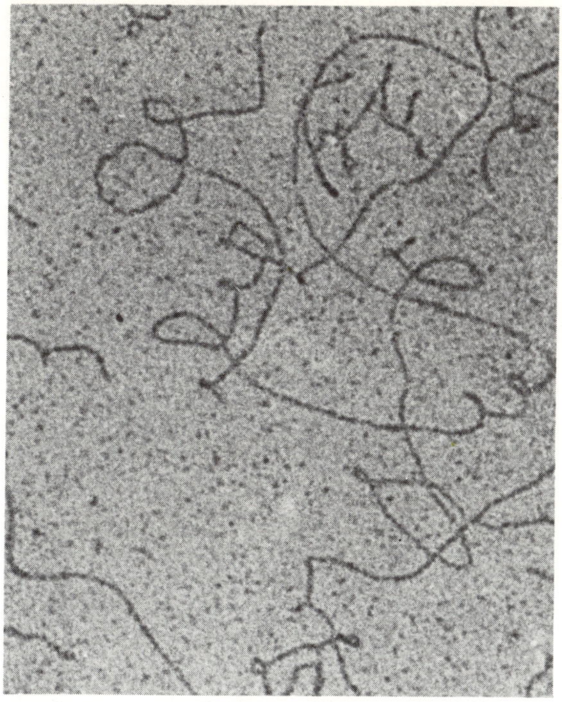

FIGURE 35. Calf thymus chromatin treated with 0.6 *M* NaCl. This treatment removes virtually all F1 (I) histones and a part of the F2b (IIb2) fraction. The loss of higher organization similar to that in 4 *M* urea (Figure 34).[946]

homologous chromosomes and contains the bulk of the nuclear satellite DNA.[745,746,974-979] The use of in situ hybridization experiments has revealed that constitutive heterochromatin containing satellite DNA is mostly accumulated in the heterochromatic regions surrounding the centromeres and the nucleolar organizer.[974,978-982] The constitutive heterochromatin, which represents a considerable portion of the genome, is inactive in RNA transcription.[974,983-987] The centromeric concentration of the repetitive DNA sequences, as distinguished by the in situ hybridization of their complementary RNA, also occurs in the polytene chromosomes of *Diptera*[998-991] (i.e., in α and β heterochromatin). The RNA complementary to the repetitive DNA sequences of mouse heterochromatin did not hybridize to similar regions on *Drosophila* chromosomes and vice versa, indicating a strict species specificity of the repetitive DNA sequences.[989]

Euchromatin and heterochromatin do not appear to be different when viewed with the electron microscope. The extended transcriptionally active euchromatin corresponds to the structures in Figures 32 and 33, while the condensed heterochromatin is transcriptionally inactive or repressed[705,834,974,983-985] and has a structure reminiscent of an aggregated form of the extended fiber elements. The condensed chromatin in the pyknotic nuclei of salamander, frog, and bird erythrocytes, as well as the chromocenters and heterochromatin blocks of a number of other cells, have been found to consist of tightly packed fibrils of 100 to 200 A thickness. Experimental disruption of heterochromatin produces the same varieties of fibril diameters as observed in diffuse chromatin.[992]

The substantial hydrodynamic difference between eu- and heterochromatin permits their fractional separation. In a method devised by Frenster et al.,[705] isolated chromatin is subjected to shear forces produced by ultrasound and the two density fractions of the chromatin (dispersed and condensed) are separated by centrifugation in sucrose solutions. This method has been frequently employed by many investigators. The yield of diffuse chromatin is, however, somewhat dependent on variations in the shear force employed. Shearing by homogenization in a blender-type Virtis homogenizer has been recommended by Duerksen and McCarthy,[993] since it offers better control of the shearing force and, hence, more reproducible results. Euchromatin and heterochromatin differ in their electron density, repetitive DNA content, templating efficiency, etc. In addition, heterochromatin is practically insoluble in water, although it does dissociate in concentrated salt solutions.[993a]

Heterochromatin and euchromatin are known to contain both histones and nonhistone proteins. Because the DNA of the two chromatin forms showed similar base compositions, it was suggested that the different structural states could be explained by the presence of non-DNA "accessory material" in euchromatin to abolish persistent condensation or in heterochromatin to cause it. Since the histones were often implicated in the supercoiling of DNA and the further aggregation of these supercoils into the mitotic chromosome structure, differences were anticipated for the histone contents of euchromatin and heterochromatin. Indeed, cytochemical staining methods specific for the basic groups of proteins (presumed to be histones) evidenced a greater content of such basic groups in condensed chromatin and thus seemed to support the above anticipation.[994,995] However, the histone content and distribution of the individual fractions were found to be the same in both types of chromatin.[985,994,996-998] As was reported by Frenster,[999] the extent of histone acetylation also appears to be nearly equal in both the heterochromatin and euchromatin. Interestingly, considerable differences have been found in the nonhistone protein, phospholipid, RNA, and phosphoprotein phosphorus contents; all these molecular species are significantly lower in heterochromatin (Table 15). On the other hand no such increase of nonhistone proteins has been detected in the euchromatin of female mealy bugs.[998]

While it is presently uncertain how much of the increased nonhistone content in euchromatin represents in situ elements of biosynthetic activity, a definite pattern concerning the role of the histone in chromatin has begun to emerge. It appears to be the nature of the interactions between the charged residues on histones, other proteins, and DNA which determines the structural features of chromatin. The partial elimination of such interactions and not the removal of histones, produces a dissociation of chromatin superstructures into the basic units.[946] The same histones are present in transcriptionally active and inactive parts of nuclear chromatin. Similarly, histones are uniformly distributed along the polytene chromosomes (with resepct to the DNA content) both in the bands and puff regions.[126,1000] In contrast to the histones, a considerable increase in the content of the nonhistone proteins can be noticed in transcriptionally active chromatin fractions and the puffs of polytene chromosomes. It is not unlikely that a local accumulation of nonhistone proteins serves to weaken the histone-histone and histone-DNA interactions in chromatin and thereby exposes previously buried parts to the polymerase enzyme. A significant removal of histones from the DNA does not occur even during the very active RNA synthesis in polytene chromosome puffs. For the specificity of DNA transcription in chromatin, it is perhaps not as important where the histones are on the DNA as where they cannot get. In other words, some nonhistone proteins may be able to keep certain regions of the DNA free of histones. The regional pattern is most likely established during differentiation. Although the regulation of DNA transcription will be discussed later, it should be mentioned that among the first changes accompanying the activation of previously inactive chromatin is its progessive decondensation and increased binding capacity of DNA for actinomycin D and other small molecules.[713-718,1001-1003]

During the preparation of cells for mitosis, the chromatin present in the cell nucleus condenses into individually discrete chromosomes. During this time, the relatively simple structure of the basic chromatin fibers (average thickness 100 A) becomes supercoiled and folds into rope-like structures. A diagram of such an action, as suggested by DuPraw,[1004-1006] is shown in Figure 36. The length of the basic DNA fibers often exceeds 1 mm and is, therefore, considerable.[1007] The DNA in this material is packed and protected by envelopes of histone and nonhistone proteins. In addition, as can be seen in Table 16, isolated chromosomes contain considerably more protein than isolated chromatin (Table 14). Whether this additional protein, in which the DNA

TABLE 15

Chemical Composition and Properties of Chromatin[985]

	Heterochromatin	Euchromatin
DNA	100.0	100.0
Histones (total)	101.0 ± 9.1	90.7 ± 7.7
Nonhistone proteins	54.9 ± 5.7	109.0 ± 6.1
Phosphoprotein (as P)	0.117 ± 0.019	0.418 ± 0.028
Histone *N*-acetyl	0.235 ± 0.035	0.385 ± 0.035
RNA (total)	1.8 ± 0.7	9.0 ± 4.7
Phospholipids (total)	3.7 ± 1.3	17.0 ± 3.6
Template activity for DNA synthesis (as % of unfractionated chromatin)	28%	470%

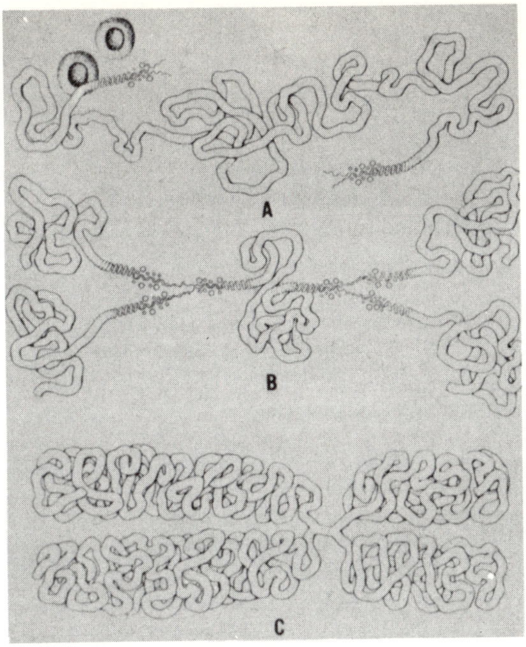

FIGURE 36. Folded fiber model of chromosome structure.[1004] (A) Each unreplicated chromosome (unit chromatid) is essentially a single 200 to 500 A fiber, which contains a DNA double helix in supercoiled configuration. (B) Replication of the chromosome occurs at several sites along the length of the fiber, where DNA polymerase replicates the DNA at fork configurations. (C) The late replicating segments of the fiber at the centromere and elsewhere serve to hold together the sister chromatids.

and histones appear to be buried, is an essential part of the chromosomes or whether it represents contamination adsorbed to the chromosomes during their isolation is not certain. The studies of Zirkin and Wolfe[1020] suggest the second possibility. As stated by DuPraw,[1006] each chromatid can be regarded as a macromolecule which can assume various configurations that depend on the actual conditions in the cell during its life cycle. Although most of the information on chromosome structure suggests that each chromatid contains one long strand of DNA (unineme) which is folded and supercoiled in a complex manner, other evidence indicates the presence of two (bineme) or more (multineme) strands of genetic material in each chromatin. This controversy was recently reviewed by Wolff[1021] and by Prescott.[1022] Regardless of their strandedness, chromosomes can be partially unfolded by mechanical stress or by chemical treatments into fibers of variable thickness that approach, for the most part, the basic fiber structure of about 100 A or less found in

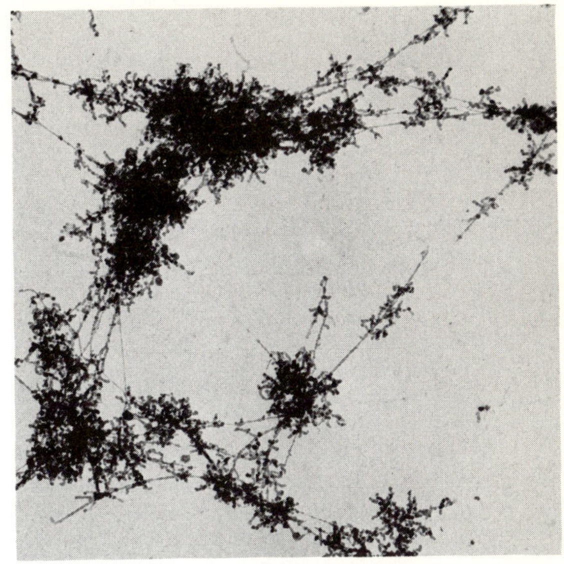

FIGURE 37. Stretched nucleohistone fibers prepared by streaking a grid over the surface of a nucleohistone solution and then staining with uranyl-magnesium acetate. Some fibers appear to have been stretched and range in thickness down to 25 A (X 57,000).[970]

isolated chromatin and nucleoprotein (Figure 37).[968,970,1004-1006,1023-1030] Histones seem to play a role only in the low level organization of the chromosome fibers, since chromatin, in the presence of urea, can be unfolded into fibers 40 A thick without any appreciable loss of histone proteins.[946] Since cations, especially Ca^{++} and Mg^{++}, profoundly affect the chromosomal architecture, local changes in their concentration may be one of the mechanisms operating in chromosome condensation and relaxation.[1008,1031] The reversible formation of disulfide linkages between the SH-groups of F3 (III) histones and some nonhistone proteins may also lead to a condensation of chromatin during chromosome formation.[227,236,298,830a,831,833] Except for the changes in the disulfide content of the F3 (III) histones, there are no substantial differences in the histone contents between interphase chromatin and metaphase chromosomes,[547,830a,831] between individual groups of chromosomes fractionated according to their size,[1032] or between the normal and supernumerary B chromosomes in rye.[1033]

In addition to studies on the structure of the interphase and metaphase chromosomes usually found in the somatic cells of eukaryotes, extensive investigations have been conducted on certain specialized forms of chromosomal structure. In

TABLE 16

The Composition of Isolated Chromosomes

Origin	Reference	DNA	RNA	Total	Protein: Acid soluble	Protein: Acid insoluble
HeLa (prepared at acid pH)	831	1.00	0.56	4.85	0.89	3.96
HeLa (prepared at neutral pH)	831	1.00	0.67	4.50	1.33	3.17
Mouse lymphoma L2	1010	1.00	1.00	5.06	–	–
HeLa	1011	1.00	0.70	3.30	0.85	2.45
HeLa	1012	1.00	0.66	4.70	2.00	2.70
HeLa	1014	1.00	0.71	4.39	2.53	1.86
Chinese hamster	1014	1.00	0.97	4.39	3.33	1.06
Syrian hamster	1014	1.00	0.90	4.09	2.95	1.14
L-cells	1014	1.00	0.78	4.09	2.05	2.05
Chinese hamster (large chromosomes A + B)	1017	1.00	0.18	2.40	–	–
Chinese hamster (medium chromosomes C)	1017	1.00	0.79	5.40	–	–
Chinese hamster (small chromosomes D)	1017	1.00	1.40	8.50	–	–
Chinese hamster	1018	1.00	0.17	2.2	–	–

Methods for the isolation of metaphase chromosomes are discussed in References 1009 – 1019. As compared with the composition of chromatin (Table 14) isolated chromosomes contain significantly more protein and RNA. This contamination appears to be chiefly ribosomal[1011] and increases inversely to the chromosomal size.[1017]

particular, the results of studies on the structure and composition of the giant polytene chromosomes and the expanded loop-like lampbrush chromosomes in oocytes have become quite well known and have proven of great interest despite the complications added by their high metabolic activity. Both types of chromosomes contain regions of loosely packed nucleoprotein threads that extend in the lampbrush chromosomes into very fine fibers which probably consist of individual DNA molecules. These extended parts are the sites of very active RNA synthesis. The ratio of DNA to histone, as determined cytophotometrically, remains constant throughout the entire length of the polytene chromosome which indicates that the complete removal of histones from DNA is not essential for active transcription.[126,1000] Histones may, however, be partially dissociated from the DNA or neutralized by the products of the transcriptional or other biosynthetic processes carried out in the puffs. In the lampbrush chromosomes, DNA loops occur at almost all chromomeres and the DNA is very spatially diluted. This makes the histochemical localization of histones much more difficult. The situation is further complicated by the great amounts of nonhistone proteins present in lampbrush chromosomes (e.g., the lampbrush chromosomes of the newt oocytes contain about 200 times the amount of protein found in the liver chromosomes of the same animal[1034]). An interesting report has been made by Izawa et al.[1035] in which arginine-rich histones F3 (III), when added to the lampbrush chromosomes in vitro, produced retraction of the loops (condensation) within 3 min. Actinomycin D and polylysine have a similar effect, whereas the F1 (I) histones appear to be almost inactive. This observation was interpreted to indicate the possible role of the F3 (III) histones in the condensation of chromatin during the formation of chromosomes.

The molecular structure of the polytene

chromosomes has been studied by many investigators. The more recent reports indicate that they are made of bundles of longitudinal fibers between 65 and 250 A in diameter, granular structures of 250 to 800 A diameter with transverse orientation at chromomeric regions, and an amorphous matrix.[1036-1039] The 230 to 250 A fibers correspond in diameter to normal diploid chromatin fibers from neuroblast nuclei[1036] and the other smaller diameters probably represent lower orders of fiber coiling (similar to mammalian chromatin). Exposure of the polytene chromosomes to bases and proteolytic enzymes disperses them. Microspectrophotometry of silkworm larvae (*Bombyx mori*) salivary gland polytene chromosomes indicates a DNA content 200,000 times greater than a diploid nucleus.[1040] This is, however, an extreme value, since most polytene chromosomes are made of 200 to 4,000 repeating DNA strands or chromatids.

Polytene chromosomes contain histones, as has been shown by staining procedures,[126,1000] whose ratio to the DNA does not change between condensed regions and the puffs. The nuclear histones extracted from the chromatin of adult *Drosophila* flies are very similar electrophoretically to calf thymus histones except that *Drosophila* lacks the typical F1 (I) histone fractions.[1041] Instead, they have a slower migrating protein, originally thought to be an aggregate. Panyim et al.[240] also reported this slowly migrating histone fraction in *Drosophila*. Treatment of *Drosophila* histones with 0.5 *M* β-mercaptoethanol reduced the F3 (III) dimer to monomer, but had no effect on the slowly migrating histone. This was interpreted to indicate that the "F1 (I)" histone is not an aggregate produced by disulfide linkages. More recently, Cohen and Gotchel[251] analyzed the histones from *Drosophila melanogaster* salivary polytene nuclei and the diploid nuclei from larval imaginal discs. They found no differences between the histones from these two sources. However, the *Drosophila* histones differed electrophoretically from those in sea urchin gastrula or calf thymus nuclei. Again, the most striking was the slow migrating very lysine-rich histone F1 (I) fraction specific for *Drosophila*. Amino acid analysis identifies this fraction as lysine-rich, i.e., corresponding to the F1 (I) histones. Unlike calf thymus F1 (I) histone, this protein in *Drosophila* is considerably less basic and differs in the contents of several other amino acids. However, this histone is not specific for the polytene chromosomes, since it occurs in other *Drosophila* tissues.

Although the histone to DNA ratio remains constant along the entire polytene chromosome length regardless of the puffs, Roberts and Kroeger[1042] suggested that histones in the puffed regions were bound to the DNA differently than in the nonpuffed parts of the chromosomes. Their conclusion was based on an increased sensitivity of histones to trypsin in the puffs. A short-term exposure to trypsin increased RNA synthesis threefold in the puffed regions but not in other areas. It was possible that, in the puffed regions of polytene chromosomes, the histone-DNA interactions were weakened to allow increased RNA synthesis. This configuration made them more susceptible to trypsin than areas inactive as genetic templates. It could be argued, however, that because of the uncoiled nature of the chromatin in the puffs, the trypsin could penetrate into the chromatin more easily than in the condensed and inactive regions. The rate of histone digestion would then be comparable to the extent of trypsin penetration into the chromatin. Although the chromosomal proteins in polytene chromosomes were able to be acetylated, the distribution of the incorporated acetate showed no accumulation in the puffs.[706]

DNA-Histone Interactions in Chromatin

Many physicochemical and biological properties of chromatin or nucleoprotein are the direct consequence of DNA interaction with histones. It is clear that such interactions are essential to the very existence of chromatin. The other macromolecular components frequently found in chromatin (e.g., nonhistone proteins, RNA, etc.) contribute refinements and individuality to the basic structure, as furniture creates the individuality of a house. The electrostatic nature of the interactions permits partial or complete dissociation of the DNA and histones in chromatin. Hence, studies based on gradual dissociation can determine what structural and other constraints histones impose upon the DNA in chromatin.

It was realized relatively early that exposure of nucleoproteins to increasing concentrations of salts causes a progressive dissociation of histones which results in a partial fractionation of both the histones and DNA.[24,25,35,863,867-869,1043] Refinements in the identification of dissociated histones, mainly by polyacrylamide gel electrophoresis, further increased the interest in probing into chromatin structure by selectively removing certain histone fractions and studying the remaining complexes. The separation of DNA from the dissociated histones can be obtained by ultracentrifugation[176,178,182,184,209,949] or exclusion chromatography.[104,177,179,1044-1047] The dissociation is cooperative[952] and rapidly reaches an equilibrium. The extent of dissociation depends on both the cations and anions constituting the salt solution.[23,176,182,184,895] The efficiency decreases in the order: deoxycholate $>$ ClO_4^- $>$ Cl^- $>$ SCN^- $>$ F^-. The fluoride anion, similar to the $HPO_4^=$ ion, is virtually inactive, while the sulfate ($SO_4^=$) is less active than chloride (Cl^-).[182,895] Since guanidine hydrochloride has the same effect as NaCl, the protein denaturing and hydrogen bond breaking activity of the guanidyl ions do not contribute substantially to the dissociation.[895] $MgCl_2$ is more effective than NaCl because of higher ionic strength and higher charge density of the magnesium ion. Smart and Bonner[184] have suggested that the differences in the efficiency of dissociating chromatin by the various salts may be associated with a differential anion binding specificity for selected sites on proteins (histones). In aqueous salt solutions, the sequence of dissociation of individual histone fractions usually begins with F1 (I), whose binding to the DNA appears to be the weakest. As the salt concentration increases, other lysine-rich fractions are progressively dissociated with the arginine-rich fractions requiring the highest salt concentrations.[23,176,179,182,381] Although this progression is not absolute and the dissociated fractions overlap at each given molarity of salt, the dissociation offers a convenient way for preparing chromatin samples deficient in certain histone fractions. An additional selectivity of extraction, which extends the value of the dissociation method, occurs with 2% NaCl in 75% ethanol or 10% guanidine hydrochloride in a 75% ethanol.[189] Both of these solvents preferentially extract the arginine-rich F3 (III) and F2a1 (IV) histones. The dissociation of F1 (I) histones is altered if concentrated urea is present in the NaCl solutions[188] or if sodium deoxycholate is used instead of NaCl.[184] Under these conditions, the F1 (I) histones are extracted last at the highest concentrations of the dissociating salt. Otherwise, the sequence of histone removal remains as indicated, i.e., the lysine-rich fractions F2b (IIb2) and F2a2 (IIb1) followed by the arginine-rich histones F3 (III) and F2a1 (IV). The chicken erythrocyte-specific histone F2c (V) dissociates like the F1 (I) histones.[188]

Using partially dehistonized chromatin, various aspects of the stability and conformation of DNA in chromatin, as well as the distribution of proteins along the DNA strands, were able to be extensively investigated. It was first reported by Zubay and Doty[48] that nucleohistone particles in solution were less extended than free DNA. In both cases, however, the double helical configuration of DNA was well preserved.[913] From the hydrodynamic behavior of partially dehistonized chromatin, Ohba[177] concluded that the DNA decreased its length upon association with histones, i.e., the number of base pairs per unit length which are perpendicular to the axis of the molecule decreased by 60%. These constraints of the DNA appeared to be mostly due to the arginine-rich histones, i.e., to the least dissociable fractions. This shortening of the DNA and, perhaps, partial unstacking of the bases in chromatin caused by the interaction of histones and DNA has been confirmed by optical studies on partially dehistonized chromatin samples.

A deformation of the DNA conformation associated with polylysine binding has been demonstrated by Cohen and Kidson.[890] Optical studies on isolated chromatin also indicate that the DNA in chromatin is deformed.[446] In this case, there apparently occurs a greater tilting of the bases, which is similar to the C conformation of DNA as suggested by Shih and Fasman.[1048] The extent of the conformational changes of DNA varies with the individual proteins participating in the complex formation.[1049] Circular dichroism and optical rotatory dispersion measurements of chromatin samples after the gradual removal (dissociation) of histones show their involvement in constraining the DNA helix into the form which it assumes in chromatin.[446,1050,1051] Increasing the ionic strength does not change the circular dichroic properties of pure DNA, but the spectrum of DNA in nucleohistone changes to that of free DNA in solution as soon as the slightly lysine-rich histone fractions are removed at salt concentrations of 0.6 to 1.2 *M* NaCl.[1050] However, according to Wagner and Spelsberg,[1052] the arginine-rich histone F2a1 (IV) is solely responsible for the conformational changes of DNA in chromatin. The very same histone fraction (F2al or IV), however, if reconstituted with the DNA in vitro had no detectable effect on the circular dichroism spectrum of DNA.[1053] Even more controversial are the findings of Pemogorov et al.,[1054] who could not find significant differences in the circular dichroism spectra of undissociated chromatin, free DNA, or chromatin in 2.0 *M* NaCl. The reasons for these discrepancies are not clear.

Removal of the very lysine-rich histones F1 (I) did not change the DNA conformation in chromatin.[1050-1052] This was in agreement with the studies of other investigators who also found that the main histone fractions affecting the conformation of DNA in chromatin were the moderately lysine-rich histones F2 (F2a2 or IIb1 and F2b or IIb2),[176,466,895] while the arginine-rich histones contributed only to a lesser extent.[892,1046,1055-1057] An opposite view was recently expressed by Smart and Bonner,[1058] who used sodium deoxycholate to dissociate histone from the DNA in chromatin. As was already mentioned, sodium deoxycholate reverses the dissociation schedule of the F1 (I) histones, and chromatin preparations devoid of all but F1 (I) histones can be obtained. When such "lysine-rich chromatins" were examined, it was found that, regardless of its chemical composition, the deformation of the DNA molecule in chromatin was very sensitive to the presence of histone. The authors concluded that the conformation of the DNA in chromatin depends on the presence of the less dissociable histones which, in the case of the sodium deoxycholate dissociation, was the F1 (I) fraction. Based on the studies of flow dichroism by Ohba[177] and themselves, Smart and Bonner[1058] concluded that a supercoiled structure which is 80 A in diameter with a repeat distance of 110 A would effectively shorten the length of the molecule by about 40%, i.e., the figure obtained from flow dichroism data. In this model, the plane of the bases would be tilted 35° with respect to the molecular axis of the chromatin strand. This agreed well with the electron microscopic observations and x-ray diffraction studies on isolated chromatin fibers.[976,1059] As could be anticipated, the circular dichroism spectrum of DNA also became altered upon its association with polylysine or polyarginine.[890,1056,1060] According to Zama and Ichimura,[1060] polylysine was much more efficient in changing the DNA conformation than polyarginine. The differences were attributed to the various effects of these two polyamino acids on the hydration of DNA. Hence, it appeared that all the histone fractions could contribute to the constriction of DNA and their individual contributions depended on the nature of their binding to the DNA in chromatin.

Another way of probing the structure of chromatin and the ways in which histones associate with the DNA is thermal denaturation studies. The melting temperature of the DNA in association with histones increases with the A + T content of DNA, indicating a preferential stabilization of A + T base pairs by the histones.[881] Partially denatured deoxyribonucleoprotein sediments on ultracentrifugation as two distinct boundaries with one having the sedimentation coefficient of completely denatured material.[1061] This can be interpreted to indicate that the thermal denaturation of nucleohistone is an all-or-none phenomenon. The melting profile of isolated chromatin is multiphasic,[176,179,186,898,902,1045,1055,1058,1062,1063] which suggests that certain histone complexes may be distributed irregularly along the DNA double helix. On the other hand, the two step nature of the melting profile may, according to Henson and Walker,[1045,1057] indicate the migration of proteins away from the denatured parts of the chromatin while the native material remains complexed with these proteins in a highly specific way. Presumably, the histones are regularly distributed along the DNA before melting. At the melting temperature of the lower step the deoxyribonucleoprotein appears to consist of denatured molecules devoid of migratory proteins, native molecules associated with the migratory proteins to various degrees, and native molecules complexed with a full complement of these proteins. Although this proposal requires further experimental support, it is unlikely that the migrating proteins are histones since it was shown by Matsuyama and Nagata[866] that histones remain associated with each single strand of the DNA molecule when subjected to thermal denaturation. In addition, it may be recalled here that Smart and Bonner[1058] have reported that histones stabilize the DNA against thermal denaturation more than can be expected on the basis of charge neutralization. The removal of about 50% of all positive charges by salt dissociation only decreases the thermal denaturation temperature (Tm) of chromatin by about 30%. To produce a similar change in the melting temperature of pure DNA stabilized to the same extent by Na^+ ions, the removal of about 75% of the Na^+ ions from the melting medium would be necessary. This suggests that individual histone molecules may partially interlock the DNA double helix and impede the separation of both individual DNA strands. A similar interference in the strand separation during thermal denaturation can be produced by cross-linking the histones in chromatin by treatment with formaldehyde.[1064]

Smart and Bonner[1058] also investigated the influence of histone fractions on the cross-linking of individual chromatin molecules. Using the selective dissociation of histones in sodium deoxycholate solutions followed by thermal denaturation, flow dichroism, and sedimentation analysis of the partially dehistonized chromatin, they found that no particular histone fraction specifically contributed to the chromatin superstructure. As the result of the earlier investigations by Littau et al.[834,904] and Chalkley and Jensen,[907,908] the very lysine-rich histone F1 (I) had been implicated in cross-linking the nucleoprotein strands in chromatin. The selective dissociation of F1 (I) histones in dilute salt or acid solutions resulted in a collapse of the crosslinked structure and the disappearance of highly aggregated, fast sedimenting chromatin complexes. The heterogeneity of formaldehyde-fixed chromatin also increased after the selective removal of F1 (I) histones, as was evidenced by centrifugation of the fixed chromatins in CsCl gradients.[1065] On the other hand, Smart and Bonner[1058] found that if the group II histones (mixture of F2a2 or IIb1, and F2b or IIb2) were removed selectively by 0.015 *M* deoxycholate, the cross-linked structure of the fast sedimenting (130s) chromatin was irreversibly eliminated. Although there were no data available concerning the effect of arginine-rich histone removal (F3 or III and F2a1 or IV) on chromatin structure, it could be argued that it was not any particular histone fraction which held the chromatin complexes together, but the sum of the electrostatic interactions of all the protein species constituting the chromatin complex. Once these interactions are weakened by the addition of sufficient amounts of salt, the structure relaxes and disappears.[907,946] As discussed previously, experiments with chromatin in urea solutions showed that no histone need be removed to achieve a complete disaggregation of chromatin into a network of 40 A diameter fibers. This process was, however, irreversible.[907]

The thermal denaturation profiles of chromatins from which individual histone fractions have been selectively removed provide valuable information on histone-DNA interactions and complement the data collected by optical rotatory dispersion and circular dichroism measurements. A particularly revealing analysis of the thermal denaturation profiles is provided by comparing temperature derivative plots, since small stability transitions can be detected in this way. In DNA complexed with purified histone fractions, at least two steps in transition can be seen. The first step corresponds to the denaturation of free DNA, while the second step is that of the DNA-histone complex.[902] As Shih and Bonner have reported,[902] the transition temperature of the second step is characteristic for each DNA-basic protein complex and depends on the composition of the histone fraction. The two-dimensional coordinate area of the fraction of hyperchromicity at the first transition can be used as a quantitative measure of free DNA in the complex, since it is linearly related to the DNA-histone ratio (Figure 38). Due to the turbidity changes which occur during the temperature increase, a similar relationship cannot be used to estimate the amount of complexed DNA.

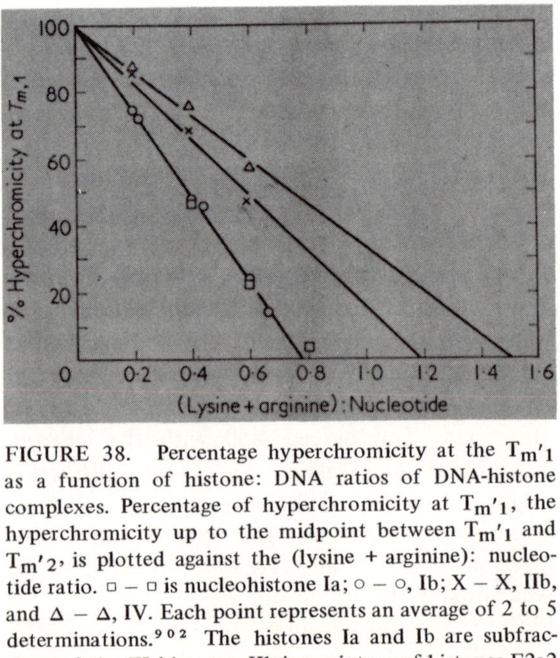

FIGURE 38. Percentage hyperchromicity at the $T_{m}'_1$ as a function of histone: DNA ratios of DNA-histone complexes. Percentage of hyperchromicity at $T_{m}'_1$, the hyperchromicity up to the midpoint between $T_{m}'_1$ and $T_{m}'_2$, is plotted against the (lysine + arginine): nucleotide ratio. □ – □ is nucleohistone Ia; ○ – ○, Ib; X – X, IIb, and Δ – Δ, IV. Each point represents an average of 2 to 5 determinations.[902] The histones Ia and Ib are subfractions of the FI histones, IIb is a mixture of histones F2a2 and F2b, and IV is histone F2a1.

A further refinement can be obtained if thermal denaturation is performed in the presence of urea. According to Ansevin and Brown,[903] DNA complexes with any calf thymus histone fraction other than the F2b (IIb2) histone increases its ultraviolet absorbance in at least three or more steps corresponding to the temperature transitions of free DNA and two or more DNA-histone complexes having different binding energies (Figure 39). The high sensitivity of the derivative method of profile analysis is demonstrated in Figure 40, which shows the transition temperatures of a mixture of three bacterial DNA species. The complexity of the denaturation profiles of the DNA-histone complexes appears to depend on how the components are reconstituted. If histones and DNA are reconstituted by slow dialysis from a concentrated salt solution, a biphasic denaturation profile was obtained.[902] Complexes made by very slowly mixing DNA and histones at low ionic strength in the presence of 3.6 *M* urea show several steps of transition.[903]

Since studies on artificial DNA-histone complexes are quite removed from the situation in vivo, it was of interest to examine the derivative denaturation profiles of both native chromatin and chromatin which was devoid of one or more histone fractions. In general, the thermal denaturation profiles of natural chromatin preparations were much more complex than those of reconstituted nucleohistones and appeared to be composed of a series of overlapping transitions too complicated to be explained by a simple two-state theory of nucleohistone association. Some chromatins could be distinguished on the basis of differences in the shape of their profiles, as shown in Figure 41. No distinction could be drawn between liver and thymus from the same organism, however. Individual maxima in rat chromatin appeared to represent different levels of stability since mild digestion with trypsin (Figure 42) or autodigestion by standing (Figure 43) produced a marked increase of the low temperature transitions and decreased high temperature regions (Figures 42 and 43). Breaking chromatin fibers either by spray fragmentation or by ultrasonic shearing into fragments of approximately 2 million daltons mol wt did not change its thermal denaturation properties (Figure 44). Similarly, freezing and thawing had no appreciable effect. The selective dissociation of histone fractions led to a decrease in the more stabilized regions of thermal transitions accompanied by an increase of low stability areas and the appearance of a melting profile peak corresponding to that of very loosely complexed DNA (Figures 45 and 46).[966,1058] The derivative thermal denaturation profile analysis of various isolated chromatins did not show any appreciable amounts of free DNA in chromatin. It did evidence a considerable heterogeneity of thermal transitions associated with the removal of histones.[902,966,1058] These findings suggested the presence of DNA regions which were differentially complexed with protein. In addition, it was also shown by ultracentrifugal analysis of fragmented, formaldehyde-fixed chromatin that there was a considerable heterogeneity in protein distribution along the DNA helices.[896,897] It was suggested from these latter studies that perhaps not all of the DNA in chromatin was associated with histones. Hopefully, further improvements in the highly sensitive derivative thermal denaturation technique might provide information about the nature and specificity of the binding between histone fractions and the individual portions of the DNA molecules.

Chromatin in its isolated form is negatively charged and is capable of binding positively charged molecules in addition to its usual complement of histones. A part of the negative charge of chromatin is contributed by the free COOH groups of the dicarboxylic amino acid residues in proteins and by the free phosphate groups of phosphorylated amino acid residues which reside in the chromosomal phosphoproteins. In addition, cationic dye binding studies show that in native nucleohistone over 50% of the DNA phosphate groups are not bound to the histones.[926,1066-1068] The dye, usually toluidine blue, binds to about 50% of the DNA phosphates in chromatin without an appreciable displacement of histones. Another cationic molecule, cetylpyridinium chloride, binds to DNA in an amount corresponding to about 40% of the DNA phosphate content.[1068] The protein-free regions of the DNA in chromatin appear to be double stranded[1069] and rather small. They are probably buried beneath protein.[1068] Richards and Pardon,[1070] in their model building studies illustrating the F2a1 (IV) histone-DNA interactions, have concluded that not more than 60% of the DNA phosphate groups in the F2a1 (IV) nucleohistone can be neutralized by the associated protein. This agrees well with the 40 to 50% of

FIGURE 39A.

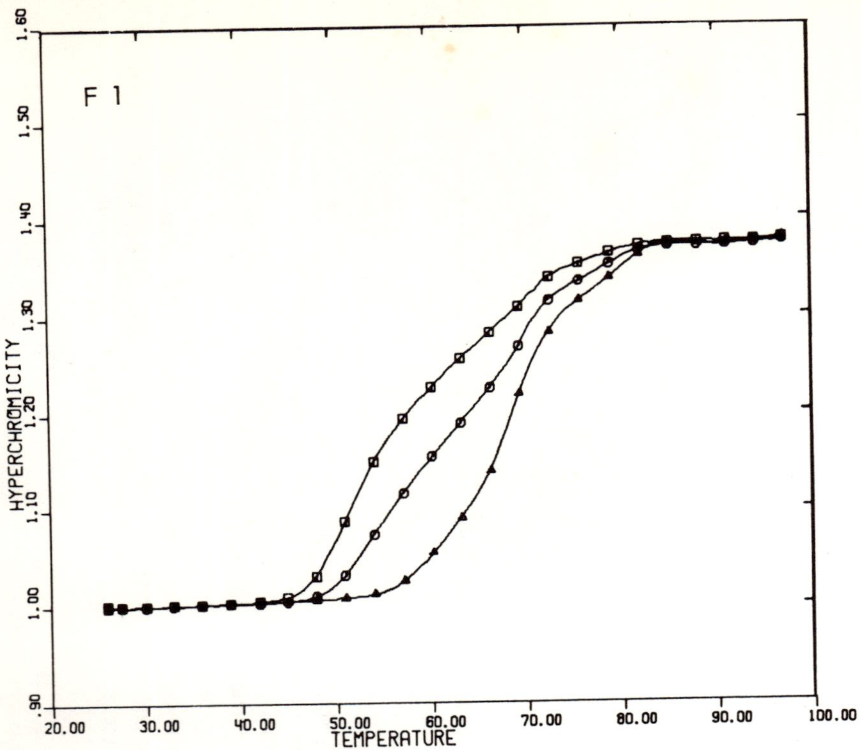

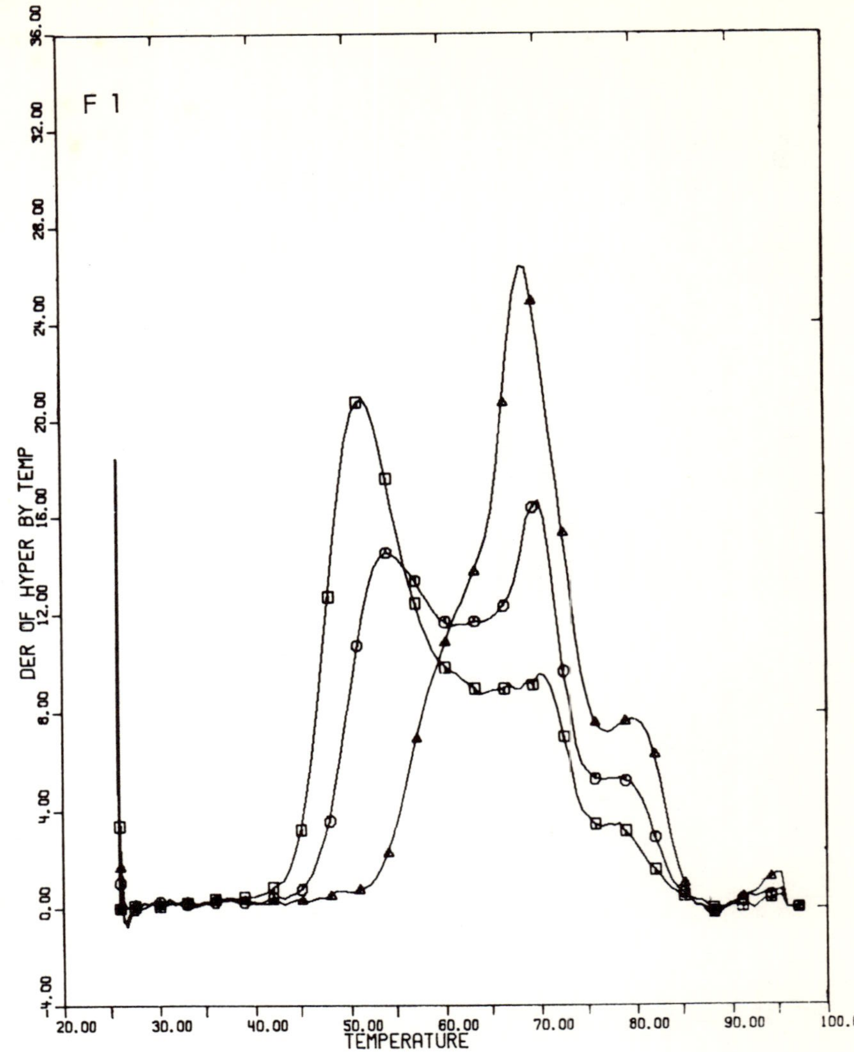

FIGURE 39. Corresponding integral (left figure) and derivative (right figure) thermal denaturation profiles of nucleoproteins formed from rat DNA and five different purified fractions of calf thymus histone at three levels of reconstitution; nominal extent of reconstitution is indicated by the value of the ratio, weight of protein/weight of DNA: □ – □, 0.25; ○ – ○, 0.50; Δ – Δ, 0.75. The type of histone used is recorded at the top of each graph. The thermal denaturation profile of histone-free DNA is shown as a dashed line with filled circles in part b. Denaturation was produced by continuous heating (programmed increase of 0.5°/min) in low ionic strength urea-containing buffer at pH 7; data were recorded and processed automatically. Only every *fifth* point in all graphs is indicated by a symbol, but all observations are used in drawing the continuous curves. The ordinate for derivative profiles is the temperature derivative of hyperchromicity at 260 nm times a factor of 1,000. From the position of the DNA curve shown in part b, it can be noted that the first peak in the derivative denaturation profile of each nucleoprotein, except 0.75 F1, should correspond to uncomplexed or weakly complexed DNA and that peaks occurring at higher temperatures are due to histone-stabilized DNA. Records of turbidity at 320 nm (not shown here) indicated that no prominent peak is produced merely by an increase in scattered light. Reconstitution was by the slow addition of individual histone fractions to dehistonized rat thymus DNA in a low ionic strength buffer containing 3.6 *M* urea at pH 7.[903]

FIGURE 39B.

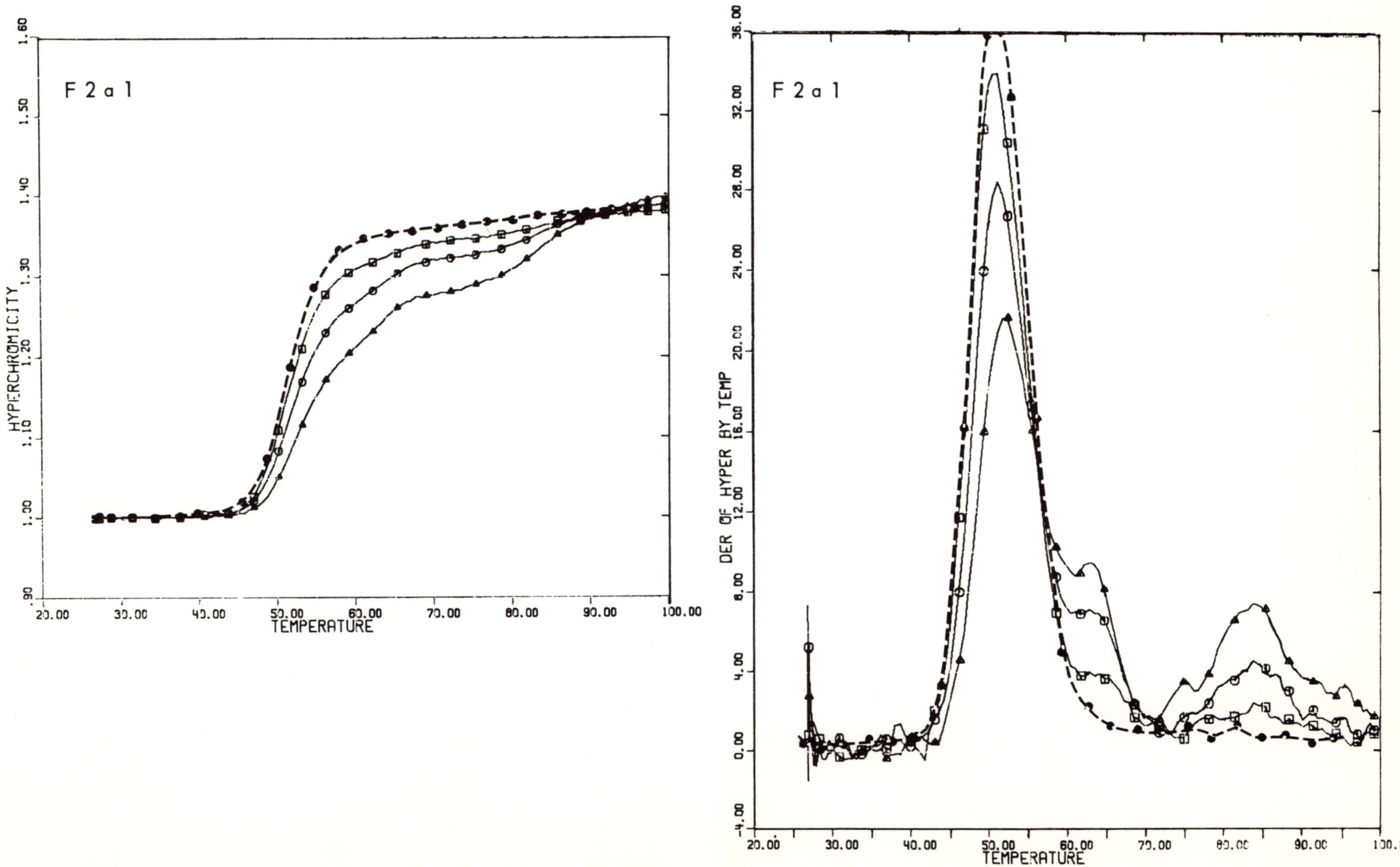

F 2 a 1
HYPERCHROMICITY
TEMPERATURE
F 2 a 1
DER OF HYPER BY TEMP
TEMPERATURE

FIGURE 39C.

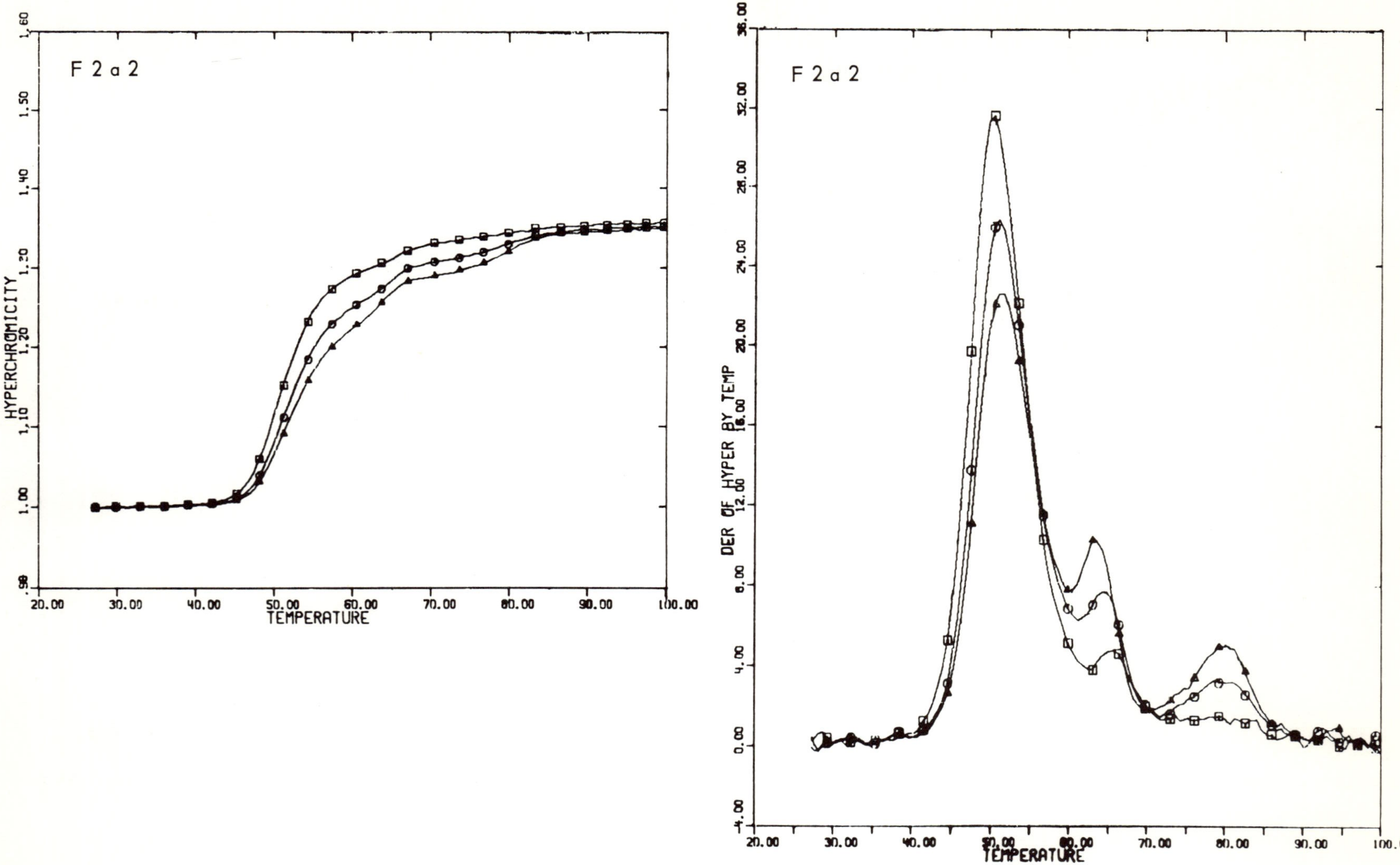
F 2 a 2
HYPERCHROMICITY
TEMPERATURE
F 2 a 2
DER OF HYPER BY TEMP
TEMPERATURE

FIGURE 39D.

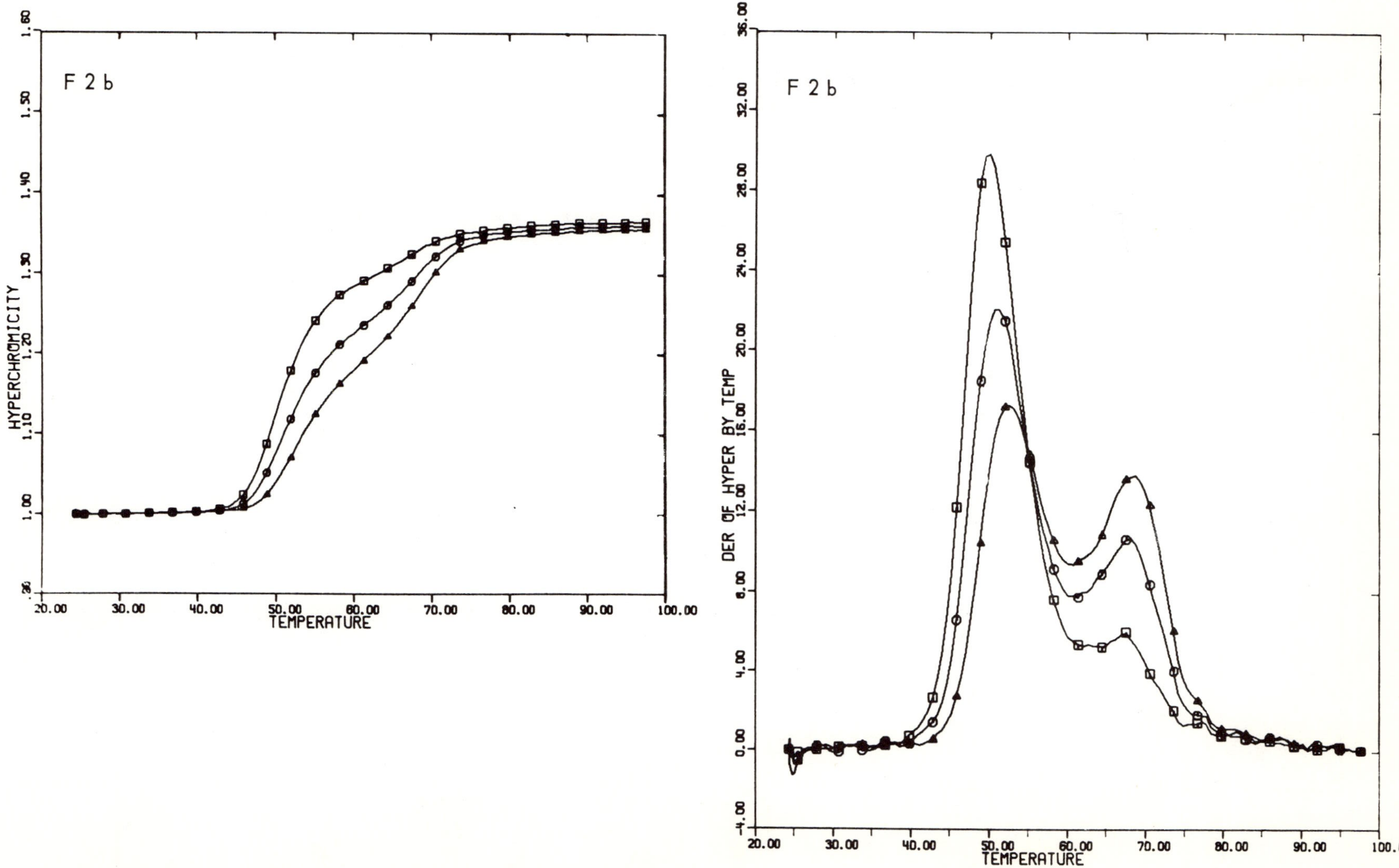
F 2 b
HYPERCHROMICITY
TEMPERATURE
F 2 b
DER OF HYPER BY TEMP
TEMPERATURE

FIGURE 39E.

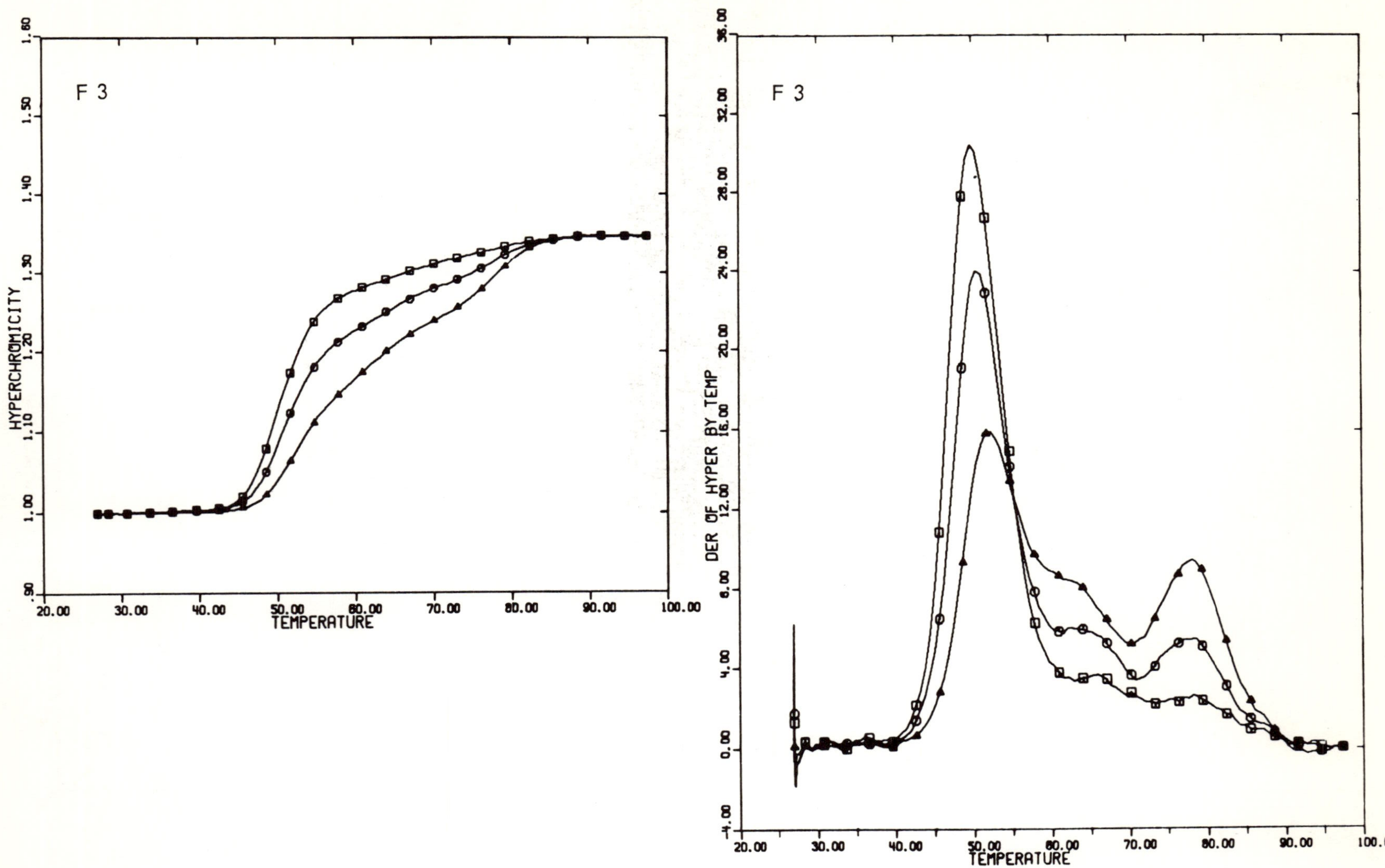
F 3
HYPERCHROMICITY
TEMPERATURE
.90
1.00
1.10
1.20
1.30
1.40
1.50
1.60
20.00
30.00
40.00
50.00
60.00
70.00
80.00
90.00
100.00
F 3
DER OF HYPER BY TEMP
TEMPERATURE
-4.00
0.00
4.00
8.00
12.00
16.00
20.00
24.00
28.00
32.00
36.00
20.00
30.00
40.00
50.00
60.00
70.00
80.00
90.00
100.00

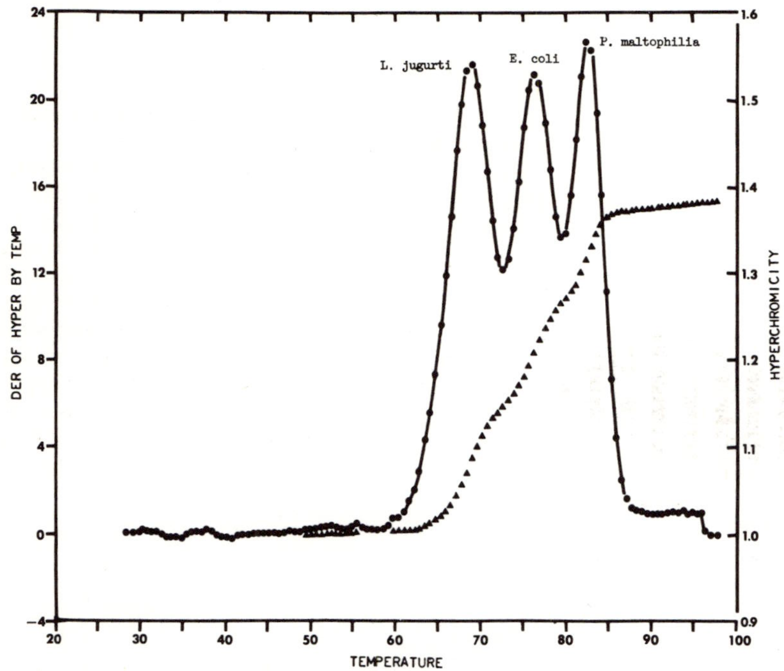

FIGURE 40. Superimposed integral and derivative thermal denaturation profiles for a mixture of roughly equal parts of three different bacterial DNAs dissolved in 0.10 SSC. The left ordinate and plotted circles refer to the temperature derivative of hyperchromicity at 260 nm multiplied by a factor of 1,000 (derivative profile). Hyperchromicity was found from the ratio of absorbance at any elevated temperature to that at room temperature, after correcting for the thermal expansion of water. The right ordinate and triangles refer to the hyperchromicity curve (integral profile) of the DNA mixture. All observed points have been plotted. Data were recorded automatically during continuous heating of the sample by a bath having a programmed temperature rise of 0.5°/min.[903]

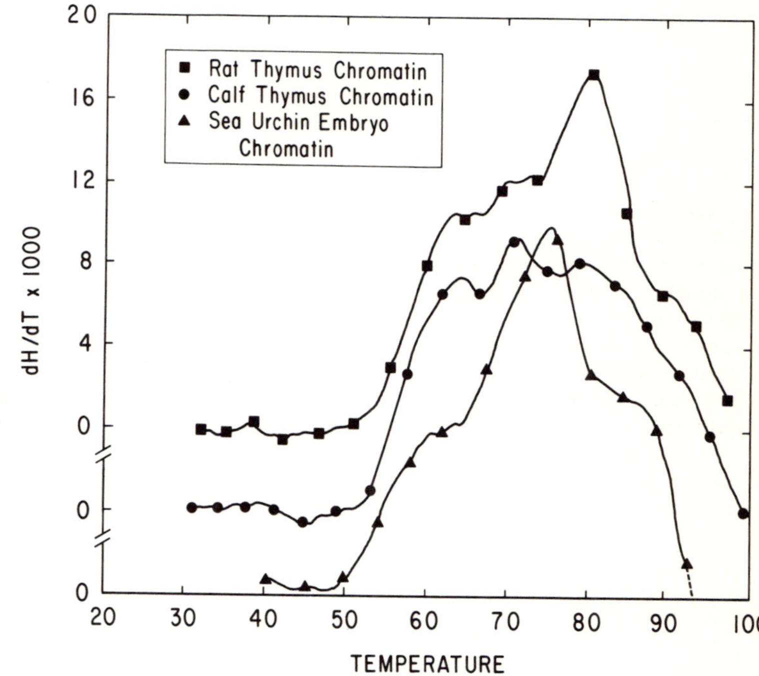

FIGURE 41. Representative thermal denaturation profiles of chromatin preparations from three sources, recorded in a medium containing 3.6 *M* urea, 0.15 mM versene, and 5 mM sodium cacodylate buffer (pH 7). The temperature derivative of hyperchromicity (dH/dT) was calculated by computer from absorbance and temperature data recorded in continuously heated samples; all points are plotted although the lines are identified by symbols located at every fifth point.

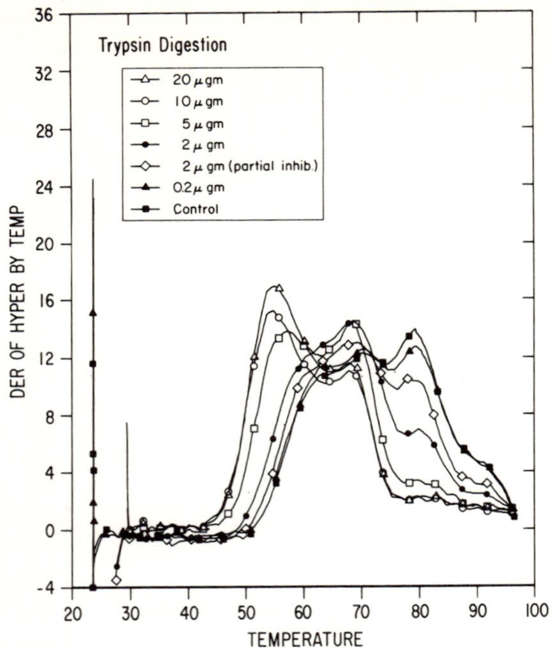

FIGURE 42. Derivative thermal denaturation profiles of trypsin-treated rat thymus chromatin. Samples were held at room temperature for about 1.5 hr following the addition of various amounts of trypsin (or no trypsin, in the control), as indicated on the figure inset in terms of micrograms per milligram of chromatin DNA. One of the 2 μg samples was partially inhibited by the addition of soybean trypsin inhibitor immediately following the trypsin. Denaturation profiles were recorded in urea-cacodylate buffer at a heating rate of 1.0°/min. Modification of a characteristic stair-step profile for fresh chromatin proceeded with greater enzyme action, showing a progressive depletion of highly stabilized areas and an increasing exposure of unprotected DNA.[966]

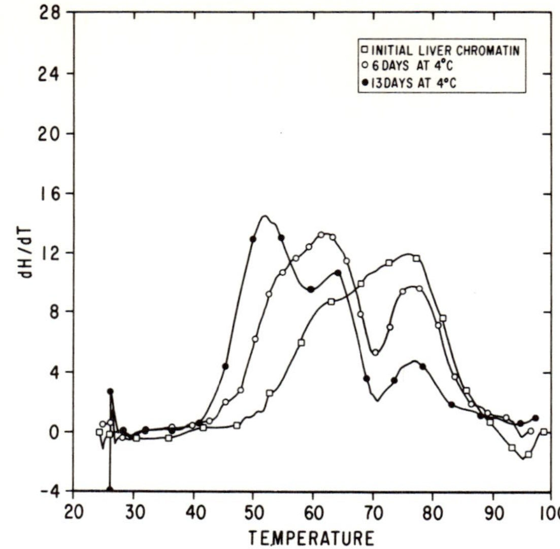

FIGURE 43. Derivative thermal denaturation profiles of rat liver chromatin held in a thawed condition for 0, 6, and 13 days. Denaturation was observed in urea-cacodylate buffer at a heating rate of 0.5°/min. Drastic changes in profiles are attributed to the autolysis of proteins and accompanying exposure of free DNA.[966]

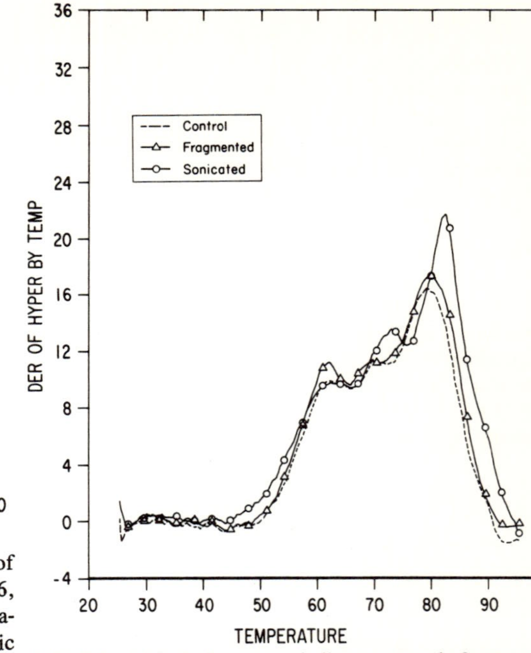

FIGURE 44. Influence of linear strand fracture on thermal denaturation of rat thymus chromatin. The "fragmented" sample was broken to strands having a modal DNA size of 2 million daltons by a gentle spraying method, as described in the experimental section; the control was the initial unmanipulated preparation of chromatin. All samples were denatured in urea-cacodylate buffer at a heating rate of 0.5°/min. The gently fragmented sample does not differ significantly from the control; the "sonicated" sample, in this and repeated experiments, differed from the control to a small but significant extent.[966]

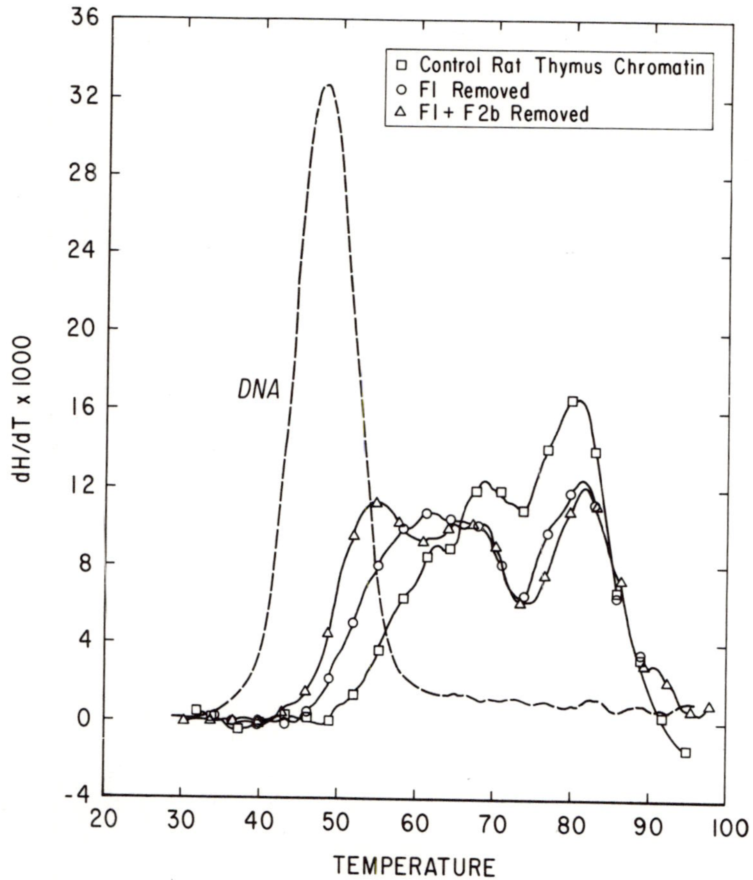

FIGURE 45. Preferential removal of lysine rich histones from rat chromatin by extraction with salts. Fraction F1 was fully removed from chromatin by extraction with 0.6 *M* sodium chloride, while only small quantities of other fractions were depleted. Histone fractions F1 and F2b were selectively extracted from chromatin with 1 *M* ammonium sulfate solution. Thermal denaturation profiles, obtained in 3.6 *M* urea, 0.15 mM versene, and 5 mM cacodylate buffer (pH 7), showed that the elimination of more highly stabilized thermal regions of chromatin was accompanied by a corresponding appearance of free or loosely complexed DNA. The denaturation profile of purified rat DNA is included for reference. The temperature derivative of hyperchromicity (dH/dT) was calculated by computer from absorbance and temperature data recorded in continuously heated samples; all points are plotted but the lines are identified by symbols located only at every fifth point.

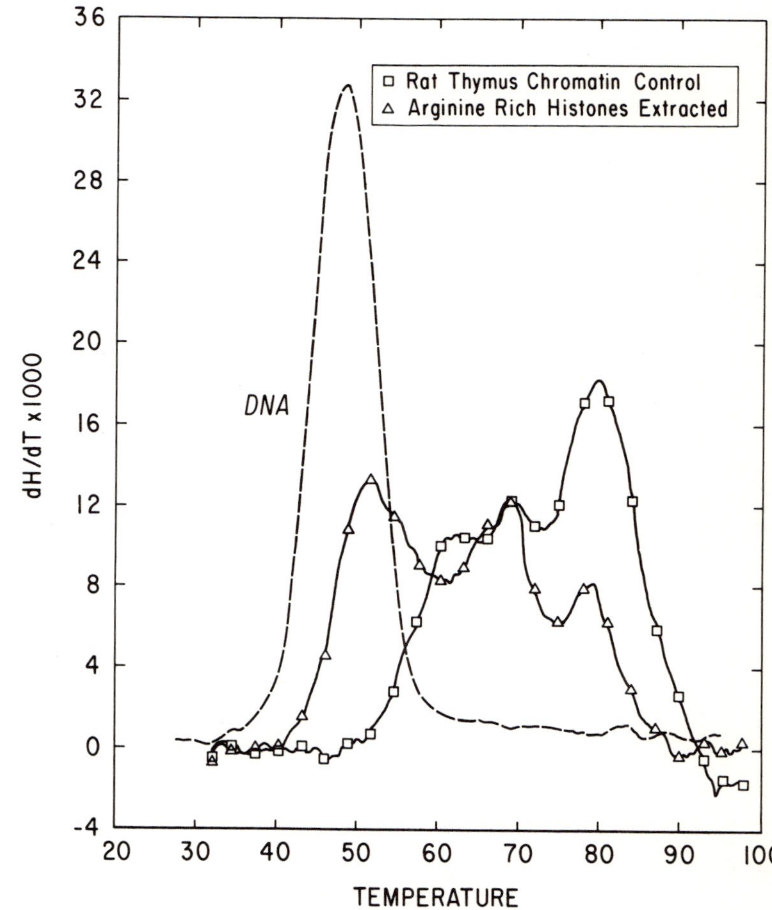

FIGURE 46. Preferential removal of arginine-rich histones from rat chromatin by extraction with an ethanolic salt solution. Histone fractions F3 and F2a1 were extracted by 10% guanidinium chloride in 75% ethanol; the control, treated with 75% ethanol alone, showed no change from untreated chromatin. The denaturation profile of rat DNA is included for reference. The sample depleted of arginine-rich histones shows a distinctive reduction of the denaturation profile in the more thermally stable regions and a corresponding increase in a region melting as free or loosely complexed DNA. The denaturation was in ureo-cacodylate medium; lines are identified by symbols placed only at every fifth point.

free phosphates detected in chromatin by dye binding experiments. As judged from luminescence characteristics of the acriflavine complexes with isolated nucleoprotein, the protein-free segments of DNA are G + C-rich.[900] Much lower estimates of protein-free DNA areas in isolated nucleoprotein are achieved by immunochemical measurements.[1071] Judging from the quantities of antigen (as DNA) required for reaction, only 1 to 5% of the DNA in native nucleoprotein (chromatin) was sterically available to react with antibody. Interestingly, this value increased to about 20% after the salt dissociation and reassociation of the chromatin. The relatively small value found by Stollar[1071] agrees well with the studies on the amounts of open template in chromatin samples from various tissues, which will be discussed later.

As a negatively charged macromolecule, isolated chromatin can bind appreciable amounts of free histones.[1072-1074] This binding is rather weak, however, and the excessive histone can be easily dissociated by dilute (10 mM) HCl or by 0.3 *M* NaCl.[1072,1074] In many ways, the binding of excessive histone resembles that of the F1 (I) histone in native chromatin.[1074] Free DNA will also accept more histone than dictated by the stoichiometric binding ratio of 1.35. The excess histone not taken up by the DNA from an unfractionated mixture of the histone fractions is enriched in the lysine-rich histones, especially the F1 (I). Interestingly, thermal denaturation studies on chromatin samples containing additional histones show them to have an increase in the highly stabilized regions of the denaturation profile. An analogous increase in the thermal stability of chromatin can be produced by the addition of polyamines or polylysine.[1074]

Similar to the cationic dyes or histones, isolated chromatin was found to bind significant quantities of polylysine.[1068,1074-1076] According to Itzhaki,[1075] at the precipitation point of isolated nucleoprotein in the presence of excessive amounts of polylysine, 100% of the purified DNA phosphates and 38% of the nucleoprotein phosphates were available for binding with polylysine. Therefore, it appeared that the stretches of free DNA in isolated nucleoprotein were large enough to accomodate polylysine molecules of an average molecular weight of 7,790 daltons, although some binding to the nonhistone proteins cannot be excluded.[1068] This agreed well with the observation that excessive histones were bound to isolated chromatin only weakly, while a smaller protamine molecule exhibited a much stronger binding, which probably occurred through the free phosphates of the DNA.[1072] With the larger polylysine molecules (62,000 daltons), cross-linking of the free DNA phosphates on different chromatin fibers might occur.[1068]

Observations indicating a considerable free phosphate content in chromatin DNA were also reported by Clark and Felsenfeld.[1076] A titration of chromatin with polylysine (degree of polymerization 670 or 50) reached saturation at a ratio of 0.5 lysine residues per DNA phosphate (nucleotide), which indicated that about one half of all the phosphate residues in chromatin were free to interact with this polyamino acid. Given the conditions of the experiment, the polylysine binding to the DNA was irreversible and noncooperative. The titration of chromatin with polylysine did not dissociate any histones up to the point of complete precipitation. On the other hand, an excess of polylysine led to the dissociation of some histones [mainly the F1 (I) fractions].[1074,1076] Clark and Felsenfeld[1076] supported their polylysine titration data by studies on chromatin digestion either with staphylococcal nuclease or by DNase I. Only about 50% of the DNA in the chromatin from calf thymus, goose erythrocytes, rabbit liver, and rat hepatoma was digested by these enzymes. Fixation of the calf thymus chromatin with formaldehyde, which binds the DNA and proteins covalently, did not change the extent of DNase digestion. Very little protein was released from chromatin during its digestion. This was probably due to the formation of insoluble complexes between the histones and nonhistone proteins released by the DNase treatment. The DNA fragments were heterogeneous and had a weight average size of about 175 base pairs. Free DNA did not produce such large fragments upon digestion with DNase, which suggested that in chromatin these lengths of DNA molecules were protected from the enzyme. This was probably due to their association with proteins, especially histones. The average base composition of the protected DNA regions did not differ from that of the total DNA (58% A + T). They also behaved like the DNA of their origin during annealing experiments using conditions described by Britten and Kohne.[1077] Therefore these protected regions were not enriched in repetitious DNA sequences. Slightly smaller fragments were obtained when poly-D-lysine rather than poly-L-lysine was used for the saturation, followed by digestion of the histones with pronase (poly-D-lysine is resistant to this enzyme) and the newly exposed DNA regions hydrolyzed by DNase. In this way, the insoluble precipitate contained DNA segments which were free of histones from the original chromatin. The average length of such fragments was about 140 nucleotide pairs. The authors also found that even at low ionic concentrations (1×10^{-3} *M* Ca^{++} or 0.6 *M* NaCl) a considerable part of the histones (more than 30%) was freely exchangeable between individual DNA molecules. Such a rearrangement of histones was first suggested by Jensen and Chalkley[908] for the F1 (I) fractions and is of considerable importance for the interpretation of experiments describing the attachment of histones to specific regions of the DNA.

The digestion of free DNA segments by nucleases may provide a new tool for studies on the structure of chromatin and chromosomes. Unfortunately, this issue is controversial at the present time, since Itzhaki[1068,1078] reports that at a sufficiently high DNA concentration the nuclease action does not stop at the 50% level, but continues until about 90% of the DNA is digested. Similar results have been obtained in this laboratory, i.e., only 10 to 20% of the DNA in isolated chromatin is not digestible by DNase I under the conditions described by Clark and Felsenfeld.[1076] The undigested portions of DNA are not dialysable and do not increase their hyperchromicity upon heating. This may indicate that the indigestible fragments are either denatured or too short to form a double helix in the buffer used for the thermal denaturation studies.[903] According to Itzhaki,[1078] very little if any DNA in isolated chromatin is not covered with protein. An essentially identical conclusion was recently reached by Mirsky[1079] who found that the extent of chromatin digestion by DNase is proportional to the concentration of the enzyme and DNA. Although estimates of 50% free phosphate have been reported, a lower content (approximately 35%) of protein-free DNA segments in chromatin has been estimated by acridine orange binding studies,[900] and from the luminescence quenching it has been concluded that these protein-free segments are G + C-rich. Clark and Felsenfeld[1076] found, on the other hand, that the DNase resistant DNA segments, as well as those free of protein, are not different in their base composition and resemble that of the total DNA (58% A + T). Since both groups of investigators used calf thymus chromatin, it is possible that the DNase digestion produces random cleavages of DNA at some of the many accessible nucleotides in chromatin.

The use of nucleases to study the structure of chromatin is further complicated by the presence of small amounts of proteolytic enzymes in most deoxyribonuclease preparations. As was pointed out by Phillips[1079a] these proteases are not inactivated by diisopropyl fluorophosphate and therefore their degradation of histones during chromatin digestion with nucleases makes the interpretation of such experiments quite difficult.

The significance of the free phosphate groups in DNA is not clear. The very high proportion (about 50%) indicates that free phosphate groups may have very little meaning for regulating the transcriptional specificity of chromatin DNA. Indeed, by DNase digestion, chromatin samples from transcriptionally active, and inactive tissues behaved very much alike.[1076] Since only about 10% of the DNA can be transcribed in isolated chromatin in vitro, this indicates that there is no correlation between the amount of free DNA phosphates in chromatin and its efficiency as a template for RNA synthesis.

Chapter 7

HISTONE INTERACTIONS WITH OTHER MOLECULES

The firm association of histones with DNA in chromatin leaves little doubt that this is the natural existence of these two macromolecular species. Since the binding of histones and DNA is primarily electrostatic, it is difficult to understand why histones do not associate in similar complexes with RNA or nonhistone proteins. The possibility of histones entering into such interactions has been reported by many investigators, some of whom suggest that this may be a mechanism for the local dissociation of histones from the DNA, which would then result in the active transcription of the denuded genes.[985,999,1080-1084]

Interactions with Proteins, RNA, and Other Polyanions

In addition to the extensive associations with each other that result in the formation of macromolecular aggregates, histones or protamines can interact with a considerable number of other proteins. Kossel[3] first recognized that the mechanism for such interactions is a simple electrostatic attraction. The mechanism has been more recently studied by several investigators.[1085-1089] The interactions are mostly nonspecific and large aggregates or a precipitate result from the neutralization of the charges on the interacting proteins. Small quantities of acidic proteins may associate with histones without precipitating them. Such complexes frequently change the behavior of histones in ion exchange chromatography.[1089]

Since the nonhistone proteins, although present in lesser amounts than histones, are an essential part of chromatin (Table 14), there is a distinct possibility that both the nonhistone proteins and histones interact to a various extent. The recent observation by Itzhaki,[1078] who found that the nonhistone proteins are not attached directly to the DNA in chromatin, further strengthens this possibility. Obviously, if extensive, such an interaction can weaken the histone-DNA binding and thereby affect the chromatin structure and function. The first experimental evidence indicating that a special group of nuclear nonhistones, the phosphoproteins, can decrease the inhibitory effect of histones on in vitro RNA synthesis is attributed to Langan.[224] Although the phosphoproteins are unable to displace the histones once they become attached to the DNA in vitro, the interaction of the phosphoproteins with histones prior to the attachment of the latter to the DNA results in a weaker histone-DNA binding. The phosphoprotein-histone complex with DNA supports in vitro RNA synthesis much better than a similar complex without the phosphoprotein.[1090] The possibility of histones complexing with the nuclear nonhistone proteins has also been demonstrated by Wang,[1091,1092] who found that some of the nonhistone proteins which form strong complexes with histones are phosphoproteins.[1093]

The various interactions of histones with certain chromatin nonhistone proteins were found to be selective in that the nature of the complexes depended on the individual nonhistone and histone protein fractions.[1094] At a ratio of histone to nonhistone protein near unity, the complexes were insoluble in neutral buffer solutions, as could be expected from the mutual charge neutralization of both components. In contrast, an excess of nonhistone protein over histone increased the solubility of the complexes. It was at this high nonhistone to histone ratio where the most selective interactions of the individual histone fractions with the nonhistone proteins were observed. The complexes did not dissociate in 2.0 *M* NaCl, which indicated a possible involvement of nonelectrostatic interactions. The nonhistone proteins, as well as their complexes with histones, were soluble in 0.25 *N* HCl. Apparently, the nonhistone protein fractions studied by Wang and Johns[1094] at least partially represented the acid soluble proteins contaminating the histones extracted from nuclei with 0.25 *N* HCl. Because the histone-nonhistone complexes in vitro were not soluble (dissociable) in 2.0 *M* NaCl, it is unlikely that they exist in chromatin since 2.0 *M* NaCl removes practically all the histones from chromatin.

Similar to the nonhistone proteins, RNA can complex with histones under physiological ionic conditions.[1095-1101] These complexes are unstable and dissociate in the presence of DNA, which competes with the RNA for the dissociated histones.[1097-1101] This makes quite improbable the possibility that the RNA-histone interaction can remove these proteins from the DNA in chromatin. Both the nonhistone proteins and RNA limit the inhibitory effect of histones on in vitro RNA synthesis only if added to the histones before these can interact with DNA. As shown by several investigators,[1096-1098,1102,1103] RNA and acidic (nonhistone) nuclear proteins can influence each other in their interactions with histones. Interestingly, free histones will form nonspecific complexes with almost any polyanionic molecule.[224,1090,1096,1097] These complexes are frequently insoluble in buffers between pH 4 and 8 and do not dissociate in concentrated salt solutions.[1094] The order in which the individual components are allowed to interact with the DNA considerably affects the solubility of the resulting complexes and the extent of histone inhibition of the in vitro RNA synthesis. Wang[1104] found that a particular fraction of the chromatin nonhistone proteins can reverse the inhibition by histones of the in vitro RNA synthesis. This reversal occurs even after the histones are complexed with the DNA. The mechanism of this reversal is not specific for the RNA polymerase reaction, since similar results can be obtained with a histone-inhibited DNA polymerase system.[1105] The nonspecificity of the action of this protein fraction indicates a possible displacement of the histones from the DNA. It is necessary to realize, however, that all the experiments discussed here have been carried out with artificial complexes between the DNA, nonhistone proteins, RNA, and histones which are made in vitro by mixing the individual components in various orders. There is no doubt that the resulting aggregates do not, by any means, approximate the complex situation in chromatin.

More realistic conclusions concerning the effects of various polyanions can be reached if isolated chromatin is used instead of the in vitro complexes. Working with isolated chromatin, which approximates more closely the situation in vivo, Teng and Hamilton[1106] found that if the templating efficiency of isolated chromatin was inhibited by added histones, this inhibition was reversible in the presence of a nuclear nonhistone protein fraction. A similar activation of template activity in isolated chromatin by a fraction of nuclear nonhistone proteins was reported by Wang[1107] and by Kamiyama and Wang.[1108] In addition to an increased template efficiency of chromatin treated with the nonhistone protein fraction, there was also a change in the base composition and hybridization properties of RNA templated by this stimulated chromatin. The RNA synthesized in vitro by the activated system also coded for the in vitro synthesis of polypeptides different from those coded by RNA templated by the control chromatin. While the findings of Teng and Hamilton[1106] may be attributed to the removal of the excess, additional histone molecules which have been shown by other investigators to associate only weakly with chromatin in vitro, the observations of Kamiyama and Wang[1108] point to a possible activation of previously restricted DNA sites on chromatin exposed to certain nonhistone proteins. According to Kamiyama et al.,[1109,1110] the active nonhistone protein fraction contained phosphoprotein kinase activity which appeared to be correlated to the stimulation effect on RNA transcription in vitro.

An activation of chromatin by a variety of polyanions was described by Coffey and coworkers.[1111-1114] In these studies, isolated nuclei or chromatin exposed to heparin, dextran sulfate, or polyaspartic acid exhibited active DNA replication in vitro when exogenous DNA polymerase was provided. The release of the template restriction was accompanied by a considerable swelling of the isolated nuclei. There was some selectivity to the activation since polyanions such as chondroitin sulfate, casein, polygalacturonic acid, or hyaluronic acid were ineffective. A number of ribonucleotide polymers – as well as RNA preparations – activated in vitro DNA synthesis when added to isolated chromatin or nuclei in the presence of exogenous DNA polymerase.[1113,1114] The exposure of isolated nuclei to synthetic polyribonucleotides resulted in their increased incorporation of labeled dTTP even in the absence of exogenous DNA polymerase. This increase was considerably lower than in experiments with the exogenous enzyme and was probably due to a limited supply of the nuclear polymerase enzyme. As observed by Ansevin et al.,[1115] treatment of isolated chromatin with polyanions such as polyglutamic, polyguanylic, or polycytidylic acid, heparin, or dextran sulfate also increased its ability to serve as a template for RNA polymerase. There were considerable differences in the effects of the individual polyanions on the in vitro RNA synthesis supported by isolated chromatin. That these differences reflected a substantial decrease in the binding strength of histones to the DNA was indicated by thermal denaturation experiments (Figures 47 and 48). As suggested by the appearance of the thermal transition maxima corresponding to that of pure DNA in chromatins treated with these polyanions, the increased template efficiency of chromatin-polyanion complexes could be explained by the partial dissociation of histones and exposure to the enzyme of DNA segments previously covered by these proteins. That heparin can dissociate measurable amounts of histones from isolated nucleohistone was first shown by Kent et al.[1116] The administration of polyanions to cells in vivo caused a marked decondensation of chromatin accompanied by an increased biosynthetic activity of the affected cells.[1117-1119] Since acid polysaccharides and other polyanions abound in cellular structures, these molecules must be taken into account when constructing models for the derepression of DNA in response to various metabolic stimuli in living cells. It was shown by Meisler and McCluer[1120] that a molecule natural to many cells, i.e., the sialic acid containing glycosphingolipids (known as gangliosides), can bind histones even if these are associated with the DNA in chromatin. The interaction was proportional to the sialic acid content of the gangliosides and resulted in decreased thermal denaturation temperatures for the complexed nucleohistones which resembled those of the free DNA. Since glycosphingolipids appeared to be present in many

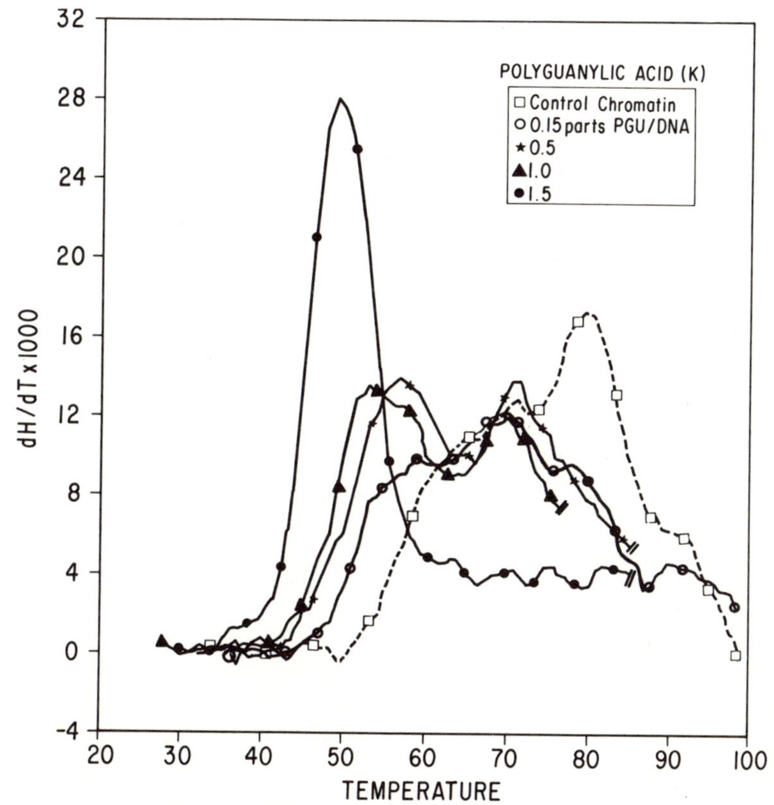

FIGURE 47. Destabilization of rat thymus chromatin by polyguanylic acid (potassium salt) added to the denaturation solvent. Figure inset shows parts by weight of polyguanylate added to one part DNA (within chromatin). Virtually all the DNA was found in an uncomplexed state at the highest concentration of polyguanylate during denaturation in a solvent containing 3.6 *M* urea, 0.15 mM versene, and 5 mM sodium cacodylate buffer (pH 7); plots terminated by parallel bars were anomalous at higher temperatures because of the formation of aggregates. The temperature derivative of hyperchromicity (dH/dT) was calculated by computer from absorbance and temperature data recorded in continuously heated samples; all points are plotted although the lines are identified by symbols located only at every fifth point.

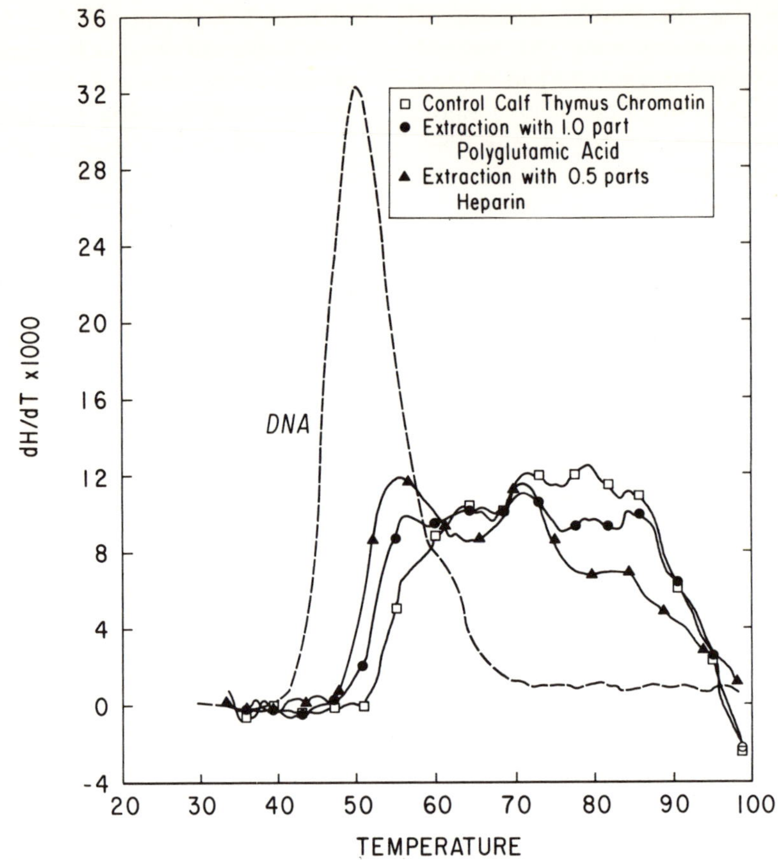

FIGURE 48. Destabilization of calf thymus chromatin following extraction of the chromatin by heparin (sodium salt). Samples of chromatin were treated in 1.5 mM sodium chloride: 0.15 mM trisodium citrate solution for 2 hr with 0.5 and 1.0 parts by weight of heparin to 1.0 part of DNA (within chromatin); the chromatin was then made 0.15 *M* in sodium chloride and recovered by centrifugation. Thermal denaturation was conducted in a solvent containing 3.6 *M* urea, 0.15 mM versene, and 5 mM sodium cacodylate buffer (pH 7). The temperature derivative of hyperchromicity (dH/dT) was calculated by computer from absorbance and temperature data recorded in continuously heated samples; all points are plotted although the lines are identified by symbols located only at every fifth point. The thermal denaturation profile for pure calf thymus DNA in the same solvent is included for reference.

cellular membranes, their interaction with histones raised the question as to how the membranes were protected from histones during the cytoplasmic synthesis and transport of these proteins into the cell nucleus.

Histone Interactions with Steroid Hormones and Carcinogens

Considerable experimental evidence indicated that hormone administration activated the transcription of DNA templates in target tissue chromatin. In addition to an increased biosynthesis of the RNA species usually present in the unstimulated tissue, hormone administration also resulted in the appearance of new RNA species previously absent.[729,730,1121] Since histones effectively inhibit the ability of DNA to serve as a template for RNA synthesis, it was suggested that the interaction of hormones with histones may influence the histone-DNA interactions in chromatin and, thereby, result in partial genetic derepression.[725,1122-1124] In support of this possibility, a labeled corticosteroid hormone which had been administered to experimental animals was found to accumulate in the nuclei of liver cells.[1125-1127] According to Sekeris and Lang,[1122] the nuclear accumulation of the hormone reached a maximum within 30 min and, during the first 3 hr, a considerable portion of the radioactivity was associated with the histones. Tritiated testosterone was bound to the liver histones to a much lesser extent. The in vivo binding of hydrocortisone to histones was confirmed by several investigators;[1128-1130] Sluyser[1123] found the arginine-rich fraction F3 (III) to be labeled with the hormone most readily. Similarly, tritium labeled testosterone injected into the rats was found to become associated with the histones of the prostate and levator ani muscle and only to a much lesser extent with the liver or spleen histones. In contrast to hydrocortisone, labeled testosterone was bound preferentially to the F1 (I) histone in rat liver, prostate, and levator ani muscle. In spleen, the F3 (III) histone accepted more labeled testosterone initially and retained it longer than the other histone fractions.[1130] However, this in vivo binding of hormones to the histones may be artifactual. Tsai and Hnilica[1131] analyzed the in vivo binding of corticotrophic hormones to nuclear proteins. Using several different procedures to isolate nuclei from the livers of rats injected with labeled cortisone or hydrocortisone, these authors found that only nuclei contaminated with cytoplasmic particles (isotonic sucrose nuclei) produced histone fractions having bound hormone. It was concluded that the hormone was bound to an acid soluble impurity, probably a protein, contaminating the histones. It remained uncertain whether this impurity was cytoplasmic, perhaps representing the cytoplasmic hormone receptor protein. Holoubek and Driscoll[1132] reported recently, that this cortisol receptor protein appeared to be part of the chromatin nonhistone complex and could be isolated together with the phosphoprotein fraction of the rat liver chromatin. They also confirmed that cortisol was not bound to the histone in chromatin. Preferential in vivo binding of hydrocortisone to the nonhistone chromatin proteins in rat thymus and liver was also seen by several other investigators.[1129,1133,1134]

In addition to the in vivo experiments, histones were found to interact with corticosteroid hormones in vitro. Sluyser[1123] reported that if histones were incubated with labeled hydrocortisone, part of the hormone was eluted together with protein when subjected to gel filtration on Sephadex G-25®. The arginine-rich histones were found to bind more hormone than the lysine-rich fractions and the protein-bound hormone remained largely undissociated during dialysis.[1130] The in vitro binding of the arginine-rich and other histones to cortisol, tetrahydrocortisol, cortisone, tetrahydrocortisone, progesterone, 17 β-estradiol, testosterone, and thyroid hormone, etc., was also reported.[1131,1133,1135-1138] The binding of corticosteroid hormones was increased in the presence of phosphate ions.[1137] Histones complexed with steroid hormones formed less stable complexes with DNA, as was evidenced by comparing the thermal denaturation profiles of nucleohistones reconstituted with hormone-free and hormone-bound histone samples.[1139] Only the histones from target tissues complexed with the appropriate hormone exhibited this effect. There was no significant difference in hydrocortisone and cortisone binding between histones from rat liver and from rat and calf thymus.[1133] The interaction of histones with corticosteroids was most likely due to hydrophobic bonding.[1140]

Corticosteroid hormones are unstable in aqueous solutions, forming 21-dehydro-derivatives which can considerably increase their binding to proteins through the intermediate formation of Shiff bases.[1136,1141-1145] These derivatives react chemically with the amino-, amido-, and guanidino-groups in proteins and peptides with a definite preference for arginine.[1142] It has been suggested that in the cells, the steroids are oxidized to 21-dehydro-derivatives which can then bind chemically with proteins, perhaps histones, forming chemically stable complexes.[1142]

Isolated histone fractions, especially the arginine-rich histone F3 (III), were found to bind labeled corticosteroid hormones quite easily. Sluyser[1140] showed that the binding of hydrocortisone to the F3 (III) histone increased linearly with the incubation time. The oxidation (dimerization) of F3 (III) histone enhanced its capacity to bind hydrocortisone. The F1 (I) histones interacted poorly and the F2a histones (IIb1 and IV) accepted considerable amounts of hormone only when aggregated. Sluyser[1140] suggested that the binding of hydrocortisone to the F3 (III) histones was realized mainly through its hydrophobic amino acid region which may be exposed on the outside of this molecule. This suggested binding was of a different character than the direct chemical interaction with arginine proposed by Monder and Walker[1142] for the 21-dehydro-derivatives.

Tsai and Hnilica[1131] analyzed complexes of

unfractionated calf thymus histones with hydrocortisone. Starch gel electrophoresis and gel filtration on Sephadex G-75 confirmed that the arginine-rich F3 (III) histone was the main in vitro hormone receptor. Over 65% of the radioactivity was associated with F3 (III) histone, which represented, at most, only 20% of the total histone protein. One third of the remaining radioactivity was bound to the F1 (I) histones. Even when reacted with a large excess of hydrocortisone, the F3 (III) fraction contained less than 1 $\mu\mu$ mole of this hormone per mg of protein. Only 0.2 $\mu\mu$ moles of hydrocortisone were bound to the F1 (I) fraction. In other words, the amount of hydrocortisone bound to the F3 (III) histones represented about 15 μmole of the hormone associated with one mole of F3 (III) histone protein, assuming its molecular weight of 15,000 daltons. This is about one molecule of hormone for every 6.6 x 10^4 molecules of F3 (III) histone. Using the data presented by Sluyser,[1140] it could be calculated that only about 0.0017% of the F3 (III) histone molecules were bound to hormone. This value was quite close to the 0.003% found by Tsai and Hnilica,[1131] who interpreted this very low combining ratio to indicate histone contamination with a corticosteroid hormone receptor. This material was probably an acid soluble protein which associated tenaciously with the F3 (III) histones. An analysis of tryptic peptides from the F3 (III) histone complexed with hydrocortisone showed that a significant amount of the labeled hormone was bound to a ninhydrinpositive, trypsin resistant substance, probably a nonhistone protein.[1131]

Another group of molecules bearing structural and functional similarities to the steroid hormones are certain carcinogens. For example, the administration of carcinogens alters the genetic transcription of chromatin in experimental animals,[1146-1150] changes the chromatin composition,[1151,1152] and considerably affects the metabolism of the chromatin proteins, probably histones.[1151] The induction of the activity of several drug metabolizing enzymes following the administration of certain carcinogens suggests that selected genes are specifically derepressed.[1153-1158] One of the enzymes increasing its activity is the DNA polymerase. In rat liver, its activity nearly doubles during the early stages of azo dye carcinogenesis, returning to nearly normal values at the appearance of tumors.[1159] Many carcinogens interact directly with both the DNA or RNA.[1160-1168] Since a number of carcinogens are known to bind to proteins and histones,[1169-1178] it has been suggested that there may be a casual link between the observed activation of DNA transcription in the chromatin of treated animals and the appearance of tumors. Possibly these newly derepressed sites (genes) may be essential for the cellular proliferation characteristic of the neoplastic growth.

Sluyser[1169] demonstrated that both labeled *N,N*-dimethyl-4-amino azobenzene and benzopyrene became associated with histones during in vitro incubation in 2 mM NaCl for 15 min. The binding varied with the composition of the histone fractions and showed some preference for the arginine-rich histones. More interestingly, it was tissue specific and showed differences between the histones isolated from rabbit liver and spleen. The labeled benzopyrene was bound to associate preferentially with the trypsin resistant "core" of the F3 (III) histones. This led Sluyser[1169] to the conclusion that benzopyrene binds preferentially to the non-basic, hydrophobic regions of the histone molecules. A more extensive in vivo binding of *N*-2-fluorenylacetamide-9-C^{14} metabolites to the arginine-rich histones than to the rest of the basic nuclear proteins was also reported by Barry et al.[1170] It was interesting that these findings and conclusions resembled those on the binding of steroid hormones to histones, especially the arginine-rich histones F3 (III).[1140] It was shown[1131] that the corticosteroid hormone binds only to very small portions of the F3 (III) histones, probably to some nonhistone protein(s) contaminating the F3 (III) fraction. This might also be true for the binding of carcinogens to histones since these proteins bind considerably less carcinogen than cortisol. Although the carcinogen binding was tissue specific while the arginine-rich histones are not, most of the bound carcinogen was found with the tryptic "core" which would be expected to contain the contaminating nonhistone proteins. Moreover, while the histone-corticosteroid hormone complexes made in vitro were stable, the histone-carcinogen complexes were dissociated by organic solvents.[1171]

The binding of carcinogens to the nonhistone proteins in chromatin was demonstrated by several investigators. Lesko et al.[1172] found only a small fraction of administered 3,4-benzopyrene associated with histones. Most of the nuclear binding was to the chromatin nonhistone proteins. Similarly, the carcinogens 3,4-benzopyrene and 20-methylcholanthrene were found to bind to chromatin in vivo.[1173] Again, most of the hydrocarbons could be recovered associated with the DNA-nonhistone complex and only about 10% of the total carcinogen associated with chromatin was soluble in 0.2 *N* HCl. As in the experiments of Kodama et al.,[1171] the carcinogen was not linked to the proteins by covalent bonds, since over 98% of it could be dissociated with cyclohexane. Other carcinogens were also found to bind preferentially to the nuclear nonhistone proteins in vivo.[1170,1174] Although the nuclear binding of amino azo dyes reportedly occurred mostly with the proteins of the nuclear sap,[1175,1176] the actual interaction of these carcinogens, as well as that of methylcholanthrene, appeared to occur primarily in the cytoplasm with a selected group of proteins which may be specific for the target tissues.[1158,1179-1181] There was a striking similarity in the molecular properties of the cortisol receptor proteins in liver and those of the cytoplasmic proteins which bind carcinogens in vivo in that the cytoplasmic receptor proteins for cortisol, dimethylaminoazobenzene, and, perhaps, for 3-methylcholanthrene, may be identical in rat liver.[1182-1184]

The activation of genetic transcription in target tissues by steroid hormones depends on the hormones' interaction with proteins. The hormone combines first with specific receptor proteins in the cytoplasm of the target cells.[1121,1185-1193] These proteins have been partially purified and appear to be relatively homogeneous with molecular weights of less than 100,000 daltons. The estradiol receptor complex of calf uterine cytosol exists in the form of 8s complexes which can be dissociated with salt into 4s subunits. The isoelectric points of the partially purified 8s and 4s receptor complexes are 5.8 and 6.4, respectively,[1188] whereas the cortisol receptor protein is more basic.[1189,1194] The initial interaction of steroid hormones (and perhaps of some carcinogens) with the cytosol receptor proteins is essential for their biological activity.[1195-1199] Recently, O'Malley et al.[1200] showed that progesterone extracted from in vitro labeled oviduct nuclei is bound to a protein which is indistinguishable from the oviduct progesterone cytosol receptor protein. This suggests that the nonhistone proteins found associated with labeled hormones in the chromatin of target tissues are hormone complexes with the cytoplasmic receptor proteins which became incorporated into the chromatin. The progesterone-cytoplasmic receptor protein complex is specific for the oviduct (target tissue of progesterone) and preferentially associates in vitro only with chicken oviduct chromatin.[1201] Results similar to the progesterone and oviduct experiments were also obtained with uterine chromatin in ovariectomized rats exposed to estradiol and with prostate chromatin in castrated male rats stimulated with dihydrotestosterone.[1202] Using hybrid chromatins obtained by reconstitution of the histones from one tissue to the nonhistone proteins and DNA mixture from another tissue, Spelsberg et al.[1203] demonstrated that the interaction specificity between the oviduct cytoplasmic receptor protein complex with progesterone and the oviduct chromatin depends on the origin of the nonhistone protein and DNA. The histones did not contribute any tissue specificity to this interaction. This indicates that the chromatin of the target tissue is genetically programmed to receive the receptor protein-hormone complex. The program is apparently tissue specific. Histones are probably only quantitative repressors of the DNA genes determined by cytodifferentiation to remain inactive and do not directly participate in the specific binding of the hormone-protein receptor complexes with the chromatin of the target tissues.

It should be remembered, however, that the interactions of carcinogens or steroid hormones with cytoplasmic receptor proteins and their transport to the specific sites of DNA in chromatin may represent only a part of the mechanism by which these substances interact with living cells. Several investigators showed the participation of microsomal fractions in the binding of methylcholanthrene or benzopyrene to the DNA or to cytoplasmic receptors.[1158,1204,1205] It was found recently by Williams and Rabin[1206] that many carcinogens degranulate rough surfaced cytoplasmic membranes by releasing their bound ribosomes. On the other hand, smooth (ribosome-free) cytoplasmic membranes, when incubated with a steroid hormone (estradiol) bound a considerable amount of the added ribosomes. Carcinogens such as 3,4-benzopyrene, 3-aminopyrene, etc., inhibited this binding. It was suggested that the carcinogens occupied or destroyed sites on the cellular membranes which normally interact with steroid hormones. This interaction was specific since a noncarcinogen (6-aminochrysene) did not affect the membranes, while a carcinogenic derivative (2-aminochrysene) degranulated the rough surfaced membranes.[1206] Although not all known carcinogens affect cellular membranes, it seems likely that many of these substances may affect the cells by using pathways normally designated for the action of steroid hormones.[1182-1184] Only additional research can explain how the interactions of carcinogens with the cytoplasmic protein receptors and the observed degranulation of the rough membrane reticulum in the affected cells join together in a sequence of events leading to neoplastic growth.

Chapter 8

HISTONES IN GENETIC RESTRICTION

In primitive unicellular organisms like bacteria, each cell cares only for itself. The regulation of DNA transcription in such a case is relatively simple and depends on the transcriptional sequence of the cellular genome and the presence of specific gene repressors. These repressor molecules are proteins and interact with the bacterial genome selectively.[1207-1214] The growth rates of unicellular cells are usually limited only by the supply of nutrients and the speed of waste product removal. This, together with their high mutation rate, indicates a very flexible and rapidly adaptable genetic apparatus. With the appearance of multicellular systems, however, a more coordinated and selective means of regulating the DNA genes is necessitated. The occurrence of histones in differentiated organisms, as compared with their absence in bacteria, suggests that these proteins play a role in the genetic regulation of differentiated cells. This is further substantiated by the close association of the histones with the chromatin DNA and by their amazing phylogenetic stability. The possible function of histones as gene regulators dates back to a proposal by Edgar and Ellen Stedman,[62,94] who assumed that the DNA-histone complex is genetically inactive. This proposal receives substantial support from the findings that histones inhibit in vitro DNA-dependent RNA synthesis as well as in vitro DNA replication.

Quantitative Restriction

Disregarding the qualitative effects of DNA-histone interactions on the specificity of DNA transcription, the quantitative ability of DNA to serve as a template is greatly limited by its association with histones, both in vitro and in vivo.

In vitro RNA Synthesis

Two investigations performed a decade ago considerably stimulated scientific interest in the properties and functions of histones. Working with chromatin from pea embryos, Huang and Bonner[1215] found a substantial decrease in the ability of DNA to serve as a template for the in vitro RNA transcription after the DNA became complexed with histones, i.e., the natural complex of DNA and histones (chromatin) only supported in vitro RNA synthesis to a very limited extent. Deproteinization of the chromatin resulted in a spectacularly increased RNA biosynthetic activity which approached that of pure DNA. In the reconstituted mixtures of DNA and isolated histone fractions, the decrease in templating activity of the DNA in the complexes was proportional to the amount of added histone. When unfractionated histone was reconstituted to the DNA at a ratio of 1:1, the incorporation of labeled nucleotides into the RNA practically ceased.

In a similar experiment, Allfrey et al.[645] removed part of the nuclear proteins, mostly histones, by a short exposure of the nuclei to trypsin. The treatment remarkably increased the subsequent incorporation of radioactive nucleosides into nuclear RNA. Although the specificity of trypsin was not limited to the histones, it could be anticipated that because of their high lysine and arginine content and lack of a rigid tertiary structure, these proteins were the most vulnerable to trypsin digestion. Acid extraction of trypsinized nuclei demonstrated a considerable depletion in their histone content. The base composition of the RNA synthesized in nuclei after the removal of histones by trypsin digestion was DNA-like, indicating that most of the previously repressed DNA became available for transcription by the endogenous nuclear polymerase. The addition of histones to the trypsinized nuclei substantially decreased their ability to biosynthesize RNA. On a weight basis, the arginine-rich histones were more inhibitory than the lysine-rich histones.

These observations supported the concept that the function of the histones in chromatin was to restrict those genes designated to be inactive. Although the approaches were diametrically designed, i.e., adding histones to the DNA in vitro or removing them from isolated nuclei, both experiments offered the same conclusion about the inhibitory activity of histones. Additionally, the RNA synthesis was inhibited in assays using either exogenous (bacterial) or endogenous (nuclear) RNA polymerase. Quantitative differences in the inhibition of the in vitro RNA synthesis directed by DNA complexed with various histone fractions were also observed by Huang et al.[898] However, in contrast to the report by

Allfrey et al.,[645] these authors found the lysine-rich histones to be more inhibitory than the arginine-rich fractions. In the system using exogenously supplied *E. coli* polymerase, the inhibition differences attributed to the individual fractions were more obvious when the DNA and histone were reconstituted by a slow decrease of the salt concentration during dialysis than when the DNA and histones were mixed directly in low salt media. The higher efficiency of the lysine-rich histones, especially the F1 (I) fraction, in inhibiting the DNA dependent RNA synthesis in vitro was confirmed by several investigators.[664,1216,1217] On the other hand, Hindley[1218] obtained exactly the opposite results, i.e., the arginine-rich histones were more inhibitory than the F1 (I) fraction. A significantly higher inhibition by the arginine-rich histones, as compared with the lysine-rich ones, was also observed by Skalka et al.[1219]

Several investigators have pointed out that the association of DNA with histones decreases the solubility of the complexes and thereby impairs the templating activity of the DNA.[332,899,1220,1221] Since the very lysine-rich histones F1 (I) are much more efficient in precipitating the DNA during the reconstitution of DNA complexes with histones, their greater inhibitory effect is most likely due to the formation of highly cross-linked reconstituted nucleohistones which have their DNA inaccessible to the polymerase enzyme.[181,332,899,910,1222] If such complexes are fragmented, their templating activity increases considerably.[1220,1221,1223] It is possible to obtain soluble DNA-histone complexes either by a slow mixing of both components[903] or by a slow removal of salt[898] and limiting the Mg^{++} and Mn^{++} concentration in the enzyme assay mixture,[1224] i.e., the manner in which the individual components are brought together affects the solubility of the final complex. Thus, the differences in templating inhibition by the very lysine-rich and arginine-rich histones may be due to their relative abilities to crosslink and precipitate DNA. These abilities change considerably with the variations in the ionic environment and the order in which individual components are mixed together.[910]

Under proper conditions, the template activity of DNA complexed with arginine-rich histones was proportional to the solubility of the reconstituted nucleohistone.[1225] Using polyarginine and polylysine, Shih and Bonner[882] found that if the polypeptide-DNA complexes were reconstituted in the presence of 5 *M* urea to decrease the interactions between the formed complexes and their components, the resulting products were more soluble. At an amino acid:nucleotide ratio of 0.6, there was essentially no difference in the template restriction of DNA by polylysine or polyarginine and the templating activity depended on the extent of hydration. Highly hydrated complexes were better templates than tightly aggregated complexes, which were virtually inactive as templates for the in vitro RNA synthesis. A similar situation was found for the individual histone fractions complexed to the DNA by salt gradient dialysis in the presence of 5 *M* urea.[902] Hence, those conditions which favored the formation of compact DNA-histone aggregates produced nucleohistones inactive as templates in the in vitro RNA synthesis regardless of whether the histone fractions were rich in lysine or arginine.

A certain length of the basic polypeptide chains associating with the DNA must be necessary for effective interactions, since proteolysis of the histones by trypsin abolished their inhibitory activity of RNA biosynthesis in vitro[332] or in vivo.[27,645] In a more detailed study, Moskowitz et al.[1226] investigated the effects of four clusters of basic amino acids isolated from the amino acid sequence of the F2a1 (IV) histone, i.e.,
16 20 35
-LYS(Ac)-ARG-HIS-ARG-LYS(Me)-, -ARG-ARG-
40 75
LEU-ALA-ARG-ARG-, -HIS-ALA-LYS-ARG-
79 91 95
LYS-, and LYS-ARG-GLN-GLY-ARG- on the in vitro RNA synthesis templated by DNA. Although the F2a1 (IV) histone itself was quite inhibitory, its basic amino acid clusters were without noticeable effect on this reaction, indicating that a certain minimum length of the basic peptides was required for the inhibition to take place. Additionally, Moskowitz et al.[1226a] found that two fragments of F2a1 (IV) histone produced by cyanogen bromide cleavage also differed in their inhibitory capacity on the in vitro RNA synthesis. The smaller C-terminal fragment (18 amino acids) was much less inhibitory than the larger NH_2-terminal (84 amino acids) fragment. Interestingly, minute quantities of these clusters, as well as other positively charged molecules (polyamines, histones, polyamino acids, etc.), actually enhanced the DNA dependent RNA polymerase enzyme activity. This latter effect might be caused, per-

haps, by displacing the newly synthesized RNA from the DNA template, by decreasing the product inhibition of the enzyme, by changing the conformation of the nucleohistone templates and enhancing the enzyme-DNA binding, or by other mechanisms.[1226-1230]

Since all the components of the RNA polymerase reaction mixture come into close contact and can form complexes, Spelsberg et al.[1231-1233] investigated the inhibition of RNA polymerase by histones. The arginine-rich histones were found to interact with both the free bacterial and mammalian enzyme and inhibit their activity to a much greater extent than if they were associated with the DNA. The inhibition was not simply related to the arginine content of the fractions.[1232] The RNA formed complexes with F3 (III) histone which were quite stable at ionic strengths equal to or greater than those encountered in living cells.[1233] The lysine-rich F1 (I) histones formed less stable complexes with the enzyme. These complexes dissociated at ionic strengths less than physiological or in the presence of free DNA (Figure 49). Although the biological significance of these findings cannot be assessed at the present time, the greater inhibition of RNA polymerase by the arginine-rich histones as compared with the very lysine-rich fractions (the latter are also more efficient in direct template

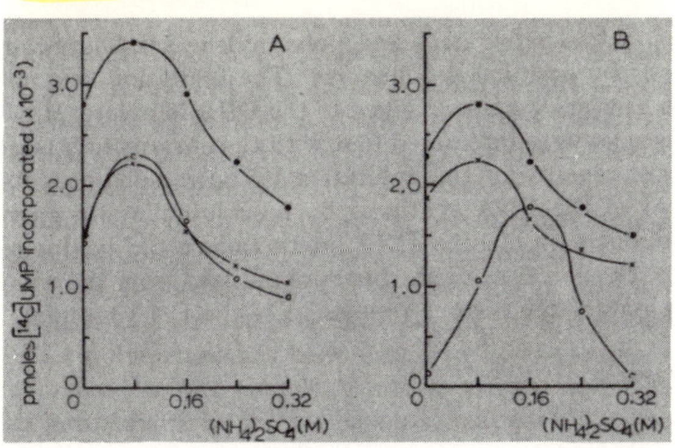

FIGURE 49. The effect of $(HN_4)_2SO_4$ concentration on the complexes of very lysine-rich F1 (I) and arginine-rich F3 (III) histones with DNA or RNA polymerase. The salt was added after the respective components were interacted. • –•, control (no histone); X – X, histone complexed to the DNA; ○ – ○, histone complexed to RNA polymerase. (A) Interactions of F1 (I) histone, (B) interactions of F3 (III) histone. Enzyme: 30 μg per reaction; histones: 20 μg per reaction; DNA: 60 μg per reaction. Notice the dissociation (activation) of the F3 histone-RNA polymerase complexes by salt (B). The F3 histone complex with DNA is not dissociated at this concentration of ammonium sulfate.[1231]

inhibition by crosslinking the DNA) did contribute to the observed controversies concerning the inhibition of the in vitro RNA synthesis by lysine-rich or arginine-rich histones.

In vitro DNA Synthesis

The role of histones in DNA replication (DNA-dependent DNA synthesis) was first investigated by Bazill and Philpot[1234] and by Billen and Hnilica.[332,1235] Isolated nuclei or nucleohistone templated very poorly for the in vitro DNA synthesis. The addition of histones to the free DNA gradually decreased its templating efficiency. The DNA polymerase reaction was almost completely inhibited at a histone:DNA ratio of unity. In a reaction mixture with *E. coli* DNA polymerase, the four histone fractions F1 (I), F2a (mixture of F2a1 or IV and F2a2 or IIb1), F2b (IIb2), and F3 (III) were all inhibitory. The F1 (I) histones suppressed the incorporation of labeled TTP into the DNA more effectively than the other fractions.[899] A mild digestion of the complexes with trypsin rapidly abolished the inhibitory activity of the histones.[332] The extent of inhibition of the in vitro DNA synthesis by individual histone fractions appeared to be closely related to their ability to cross-link and precipitate the DNA strands, since the templating efficiency decreased with the lowered solubility of the complexes.[332,899] The F1 (I) histones, which were known to cross-link the DNA in vitro most efficiently,[1222,1223] were also the most inhibitory in the DNA polymerase reaction.[898,899,1236]

Gurley et al.[1237] confirmed the histone inhibition of the in vitro DNA synthesis. Again, the F1 (I) histones were more effective than the other fractions. Interestingly, the activities of thymidine kinase and deoxyribonucleotide kinase were practically unaffected by the added histones. In a more detailed study, Wood et al.[1238] established that the high inhibitory activity of the F1 (I) histones did not result from the formation of aggregates, since their dispersion by ultrasound did not change the inhibition. Similar to the RNA polymerase reaction, small quantities of histones actually enhanced the incorporation of dATP into the DNA in vitro. Isolated pea embryo chromatin was also found to be an active template for the in vitro DNA polymerase reaction and supported the incorporation of labeled dATP into the DNA more efficiently than that of labeled ATP into the

RNA.[1239] The exposure of chromatin or isolated nuclei to polyanionic macromolecules such as heparin, polyaspartic acid, dextran sulfate, polynucleotides or natural RNA increased considerably their ability to template for DNA replication in vitro.[1111-1114] That this increase was due to the destabilization of the DNA-histone binding in chromatin was shown by detectable changes in its thermal denaturation properties.[1115]

In agreement with the in vitro data, Sluyser et al.[1236] found that the in vivo synthesis of DNA in regenerating rat liver was significantly inhibited by injections of histones or polylysine. The latter was a much more powerful inhibitor than any of the histone fractions, most likely because of the resistance of polylysine to cellular proteolytic enzymes. Although the F1 (I) histones inhibited the in vitro DNA synthesis the most of all the fractions, they did not differ from the other histones in their effects in vivo. However, the results were almost certainly complicated by a variable proteolytic degradation of the histones after their injection into the animals and the different abilities of the individual fractions to penetrate cellular membranes.

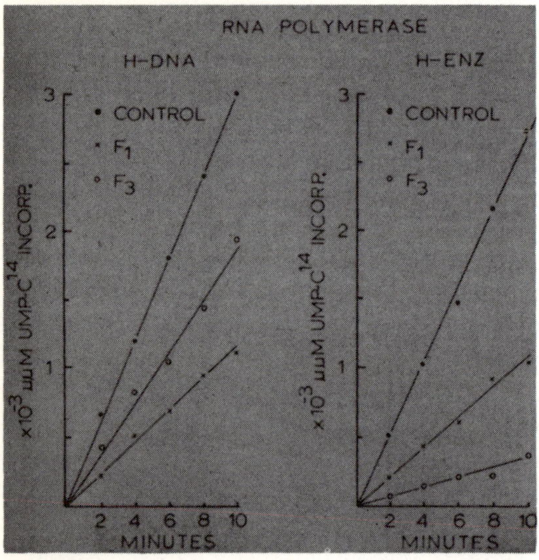

FIGURE 50. The inhibition of RNA synthesis in vitro by very lysine-rich (F1 or I) and arginine-rich (F3 or III) histones. The effect of preincubations of histones with DNA (H-DNA) and RNA polymerase (H-ENZ). The controls are reactions with no histone. Histone, 20 μg per assay; enzyme, 30 μg per assay; and DNA, 60 μg per assay. The arginine-rich histone F3 (III) inhibits the in vitro RNA synthesis much more when associated with RNA polymerase than with the DNA.[1233]

It is noteworthy that, unlike the systems synthesizing RNA in vitro, no discrepancies in the inhibition by F1 (I) and arginine-rich histones have been observed in the in vitro biosynthesis of DNA. Although the absolute amount of inhibition varied, all of the investigators have seen a much higher inhibition by the very lysine-rich F1 (I) histones as compared with the other fractions. The likely reason for this difference between the RNA and DNA polymerase systems is the insensitivity of the DNA polymerase enzyme to the inhibition by histones. As shown in Figure 50, arginine-rich histones inhibit the RNA polymerase reaction system more efficiently if first interacted with the enzyme instead of being first interacted with the DNA. DNA polymerase does not show such an interaction with arginine-rich histones (Figure 51)[1232-1233] and is inhibited only through the histone complexing with DNA templates. The higher inhibition of the DNA polymerase reaction in vitro by F1 (I) histones probably reflects their efficiency in crosslining the DNA molecules.

Chromatin as a Template for in vitro RNA Synthesis

The interactions of histones with DNA are complicated by the presence of other macromolecules such as nonhistone proteins, RNA, etc. Since most of these components are electro-

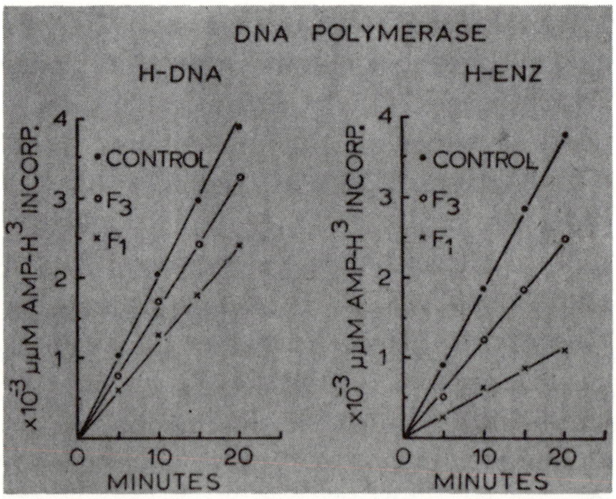

FIGURE 51. The inhibition of DNA polymerase by very lysine-rich (F1 or I) and arginine-rich (F3 or III) histones. The inhibition is not affected by the sequence in which the reaction components were interacted. H-DNA: the histones were first interacted with DNA; H-ENZ: the histones were first interacted with DNA polymerase. The controls are assays without histones. Histone, 50 μg per assay; DNA polymerase: 0.9 μg per assay; and DNA, 60 μg per assay.[1233]

negative, interactions between their charged groups and histones will most likely weaken the association of the latter with DNA. Additionally, because of the peculiarities in the individual histone types and their cross-linking of DNA strands, the histone distribution along the DNA appears to be discontinuous and irregular with parts of the DNA molecule remaining free of these proteins.[896,897,900,1050,1067-1069,1075,1076] The areas of histone-free DNA are rather small, however.[1078,1079] The irregular distribution of histones in chromatin fibers may interfere with the cooperativity of histone-DNA binding and dissociation in chromatin. Furthermore, the presence of more than one kind of histone in chromatin suggests different functions of each of the histone fractions in the template restriction and supercoiling of DNA. The extreme phylogenetic stability of the histone patterns in higher animals [with the exception of F1 (I) histones] also indicates that each of the fractions has a limited and probably specific role in maintaining the physical and biological integrity of chromatin.

The combined efforts of several investigators[23,48,705,895,930] showed that regardless of its complicated molecular structure, chromatin could be isolated from a variety of tissues in a form resembling that in the living cells. As a consequence of cytodifferentiation existant in higher living forms, parts of the DNA genome are restricted and inactive in RNA transcription.[930] Under carefully controlled enzymatic assay conditions using isolated chromatin or pure DNA as templates for the in vitro RNA synthesis,[23,895] the amount of open template (DNA genes accessible to the RNA polymerase transcription) could be estimated. For most tissues, depending on the method used for its isolation, chromatin exhibited activity in the range between 5 and 15% of open template with extreme values of less than 0.1% reported for sperm chromatin[27,360] and about 32% for growing pea cytoledons.[895] Usually, the open template determinations which employed enzyme kinetic experiments having saturating enzyme or template concentrations yielded higher values than data derived from RNA-DNA hybridization studies. This was probably because most of the hybridization experiments offered information on the amount of repetitive (or very similar) RNA species complementary to the DNA with the unique RNA species escaping detection. Since histones were known to be extremely sensitive to proteolytic enzymes, the possibility of proteolytic damage to the chromatin must be considered in the interpretation of data describing the amount of open template in isolated chromatin. Obviously, even limited proteolytic damage to the histones would result in an increase in the rate of in vitro RNA synthesis templated by isolated chromatin. In addition to proteolysis, fragmentation and conformational changes could occur during chromatin isolation which would also contribute to an increase in its templating activity.[907,1220,1221] The studies on the transcriptional activities of euchromatic and heterochromatic regions, which were discussed in more detail in the previous chapter, showed that chromatin conformation considerably influenced its transcriptional activity in vitro. The conformational differences between the euchromatin and heterochromatin appeared to be resistant to mechanical stresses such as shearing or exposure to ultrasound. This greater stability of the heterochromatin allowed its isolation.[705,974,985,993] Because of the difficulties in estimating accurately the extent of the chemical and mechanical damages incurred during the isolation of chromatin, the data describing the amount of open DNA templates in isolated chromatin must be regarded only as crude estimates. Experience with the isolation of nuclei showed the numerous difficulties in obtaining these cellular organelles in comparative purity from various tissues. It can be anticipated that a similar situation may prevail in the isolation and purification of chromatin. Obviously, the amount of impurities would affect the templating efficiency of isolated chromatin.

It was known from physicochemical studies that histones were essential for the supercoiling of DNA and, hence, the conformation of chromatin. Since the biological function of chromatin depends on its conformation, it could be expected that thc selective dissociation of the individual histone fractions would change its effectiveness to serve as a template for the DNA-dependent RNA synthesis. Hence, it came as a surprise when the selective removal of a considerable part of the F1 (I) histones from isolated chromatin (these proteins represent approximately 20% of all chromatin histones) did not much change the templating efficiency of the residual chromatin.[182,185,186,895,908] The small increase in the templating activity of chromatin[1240-1243] exposed to salt concentrations which dissociate the F1 (I) histones [0.4 to 0.5 *M* NaCl, 0.1 *M* $MgCl_2$, 0.2 *M* $(NH_4)_2SO_4$, etc.] was probably due to the activation of the endogenous RNA polymerase, which may be inhibited at low ionic strengths by arginine-rich histones or other proteins.[1231-1233] This and several other findings indicate that the F1 (I) histones may be bound to the DNA in chromatin differently from the other histones. Although the selective removal of the F1 (I) histones resulted in a profound conformational change of the chromatin, as reflected in a loosening of its ultrastructure and increased solubility,[904,907,908] the template efficiency of chromatin appeared to be little affected. According to Koslov and Georgiev,[1244] the F1 (I) histones might prevent the movement of polymerase along the DNA as was estimated from the lengths of the RNA molecules templated by free DNA, isolated chromatin, and chromatin devoid of F1 (I) histones.

The apparent differences between contribution of F1 (I) histones to the template activity and conformation of chromatin might be explained by changes which were likely to occur during the isolation of chromatin. It was suggested by Smart and Bonner[1245] that the F1 (I) histones are bound in chromatin in such a manner that they are much more exposed to the aqueous environment than the other histone fractions. This might be realized either by the F1 (I) histones cross-linking the surface of supercoils consisting of DNA complexes with other histones or by covering such complexes in a sheath-like structure with all the other histones and DNA inside the sheath. Also, the increased hydrophobic contribution of the F1 (I) histones to chromatin might be due to a loose and open binding of these proteins to DNA, assuming that all the histones were arranged linearly along the DNA molecule. Smart and Bonner[1245] supported their conclusions by pointing out that the F1 (I) histones in chromatin were much more susceptible to proteolysis than the other fractions and were the first fractions to be dissociated if the ionic strength of the chromatin solution was gradually increased. The F1 (I) histones became fixed to the DNA before the other fractions when chromatin was exposed to increasing concentrations of formaldehyde.[1064] The F1 (I) histones also have the highest net positive charge of all histone fractions (18.3 mol %).

Jensen and Chalkley[908] showed that the exposure of chromatin to 0.05 to 0.3 *M* NaCl caused its extensive aggregation. The aggregation was reversible if the ionic strength was decreased. The aggregated chromatin (130s) was less efficient in templating for the in vitro RNA synthesis as compared with the nonaggregated 30s chromatin. Rearrangement of F1 (I) histones by exposing the chromatin to 0.6 *M* NaCl followed by dialysis against 0.01 *M* Tris buffer eliminated the differences in template efficiency between the two types of chromatin and brought both to the unaggregated values.[907] Also, if the aggregation of chromatin in 0.14 *M* NaCl was attempted in the presence of free DNA or RNA, it did not take place.[908] This was thought to be due to an increase in the negative charge of the isolated nucleohistone associated with the complexing of the F1 (I) histone to the free DNA or RNA which, thereby, increased the solubility of chromatin. Low salt concentrations were reported to change the structural organization of chromatin[1247-1249] as well as its templating efficiency in vitro.[1247,1250] An increase of ionic strength from 0.001 to 0.1 of very diluted chromatin solutions brought about a collapse of the extended chromatin structure. This collapse could be prevented by a selective removal of the F1 (I) histones from chromatin.[1248]

Although the F1 (I) histones may cross-link the supercoiled DNA-histone complexes of chromatin in vivo, during the isolation of chromatin these cross-links may be disturbed and the F1 (I) histones become rearranged in a manner increasing the solubility of the product. The selective removal of the rearranged F1 (I) histones would no longer affect the template efficiency of the chromatin since this is already increased as a result of the altered chromatin conformation in which the F1 (I) histones are now occupying sites not accessible to the polymerase enzyme. However, according to Smart and Bonner,[184,1058] if sodium deoxycholate is used for the dissociation of histones from chromatin, the first histone fractions released consist of a mixture of F2b (IIb2) and F2a2 (IIb1) histones. The very lysine-rich F1 (I) fractions are dissociated last. Nevertheless, this dissociation sequence disrupts the cross-linking of chromatin in a manner similar to the selective removal of F1 (I) histones by NaCl solutions. It appears that the special conformation of chromatin resulting from the association of DNA with histones is conferred upon DNA by all the histone species. Recent reports by Clark and Felsenfeld,[1076] who found that there is considerable protein exchange on chromatin DNA at ionic strengths even lower than those used for the selective dissociation of histones, impose serious limitations on the interpretation of dissociation studies with isolated chromatin. If histones at higher ionic strengths can move freely along the DNA molecules, it is difficult to assign specific roles to the individual fractions since other histone molecules can assume positions on the DNA vacated by the dissociated species.

The removal of histones other than the F1 (I) results in a considerable increase in the template efficiency of dehistonized chromatin.[182,185,186,895,930,1063,1247,1251,1252] If practically all of the histones are removed in 2.0 *M* NaCl, the resulting material, which is also devoid of a portion (about 20%) of the nonhistone proteins present in the original sample, still exhibits some template restriction as compared with pure DNA.[182,1253] This indicates that some components other than histones may participate in quantitatively restricting the chromatin DNA.

Marushige and Bonner[930] found that because of the irreversible binding of RNA polymerase to the DNA at the initiation of transcription, the apparent Km value represents the amount of DNA required to bind irreversibly a given amount of polymerase. From double reciprocal plots expressing the effect of template (DNA) concentration on the rate of RNA synthesis in vitro, the authors concluded that the Km values of pure DNA and of isolated chromatin are the same. This indicates that both the DNA and chromatin bind the same amount of polymerase enzyme but in chromatin the histones prevent part of the bound polymerase from engaging in transcription. Although the situation is most likely more complex, the double reciprocal plots have been used by numerous investigators to compare the template activities of chromatins isolated from various sources. It was found that the template activity of chromatin can be stimulated or decreased by various polyanions,[985,1115,1241,1242] during compensatory hypertrophy of liver following partial hepatectomy,[722-724,1254-1256] during various changes in the differentiated state of cells such as proliferation,[1257] neoplasia,[939,940] development[27,388,895,937,1258-1260] senescence,[936,1261,1262] spermatogenesis,[360] hematopoiesis,[1263,1264] response to infections,[1265,1266] or drugs such as hormones,[725,726,1121,1191,1267-1276] vitamin D,[1277] and carcinogens.[1146-1150] According to a recent report, chromatin template activity also changes with the diurnal rhythm in mouse liver and perhaps other tissues.[1278] Interestingly, the diurnal variations are unaffected by changes in the diet of animals including starvation. Although it may be connected to the diurnal variations in corticosteroid hormones, the rhythm of templating activity is not completely obliterated by adrenalectomy. The templating activity of chromatin also changes during the cell cycle[547,1257] (probably because of differences in the condensation of chromatin during the G_1, S, G_2, and M periods) and in irradiated animals.

The changed properties of chromatin from old or irradiated animals most likely reflect the modifications of its composition. In chromatin from old animals the ratios of histones to the DNA are different from those found in young individuals, as was determined by chemical analysis, thermal denaturation, and dye binding experiments.[1279-1286] The exact nature of the structural and compositional changes of chromatin associated with senescence is presently unknown. Effects opposite to those associated with aging have been seen in tissues from irradiated animals. Radiation (x- or γ-rays, 600 R, or more of total body dose) results in a decreased association of histones (especially the lysine-rich fractions) with chromatin DNA.[1287-1291] This labilization of DNA-histone bonds in chromatin after irradiation, among other methods, can be easily detected by the increased sensitivity of chromatin histones to be dissociated by heparin or other polyanions.[1292-1296]

In chromatin, the DNA is partially protected from radiation damage by its close association with proteins, especially histones.[1297,1298] A partial dissociation of histones (0.4 to 0.8 *M* NaCl) results in a substantial increase in the radiation damage of chromatin as measured by the extent of the release of histones by radiation. The loosening of the histone interaction with DNA probably arises from the partial deamination of the basic amino acids.[1290] Lloyd and Peacocke[1299] have established that the dissociation of histones from DNA by γ-radiation is probably the result of a single hit process, while the degradation of DNA is a two-hit process. A combination of these two processes may explain the radiation-induced decrease in the molecular weight and in the radius of gyration of the nucleohistone. The release (dissociation) of histones from the irradiated chromatin could then occur as a result of the breakage of one of the DNA strands with a consequent disorganization of the helix in the vicinity of the break.

Qualitative Restriction

Beside the quantitative changes of the templating activity of chromatin which result from the removal of histones, alterations in the base composition of RNA newly synthesized on chromatin templates devoid of histones were reported in the literature. In general, the RNA templated by such modified chromatin resembled the DNA in its base composition.[645,664] This indicated that the sequences of nucleotides on the DNA which were previously covered by histones differed in their base ratios from those originally available for transcription. More sophisticated information indicating the existence of a specificity of DNA transcription during various physiological conditions of eucaryotic cells came mainly from studies on the hormonal stimulation of RNA synthesis in target tissues, especially polytene chromosomes. It was found that the administration of a steroid hormone, ecdysone (the molting hormone), to the larvae of certain Dipteran species induced selected parts of their polytene chromosomes to become activated with respect to RNA and protein synthesis. This activation could be directly observed under the light microscope by the appearance of extended zones or bands of acute RNA synthesis on the chromosomes (puffs). During normal life, the appearance of specific puffs was associated with the developmental schedule of the animal and any interference with this temporal sequence affected further differentiation.[1300] It was shown that the formation of certain puffs could be correlated to the appearance of specific proteins, whose biosynthesis was most likely coded by the RNA produced in these puffs.[1301] The involvement of histones in the puffing process was indicated by the unfolding of the tightly coiled chromosome regions in the related chromomeres. It was known that such an unfolding was a prerequisite for the activation of genetic loci. Since, according to the current models of chromosome structure, the supercoiling of the chromatin fibers in chromosomes most likely resulted from the association of DNA with histones, a weakening of such bonds could lead to a partial uncoiling. That a complete dissociation of histones did not take place was indicated by the nearly constant ratio of DNA and histones in the puffs and in the condensed parts of the polytene chromosomes.[126,1000] This was in agreement with the findings that chromatin could, in the presence of urea, be unfolded into the basic fiber structure of approximately 40 A diameter without a complete dissociation of histones.[946] In addition to the RNA content, considerable amounts of proteins accumulated in the puffs. The accumulation of the nonhistone proteins appeared to be the first event in the appearance of puffs induced by ecdysone.[1302] According to Berendes,[1303,1304] the proteins began to accumulate in the puff before the increase of RNA synthesis. As RNA synthesis proceeded, the protein content of the puff decreased. Labeling experiments indicated that these proteins were not synthesized in the puffs, but were transported to the puff from the site of their biosynthesis elsewhere in the cell.[1303-1305] It seems possible that the nonhistone proteins locally destabilized the supercoiled structure of the chromosome and allowed previously inaccessible DNA sites to become available for transcription. However, this mechanism would require an extremely selective interaction of protein(s) specific for each puff with the chromatin structures localized at the sites of future puffing. On the other hand, these proteins might be associated with the transport of RNA transcribed in the puff to the cytoplasm.[1306-1308]

A change in the transcriptional specificity of chromatin similar to that observed in polytene chromosomes after ecdysone administration can be produced in vertebrate tissues by certain hormones. The target tissues respond to hormonal stimulation, in part, by the biosynthesis of new species of RNA[1121,1272,1309-1317] which, in turn, code for new protein species.[1121,1272,1318-1323] If histones serve as gene repressors and reversibly inhibit RNA transcription in chromatin, the selective stimulation of the biosynthesis of specific RNA species resulting from the administration of hormones should produce changes in the biosynthesis or other properties of histones. In rat uteri, estrogen treatment of the animals causes an increase in the labeling of nonhistone proteins[1106,1319,1324] which partly occurs within 30 min after the administration of hormone.[1319] The injection of cortisone into rats induces a rapid synthesis of a nonhistone nuclear protein in the rat livers which is characterized by a single band in polyacrylamide gel electrophoresis.[1323] The enhanced nonhistone protein synthesis observed in stimulated rat uteri occurs only secondary to the RNA synthesis which occurs within 15 min after the administration of estrogen.[1322] This RNA appears to accumulate as soon as the hormone enters the nucleus. The function of this rapidly synthesized RNA is unknown. Perhaps it is the "activator RNA" necessary, according to the scheme of Britten and Davidson,[1325] for the activation of producer genes. While many investigators confirm the increased biosynthesis of nuclear nonhistone proteins to be one of the changes produced by hormones administered to the animals, histone biosynthesis appears to be unaffected[1106,1268,1326-1328] or at most decreases slightly.[1106,1329]

Barker[1330] found a slight decrease of histones in the uterine chromatin of estradiol stimulated rats that involves specifically the F3 (III) histone fraction, which decreases within 15 min after the hormone administration to 58% of the control values. The loss of F3 (III) histone appears to be caused by its complexing with a nonhistone protein which is rapidly biosynthesized after the administration of the hormone. It is not known whether the complexing of F3 (III) histones with this nonhistone protein results in the derepression of the DNA genes originally inhibited by the F3 (III) histones. This protein may be similar or identical to the fraction found in uterine chromatin by Teng and Hamilton[1106] which is capable of reversing the in vitro inhibition of RNA synthesis caused by added histones. The biosynthesis of the nonhistone protein which associates with the F3 (III) histones is resistant to inhibition with actinomycin D.[1330] It is noteworthy that Mueller et al.[1331] predicted the presence of a protein responsible for the activation of RNA synthesis in the estradiol-stimulated uterus.

An increased biosynthesis of nonhistone proteins in the chromatin of mouse salivary glands was also observed 30 min after the injection of isoproterenol.[566] The DNA and histone synthesis associated with cellular proliferation did not occur until about 20 hr after the injection indicating that the biosynthesis of certain nonhistone proteins may be necessary to activate the S period of the cell cycle. On the other hand, Salganik et al.[1332] did not find any changes in the labeling of rat liver chromatin proteins after injecting the animals with hydrocortisone. A more detailed analysis suggested a possible shift in the chromatin nonhistone proteins from heterochromatic areas to transcriptionally active euchromatic areas. The amount of diffuse chromatin not only increased after hydrocortisone injection, but the diffuse chromatin accumulated three to four times more nonhistone protein than the repressed, condensed chromatin, which actually lost some nonhistone proteins. Murthy et al.[564] and Stellwagen and Cole[526] showed that the corticosteroid hormones had no significant effect on the biosynthesis and the amount of histones in rat liver chromatin or in mammary glands. There was an exception, however, in that the very lysine-rich histone F1 (I) subfractions were found to change their biosynthetic rates disproportionally to each other in hormone stimulated mammary gland explants.[550,551] The biosynthesis of F1 (I) histones as well as the in vivo chemical modifications of these proteins were discussed in previous chapters. Since the F1 (I) histones may be involved in the maintenance of chromatin superstructure, the altered biosynthesis of F1 (I) histones resulting from the administration of hormones might indicate the necessity for a conformational modification of chromatin in the mechanism of action of certain hormones. An interesting observation was recently reported by Katz et al.[1333] who studied the stimulation of transcriptional activity in nuclei from ovariectomized rats by estradiol-progesterone or trypsin. The overall extent of the stimulation was similar under both experimental conditions. If, however, the estradiol-progesterone stimulated nuclei were exposed to trypsin, very little additional stimulation of RNA synthesis could be detected. Apparently, the chromatin DNA sites stimulated by the hormone treatment were those which were also sensitive to trypsin. These results resembled those reported by Roberts and Kroeger[1042] who found that the histones in the puffs of polytene chromosomes were more sensitive to tryptic proteolysis than the histones in the condensed, nonpuffing chromosome regions. Possibly, the apparent competition of the estrogens and trypsin for similar sites in the uterine chromatin only indicated that these sites were less condensed relative to the remaining chromatin regions.

Considerable evidence has been presented which demonstrated the formation of molecular hybrids between the DNA strands of different species as well as between single DNA strands and homologous species of RNA.[1334,1335] Although the extreme complexity of the eukaryotic genome limits the usefulness of DNA-RNA hybridization, the technique nevertheless offered a sensitive tool for studies on the biochemical nature of the genetic restriction of the DNA in chromatin. In the presence of increasing amounts of competing RNA species, it has been shown by hybridization assay that the hybridizable RNA templated from isolated chromatin is tissue specific (Figures 52 and 53).[1336] This specificity was not substantially altered by the isolation of chromatin since RNA species synthesized in vitro were indistinguishable from those produced in vivo by the corresponding tissues (Figure 54). Pure DNA served as a more efficient template for the in vitro RNA synthesis as compared with isolated chromatin and, in addition, there were many more RNA species templated by pure DNA than by any of the isolated chromatin samples. The species origin of the polymerase enzyme did not affect the specificity of transcription (Figures 52 and 53). It was concluded that the DNA in chromatin is restricted and this restriction is tissue or organ-specific.[47,739,895,1336-1339]

The restriction of DNA in chromatin is mediated by proteins, especially histones, since the removal of protein from chromatin resulted in RNA transcription very similar or identical to that templated by pure DNA.[740] It should be realized, however, that the time allowed for hybridization in all the experiments discussed here was much too short to permit hybridization of the RNA templated from the unique nucleotide sequences in DNA. This, unfortunately, limited the information obtained to the repetitive or similar RNA species which could be hybridized within the usual hybridization time. According to Paul and Gilmour,[740] the acid removal of only the histones resulted in chromatin which did not show the full template activity of DNA, although a considerable number of new RNA species could be synthesized in vitro from this material. Artificial nucleohistone formed by the in vitro reconstitution of DNA and histones was completely restricted with no templating activity. If the reconstitution was made with histones and dehistonized (acid-extracted) chromatin instead of pure DNA, the resulting

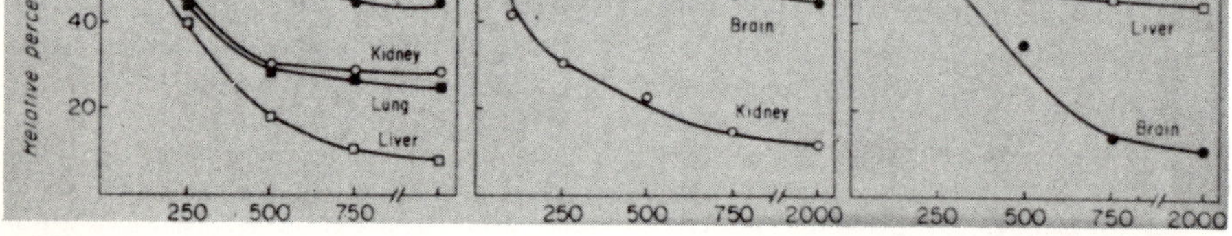

FIGURE 52. Tissue specificity of RNA synthesized from chromatin originating from various mouse organs using *E. coli* RNA polymerase. (A) RNA was synthesized using a mouse liver chromatin template and *E. coli* RNA polymerase. The reaction was run at 37° for 10 min and the RNA was then extracted with hot phenol. RNA was extracted from various tissues and used as competitor for the hybridization of the in vitro synthesized RNA to DNA. Hybridization was with 12 μg of DNA in 0.4 ml of 2 X SSC at 67° for 18 hr; 1211 cpm was hybridized in the absence of competitor. (B) RNA was synthesized using a mouse kidney chromatin template and *E. coli* RNA polymerase. The reaction was run at 37° for 10 min and the RNA was then extracted with hot phenol. RNA extracted from various tissues was used as competitor for the hybridization of the in vitro synthesized RNA to DNA. Hybridization was with 12 μg of DNA in 0.4 ml of 2 X SSC at 67° for 18 hr; 406 cpm was hybridized in the absence of competitor. (C) RNA was synthesized using a mouse brain chromatin template and *E. coli* RNA polymerase. The reaction was run at 37° for 10 min and the RNA was then extracted with hot phenol. RNA extracted from various tissues was used as competitor for the hybridization of the *in vitro* synthesized RNA to DNA. Hybridization was with 12μg of DNA in 0.4 ml 2 X SSC at 67° for 38 hr; 504 cpm was hybridized in the absence of competitor (Smith et al.[1336]).

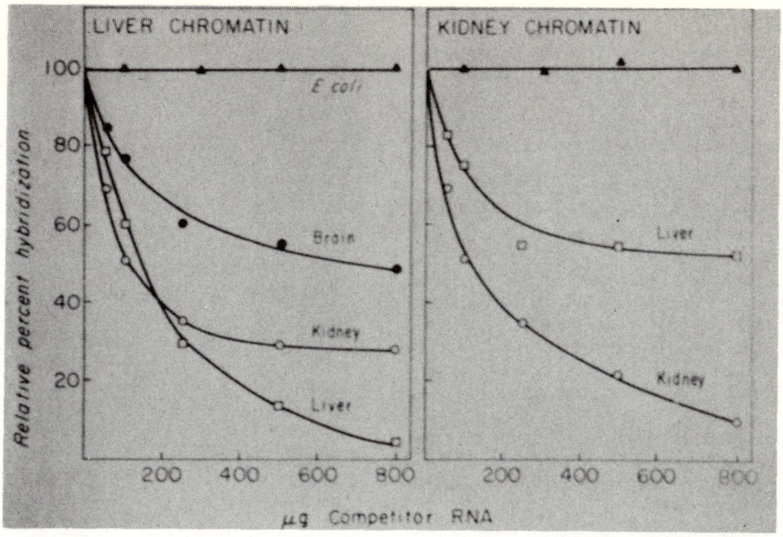

FIGURE 53. Tissue specificity of RNA synthesized from chromatin originating from liver or kidney using mouse RNA polymerase. (A) RNA was synthesized in vitro from a mouse liver chromatin template using mouse RNA polymerase. The reaction was run in 5 ml buffer at 37° for 10 min and the RNA was extracted with hot phenol. RNA isolated from various tissues was used as competitor for the hybridization of the in vitro synthesized RNA to DNA. Hybridization with 12 μg of DNA in 0.2 ml of 2 X SSC at 67° for 18 hr; 384 cpm was hybridized in the absence of competitor. (B) RNA was synthesized in vitro from a mouse liver chromatin template using mouse RNA polymerase. The reaction was run in 5 ml buffer at 37° for 10 min and the RNA was extracted with hot phenol. RNA isolated from various tissues was used as competitor for the hybridization of the in vitro synthesized RNA to DNA. Hybridization was with 12 μg of DNA in 0.2 ml of 2 X SSC at 67° for 18 hr; 336 cpm was hybridized in the absence of competitor.[1336]

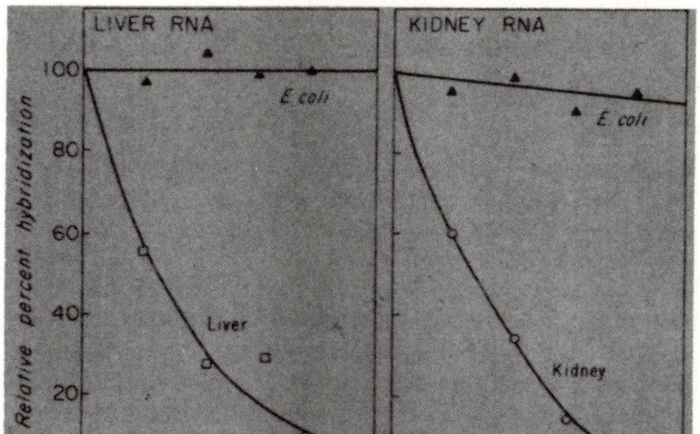

FIGURE 54. Competition of RNA synthesized in vitro from chromatin in the hybridization reaction of RNA labeled in vivo. (A) ^{32}P liver RNA (5μg) labeled in vivo was hybridized with 12 μg of DNA in the presence of increasing amounts of RNA synthesized in vitro with liver chromatin. Hybridization was in 0.2 ml of 4 X SSC for 18 hr at 67°; 520 cpm was hybridized in the absence of competitor. (B) ^{32}P kidney RNA (5 μg) labeled in vivo was hybridized with 12 μg of DNA in the presence of increasing amounts of RNA synthesized in vitro with kidney chromatin. Hybridization was in 0.2 ml of 4 X SSC for 18 hr at 67°; 620 cpm was hybridized in the absence of competitor.[1336]

complex exhibited a template activity which was as quantitatively and qualitatively restricted as was the original chromatin. Similarly, if all the proteins were dissociated from the DNA in chromatin and then reconstituted to the DNA, the reconstituted chromatin templated for RNA species very similar to those templated by the original, undissociated chromatin (Figure 55). If chromatin nonhistone protein isolated by centrifugation in 4 *M* CsCl (pellicle protein in Figure 55) was present in the histone and DNA mixture during the reconstitution, the complete restriction of DNA by histones was prevented. Paul and Gilmour[740] concluded that although histones are essential to restrict the DNA quantitatively, their effect is not specific. Other macromolecules, most likely nonhistone proteins, were necessary to obtain the ordered DNA-histone association which results in the tissue specific restriction of DNA in chromatin. That the specificity of restriction resided with the nonhistone protein complex of chromatin was made apparent when these proteins were replaced with bovine serum albumin and

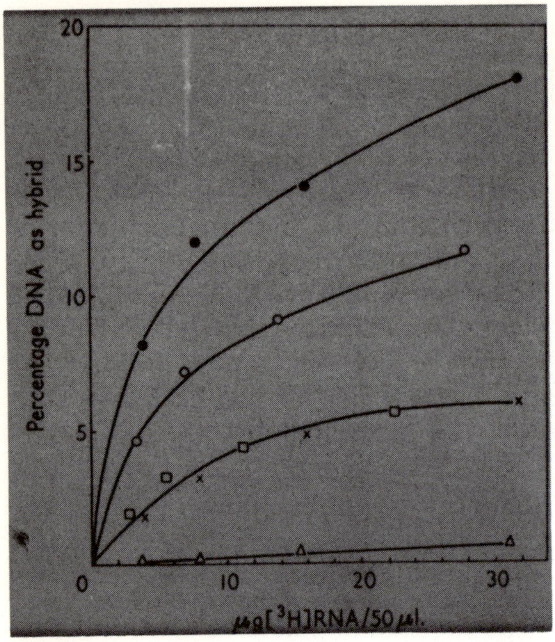

FIGURE 55. Hybridization to DNA of ^{3}H-labeled RNA preparation made from DNA (● – ●); whole chromatin (X – X); DNA + nonhistone protein fraction (○ – ○); DNA + histone (Δ – Δ); DNA + nonhistone protein + histone (□ – □). The hybridization was carried out at 67° C for 16 hr with 5 μg of denatured calf thymus DNA per filter. The chromatin, nonhistone proteins and histones were also from calf thymus.[1153]

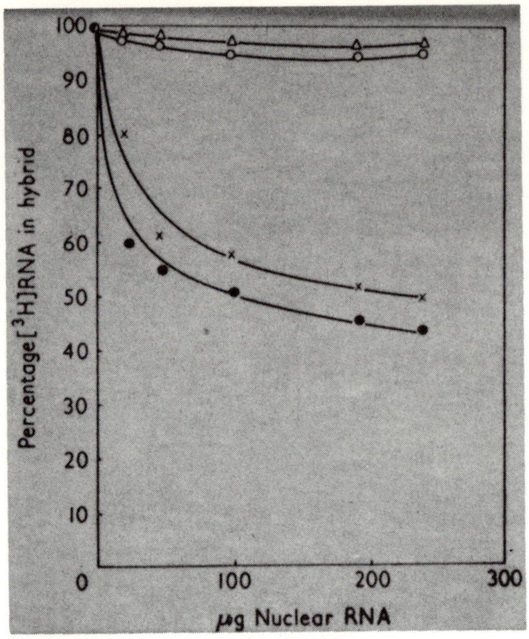

FIGURE 56. Competitive hybridization between natural calf thymus nuclear RNA and ^{3}H-labeled RNA preparations synthesized in vitro from DNA (○ – ○); natural (whole) chromatin (X – X); DNA + histone + nonhistone protein (● – ●); DNA + histone + bovine serum albumin (Δ – Δ). 0 to 240 μg unlabeled natural RNA were added to hybridization incubations containing 50 μg each of ^{3}H-labeled RNA in the presence of 5 μg denatured calf thymus DNA. The total volume of each hybridization mixture was 50 μl.[1153]

reconstituted with histones to the DNA and the resulting nucleohistone, although transcriptionally active, supported the in vitro biosynthesis of nonspecific and random RNA species (Figure 56).

The nature of in vitro RNA templating by chromatin samples devoid of certain histone fractions which had been selectively removed by salt extraction was studied by Spelsberg and Hnilica.[182,1340] As shown by both rate studies of the in vitro RNA synthesis and DNA-RNA hybridization, the dissociation of any histone fraction other than F1 (I) resulted in a partial loss of the transcriptional specificity of the dehistonized chromatin. Selective dissociation of the F1 (I) histones did not increase the template activity of isolated chromatin and did not result in the biosynthesis of new RNA species (Figure 57). This seemed to indicate that the F1 (I) histones either did not participate in the tissue specific restriction of DNA in chromatin or that the transcription of DNA segments covered by these proteins could not be detected under the conditions used in this experiment. Removal of other lysine-rich histones, mainly the F2b (IIb2) fraction, considerably increased the number of RNA species templated by the treated chromatin (Figure 57). The selective removal of the lysine-rich fractions either with NaCl or $(NH_4)_2SO_4$ solutions yielded comparable results. When the arginine-rich histones F3 (III) and F2a1 (IV) were selectively removed, additional RNA species were transcribed in vitro from the residual chromatin (Figure 58) as compared with the original sample. This observation was in agreement with the conclusion of other investigators that all the histone fractions [perhaps with the exception of F1 (I) histones] contributed to the physicochemical and biological properties of chromatin.[1245] Similar to the reports of Paul and Gilmour,[740] the removal of all of the histones did not result in a complete derepression of the DNA. Nevertheless, chromatin that was practically devoid of all histones templated for considerably more species than any of the partially dehistonized samples.[182,740,1253]

Aside from the small contribution of the

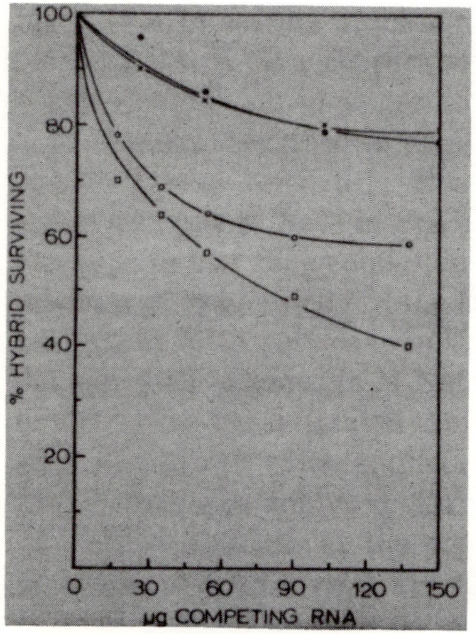

FIGURE 57A.

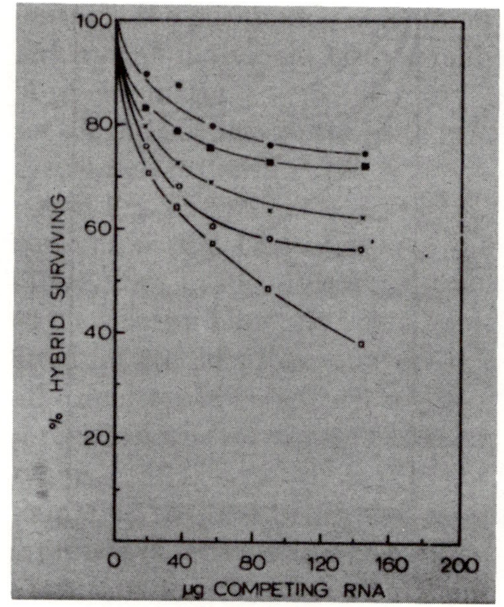

FIGURE 57B.

FIGURE 57. The effect of removal of lysine-rich histones of RNA synthesis in vitro. (A) Competitive hybridization using RNA templated from rat thymus chromatin which has been extracted with water (• – •). 0.5 *M* NaCl (X–X) and 2.0 *M* NaCl (○ – ○), and RNA templated from pure rat DNA (□ – □). [^{3}H] RNA templated from pure rat DNA was used at 50 μg per reaction with 2 μg rat DNA per filter. (B) Competitive hybridization using RNA templated from rat thymus chromatin which has been extracted with water (• – •), 0.2 *M* $(NH_4)_2SO_4$ (■ – ■), 0.3 *M* $(NH_4)_2SO_4$ (X–X), 1.0 *M* $(NH_4)_2SO_4$ (○ – ○), and RNA templated from pure rat DNA (□ – □). [^{3}H] RNA templated from pure rat DNA was used at 50 μg per reaction with 2 μg DNA per filter.[1340]

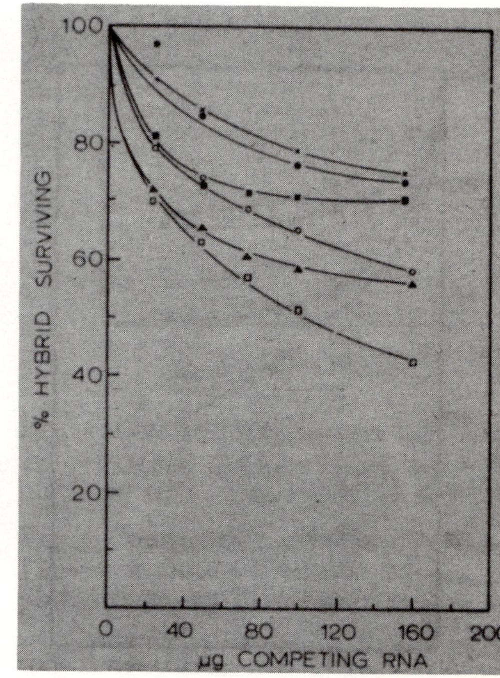

FIGURE 58. The effect of removal of arginine-rich histones on RNA synthesis in vitro. Competitive hybridization using RNA templated from rat thymus chromatin which has been extracted with water (• – •), 75% ethanol (X–X), 75% ethanol + 2% NaCl (■ – ■), 75% ethanol + 10% guanidine-HCl (○ – ○), 75% ethanol-0.25 *M* HCl (Δ – Δ), and RNA templated from pure rat DNA (□ – □). [^{3}H] RNA templated from pure rat DNA was used at 50 μg per reaction with 2 μg rat DNA per filter.[1340]

nonhistone proteins to the quantitative restriction of DNA in chromatin, it was obvious that histones were almost entirely responsible for preventing the transcription of genes inactivated by cytodifferentiation. When alone, histones combine with the DNA in a random fashion and produce nucleohistones without any tissue specificity for hybridizable RNA. Since the selective removal of histone fractions from isolated chromatin resulted in the derepression of qualitatively different segments of DNA, the in vivo specificity of the DNA-histone interactions must be directed by other molecules. In an attempt to elucidate the biochemical nature of substances participating in tissue specific DNA restriction in chromatin, Bekhor et al.[739] studied the necessity of using urea in the reconstitution of chromatin. If chromatin was dissociated in 2 *M* NaCl and reconstituted by gradient dialysis, its tissue specificity as a template for in vitro RNA synthesis was lost. Dissociation with 2 *M* NaCl in 5 *M* urea followed by a gradual removal of NaCl in the presence of urea and a subsequent removal of the urea by dialysis against urea-free buffer resulted in reconstituted chromatin samples similar to the native material (based on a comparison of the hybridization properties of the RNA species templated by native and reconstituted chromatins). The necessity for the presence of urea during the reconstitution of chromatin was confirmed by other investigators.[47,1253,1339,1341]

Although the findings of Paul and Gilmour[740,1253] point to a possible role of nonhistone proteins in maintaining the differentiated state of chromatin restriction, Bekhor et al.[739] and Huang and Huang[1339] proposed that a special class of RNA molecules, i.e., chromosomal RNA, was responsible for the tissue specific reconstitution of chromatin in the presence of 5 *M* urea. This RNA of a peculiar base composition was able to be extracted together with the proteins of chromatin during its salt dissociation. It appeared to be firmly bound to a nonhistone protein fraction,[1339] probably through the serine residues of these proteins.[1342] Bekhor et al.[739,1343] suggested that the necessity of urea for the reconstitution of native-like chromatin might indicate the involvement of the hybridization of chromosomal RNA to specific nucleotide sequences in the DNA. Although this proposed role of chromosomal RNA offered a relatively simple mechanism for the specific restriction or perhaps protection of DNA genes, it did not explain how the hybridization of the unique nucleotide sequences could be obtained in only several hours at low temperature and in the presence of large quantities of proteins.

More recently, Gilmour and Paul,[1344] using rabbit tissues, and Spelsberg et al.,[47,1341] working with rat tissues, showed that the nonhistone protein fraction of chromatin was essential for the tissue specific RNA synthesis templated by isolated chromatin in vitro. Both groups found that the specificity of the in vitro RNA synthesis detectable by DNA-RNA hybridization depended on the origin of the nonhistone proteins in isolated chromatin. In particular,[47,1341] histones were dissociated from rat liver and thymus chromatin by concentrated NaCl in the presence of 5.0 *M* urea and, after extensive ultracentrifugation of the dissociated samples at pH 6.0, resulted in pelleted sediments containing DNA and most of the nonhistone proteins (the supernatants contained practically all the histones and a small part of the nonhistone proteins). Half of the pellets were resuspended in their respective supernatant fluids for the reconstitution of homologous chromatins. The other half were resuspended with the supernatants from the other tissue, (i.e., the histone from thymus with liver DNA and nonhistone proteins and vice versa). The suspensions were reconstituted (hybrid chromatins) by a slow dialysis and used as templates for the in vitro RNA synthesis. Hybridization analysis of the newly synthesized RNA demonstrated that the nonhistone proteins remaining with the pelleted DNA determined the tissue specific pattern of DNA restriction (Figure 59). In another test, the dehistonized rat liver DNA-nonhistone protein complex was reconstituted with acid extracted, purified calf thymus histone. As shown in Figure 60, the hybrid chromatin of this mixture displayed a restriction pattern similar to a control rat liver chromatin, which suggested that the pattern of tissue specific restriction was determined by molecules sedimenting with the DNA during ultracentrifugation at pH 6.0. Polyacrylamide gel electrophoresis of the proteins contained in this pellet showed a considerably heterogeneous content and the virtual absence of histones. Since no attempt was made to remove RNA from the pelleted proteins, its participation in the

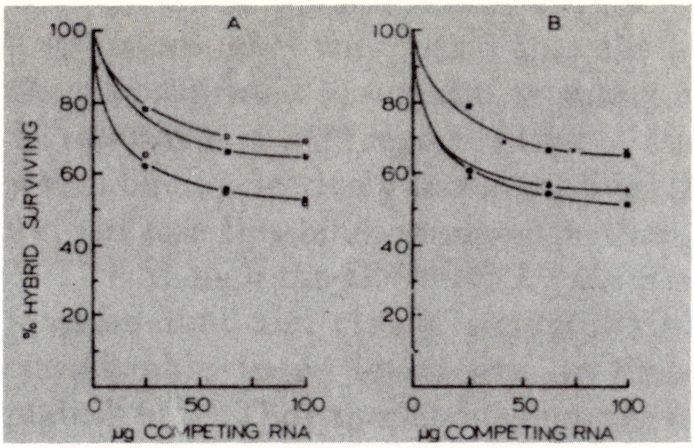

FIGURE 59A.

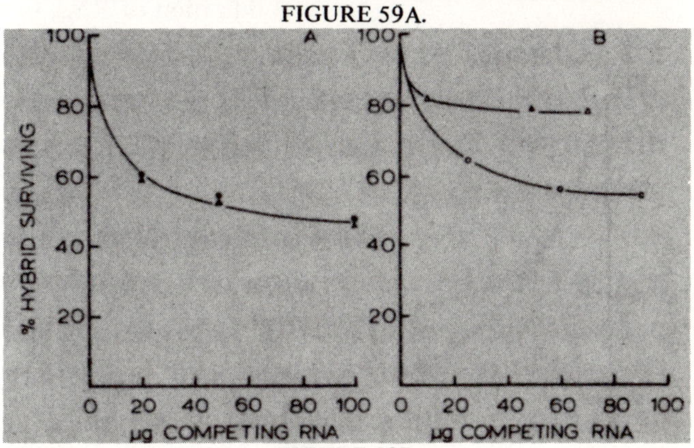

FIGURE 59B.

FIGURE 59. (A) Competition experiment of unlabeled RNA templated in vitro from ● – ●, native liver chromatin; ○ – ○, reconstituted liver chromatin; ■ – ■, native thymus chromatin; □ – □, reconstituted thymus chromatin; ▲ – ▲, hybrid chromatin composed of histone from thymus and the DNA and associated nonhistone protein from liver chromatin; and X – X, hybrid chromatin composed of histone from liver chromatin and the DNA and associated nonhistone protein from thymus chromatin. All reconstitutions were performed by dialyzing from 2.0 to 0.4 *M* NaCl in the presence of 5.0 *M* urea + 0.05 *M* sodium acetate (pH 6.0) for 16 hr followed by the thorough dialysis against 0.05 *M* sodium acetate (pH 6.0). The chromatin was further cleaned by carrying it through the chromatin isolation procedure. In A and B, the unlabeled RNA is competed against 70 µg [^{3}H] RNA templated in vitro from native rat liver chromatin. The 100% of hybridization represents 410 counts/min radioactivity with 4 µg DNA per filter. (B) Competition experiments of unlabeled RNA is templated in vitro from: ● – ●, liver chromatin devoid of all histones [extracted with 2.0 *M* NaCl + 5.0 *M* urea + 0.05 *M* sodium acetate (pH 6.0]; ▲ – ▲, thymus chromatin devoid of all histones; ○ – ○, hybrid chromatin composed of the histone of thymus chromatin and the DNA and associated nonhistone protein of liver chromatin; and △ – △, hybrid chromatin composed of histones from liver chromatin and the DNA and associated nonhistone protein of thymus chromatin. Reconstitution conditions are as described in the legend of (A). In A, the unlabeled RNA is competed against 110 µg of [^{3}H] RNA templated in vitro from dehistonized rat liver chromatin. In B, the unlabeled RNA is competed against 50 µg of [^{3}H] RNA templated in vitro from native rat liver chromatin. The 100% of hybridization represents in A, 2,200 counts/min radioactivity with 2 µg DNA per filter; and in B, 212 counts/min radioactivity with 2 µg DNA per filter.[47]

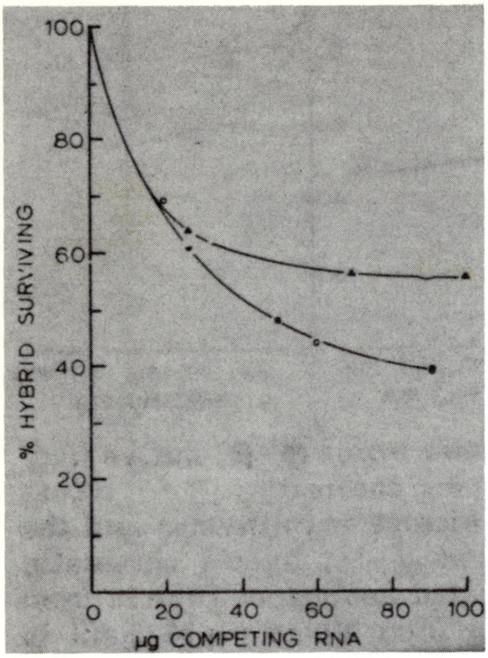

FIGURE 60. Competition experiments of unlabeled RNA templated in vitro from ● – ●, native liver chromatin; ▲ – ▲, native thymus chromatin; and ○ – ○, hybrid chromatin composed of total histone from calf thymus and the DNA and associated nonhistone protein from rat liver chromatin. The unlabeled RNA is competed against 50 µg of ^{3}H-labeled RNA templated in vitro from native liver chromatin. The 100% of hybridization represents 212 counts/min radioactivity with 2 µg DNA per filter.[47]

tissue-specific DNA restriction could not be ruled out. Perhaps a combination of both the nonhistone proteins and RNA might be necessary to obtain a functional chromatin complex. It is noteworthy that many nonhistone proteins contain firmly bound RNA (ribonucleoproteins). Experiments on the reassembly of ribosomes in vitro[1345] showed that a highly cooperative reassociation of RNA and ribosomal proteins could be obtained by reconstitution in the presence of urea and high salt concentrations. The resulting complexes were functional in in vitro protein synthesis. Significantly, the conditions for the reconstitution of ribosomes were almost identical to those used for chromatin. The cooperative and specific binding of RNA and proteins in ribosomes did not, however, postulate the presence of special classes of RNA. As was shown recently by Teng et al.,[792] a group of nuclear nonhistone proteins (phosphoproteins) can bind directly and selectively to the DNA forming transcriptionally stimulated complexes.

Studies on the chemical nature of the macromolecules restricting DNA in chromatin and especially on the biochemical nature of such restriction have been made possible by the development of techniques for the isolation of chromatin in its quasi-native state. Unfortunately, the technique of DNA-RNA hybridization, which is frequently employed to analyze the RNA species templated by isolated and experimentally altered chromatin, yields information only about selected parts of the DNA genome. The molecular hybridization of RNA to denatured DNA was first developed to analyze RNA species in bacteria where the genome is relatively small and the individual RNA species are present in sufficient concentration to form hybrids with its complementary DNA in a reasonable time span. In eukaryotes, the genome is much larger and the individual (unique) RNA species templated by such DNA are numerous and diluted, thereby extending considerably the time required for their hybridization. Although it has been shown recently by several investigators[1346,1347] that it is possible to obtain a selective hybridization of the nonrepetitive (unique) RNA species, this approach has not yet been applied to studies on the role of histones and nonhistone proteins in the restriction of chromatin DNA. The reason why any hybridization of RNA to complementary DNA can be seen in experiments with eukaryotic organisms, is the presence of reiterated DNA sequences in their chromatin. It was first found by Britten and Kohne[1084] that in higher organisms a considerable part of DNA is present in many identical or closely similar copies of nucleotide sequences, duplicated 10,000 to 100,000 or more times. Altogether, this repetitious DNA may account for 20 to 50% of the total chromosomal DNA. As has been discussed earlier in this book, the repetitious DNA is accumulated in specific chromosomal regions, especially in heterochromatin. The presence of repetitious DNA in the eucaryotic genome probably accounts for most of the DNA-RNA hybrids formed between the RNA synthesized in vitro from chromatin templates and the denatured homologous DNA under experimental conditions used for the studies on reconstituted chromatin preparations.

The hybridization data are further complicated by the nonidentity of the repetitious DNA sequences. Closely similar DNA molecules can form hybrids with DNA segments which are not exactly identical. This results in significant mismatching and cross reactions among the DNA segments and their transcribed RNA complements, decreasing the selectivity of the hybridization procedure. Although mismatching can be partly eliminated by more stringent conditions during the hybridization, RNA molecules closely similar in their base sequences are indistinguishable.[1335] Despite all these limitations, the hybridization experiments with RNA transcribed in vitro from isolated chromatin templates show that the repetitive DNA genes can be transcribed and their transcription is specifically restricted in the various tissues of the same animal. Furthermore, the "closely similar" RNA species represent a considerable part of the eukaryotic RNA. The transcription of this RNA appears to be regulated by a coordinated action of the histones and nonhistone proteins, with the latter perhaps preventing the histones from associating with DNA genes designated to remain transcriptionally active.

Chapter 9

THE NONHISTONE PROTEINS

Although a detailed discussion of the nonhistone proteins and other macromolecular components of chromatin is beyond the scope of this book, these proteins do form a functional unity with chromatin DNA and histones and should receive brief attention. For more detailed information, the available reviews of this subject should be consulted.[109,159,545,1348,1349]

Isolation of Nonhistone Proteins from Chromatin

For a long time, the nonhistone proteins of chromatin presented a considerable challenge to biochemists concerned with the characterization of nuclear proteins. These macromolecules did not yield to gentle attempts of their solubilization and were frequently described as a protein fraction soluble in dilute solutions of NaOH or KOH. Furthermore, it was only several years ago that methods for the isolation of chromatin were developed which permitted more meaningful excursions into the biochemistry of chromatin acidic proteins (nonhistone proteins).

Wang[1091,1092] analyzed the nuclear protein residue present after an extraction with 1.0 *M* NaCl which removed most of the lysine-rich histones and some DNA and nonhistone proteins.[176,182] This residue consisted of proteins soluble in 2% deoxycholate. Using isoelectric precipitation, Wang[1091,1092] obtained four fractions which differed in their amino acid composition. All four fractions contained RNA, and interacted strongly with histones.

A different approach to the isolation of nonhistone proteins was developed by Huang and Bonner,[1215] who subjected chromatin dissolved in concentrated CsCl solution to extensive ultracentrifugation. Proteins which are lighter than DNA migrated to the top of the centrifuge tube where they formed a thin layer (pellicle). This method was used by Benjamin and Gellhorn[790] for the isolation of nonhistone proteins from acid-extracted nuclei or chromatin. The nonhistone proteins from rat and mouse nuclei were acidic in amino acid composition, represented approximately 15% of the total nuclear protein content, contained significant amounts of phosphorus, and were heterogeneous in polyacrylamide gel electrophoresis. Dissociation with salt followed by ultracentrifugation was also used by other investigators to isolate the nuclear nonhistone protein fraction.[47,182,323,953,1044,1090,1106,1338,1344,1350-1356]

The salt dissociation of chromatin was further improved by the introduction of 2.0 *M* NaCl in 5.0 *M* urea.[739] At pH 8.0, the supernatant formed after prolonged ultracentrifugation contains practically all the histones and nonhistone proteins, while only about 2 to 3% of the chromatin protein sediments with the DNA.[47,739,1253,1339,1341] In addition to centrifugation, the proteins could be separated from the DNA by gel filtration on Biogel A-50® agarose columns.[1044] If the pH of the 2.0 *M* NaCl and 5.0 *M* urea solution is decreased to 6.0 or less (5.0 or 4.0), a major part of the nonhistone chromatin proteins sediments together with the DNA leaving histones in the supernatant.[47,1341,1357] According to Shaw and Huang,[1044] a practically complete dissociation of the proteins from the DNA can be obtained in 3 *M* NaCl and 7.0 *M* urea. The rapid extraction of histones with acid prior to the solubilization of residual chromatin in 3 *M* NaCl in 7.0 *M* urea did not visibly degrade the nonhistone proteins.[103]

An entirely different technique for the isolation of chromatin nonhistone proteins was developed by Gronow and Griffiths.[1358] Freshly isolated rat liver nuclei were extracted with a pH 7.6 solution of 8 *M* urea + 0.05 *M* sodium phosphate buffer containing *N*-ethylmaleimide. This treatment removed about 70% of the total nuclear protein but none of the DNA. The ethylmaleimide was included to prevent interactions of sulfhydryl-containing proteins. Subsequent electrophoresis of extracted proteins in polyacrylamide gels revealed their considerable heterogeneity; the extraction procedure was well reproducible. A representative electrophoretogram of these proteins obtained by isoelectric focusing is shown in Figure 61.

It was shown in our laboratory that this new technique is applicable to isolated chromatin. Since no detectable modifications of electrophoretic patterns resulting from the interactions of sulfhydryl-containing proteins were observed among nonhistone proteins from chromatin, the *N*-ethylmaleimide was omitted. In a typical ex-

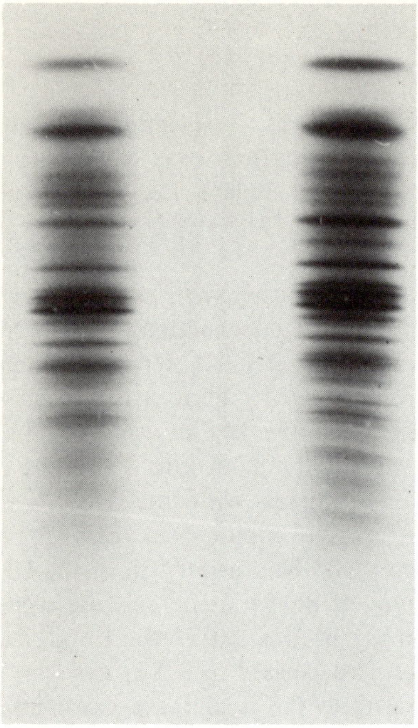

FIGURE 61. Electrophoresis of nuclear nonhistone proteins. Isoelectric focusing of nuclear urea-sodium phosphate buffer extracts in 5% polyacrylamide gel. Highest pH (approximately 9) was at the top of the gels. Each gel contains proteins from an individual rat liver. Approximately 130 μg of protein were applied to each gel and electrophoresed for 5 hr. Coomassie Brilliant Blue staining.[1358]

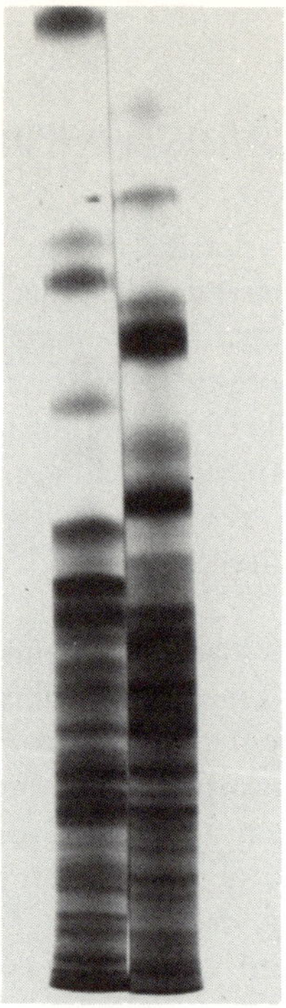

FIGURE 62. Nonhistone proteins extracted from calf thymus (CT) or rat liver (RL) chromatin with 4.0 *M* urea – 0.1 *M* sodium phosphate buffer, pH 7.0. Electrophoresis in the presence of 0.1% SDS, Coomassie Brilliant Blue staining. Origin of migration is at the bottom of the gels. (Courtesy of Carlleen M. Groves.)

traction procedure, isolated chromatin is extracted with 4 *M* urea + 0.1 *M* sodium phosphate buffer, pH 7.0. This procedure removes considerable portions of the nonhistone proteins leaving DNA and most of the histones insoluble. As shown in Figure 62, the proteins soluble in urea-phosphate are highly heterogeneous. After the extraction, histones can be solubilized from the DNA-containing residue by diluted acid.

The use of detergents for the isolation of the nonhistone chromatin proteins was introduced by Marushige et al.,[1359] who first removed histones with 0.2 *N* HCl and then solubilized the residue in buffered (pH 8.0) sodium dodecyl sulfate (SDS). After decreasing the SDS concentration by overnight dialysis, the DNA was sedimented by prolonged ultracentrifugation. Using this simple procedure, Elgin and Bonner[103] compared the nonhistone proteins from the chromatins of several animal and plant tissues and demonstrated their heterogeneity and limited tissue specificity by polyacrylamide electrophoresis. The SDS procedure has been successfully employed for the isolation of nonhistone proteins by several other investigators[104,1357,1360] and in this laboratory. Another detergent, Triton X-100® was also used for the isolation of nonhistone proteins from rat brain and liver nuclei.[1361]

Most recently, Elgin and Bonner[1362] isolated several fractions of chromatin nonhistone proteins in a purity sufficient to warrant their analysis. These authors dissociated isolated chromatin first with concentrated formic acid which was later diluted with NaCl and urea solutions to a final mixture of 25% formic acid, 8 *M* urea, and 0.2 *M* NaCl. The chromatin proteins separated from DNA by ultracentrifugation were then fractionated by chromatography on Sephadex SE C-25® resin and other media. A limited number

(10 to 15) of protein species were detected. The mean molecular weights ranged from 12,500 to 142,070. Two of these proteins were isolated as single purified proteins and designated as fractions θ_1 and ϵ. Their amino acid composition is shown in Table 17.

A special class of nuclear nonhistone proteins are the phosphoproteins. These proteins have several phosphorylated serine or threonine residues per molecule[1364] and are accumulated in the cell nucleus (nuclei of rat liver contain 20 to 25 times more phosphoprotein than cytoplasm[224,1096]). In general, the phosphoproteins can be isolated from the supernatant fluid left after the rapid reconstitution by dilution of chromatin which had been dissociated in concentrated NaCl. Since this rapid reconstitution of chromatin results in a complete loss of its specificity to template for selected species of RNA,[739] the phosphoproteins may play a decisive role in maintaining such tissue specificity. The protein-containing supernatant which remains after the reconstitution is dehistonized and the phosphoproteins recovered by adsorption on calcium phosphate gel (hydroxylapatite).[1090]

Langan[1090] characterized the phosphoprotein fraction from rat liver nuclei. It contained 3.7 to 4.5 phosphate groups per 100 amino acid residues, about 90% of which were esterified to serine residues and only about 10% to the threonine residues. A similar fraction was isolated from calf thymus nuclei by Kleinsmith and Allfrey.[1364] This protein fraction contained a kinase enzyme which, in the presence of ATP and Mg^{++}, could phosphorylate phosphoproteins in vitro. In addition to the kinase, calf thymus nuclei also contain a phosphatase enzyme capable of dephosphorylating the nuclear phosphoproteins. Kleinsmith and Allfrey[1365] suggested that the phosphoproteins may have an important role in modifying chromatin conformation by exchanging their phosphate groups and modifying the overall negative charge of chromatin. This suggestion recently received experimental support. Kamiyama and Wang[1108] isolated a fraction of nuclear nonhistone proteins which activated RNA transcription in isolated chromatin. Further analysis of this fraction indicated that it consists of phosphoproteins and phosphokinase. The activity of phosphokinase could be correlated to the stimulation of in vitro RNA synthesis templated by chromatin.[1109,1110]

TABLE 17

Amino Acid Composition of Purified Nonhistone Proteins from Chromatin

Amino acid	θ_1	ϵ	a	b
Lysine	4.2	**13.3**	6.2	9.5
Histidine	1.3	1.3	1.5	1.3
Arginine	2.7	5.3	5.6	5.2
Aspartic acid	**9.0**	**10.6**	**9.5**	**9.5**
Threonine	5.5	3.7	5.5	5.0
Serine	**13.3**	9.1	7.5	7.6
Glutamic acid	**12.8**	**17.2**	**12.4**	**14.0**
Proline	6.1	4.8	5.0	4.7
Glycine	**16.4**	9.5	7.5	9.4
Alanine	9.0	8.2	7.6	9.6
Half cystine	0.0	0.0	1.3	ND
Valine	4.2	2.5	5.6	7.3
Methionine	1.5	1.4	1.9	1.8
Isoleucine	2.7	2.8	4.3	3.4
Leucine	4.8	4.5	9.9	7.2
Tyrosine	2.6	2.8	2.6	1.9
Phenylalanine	3.9	2.9	3.9	2.4
Tryptophan	ND	ND	2.0	ND
Acidic/basic	2.7	1.4	1.64	1.46

All values are percent of total moles of amino acids recovered. ND – not determined

θ_1 Rat liver chromatin nonhistone protein fraction θ_1.[1362]
ϵ Rat liver chromatin nonhistone protein fraction ϵ.[1362]
a Homogeneous nonhistone protein from rat liver chromatin.[1359]
b Nonhistone protein released from rat liver chromatin during template activation.

The main features distinguishing each protein are printed in a boldface type.

Another method for the isolation of phosphoproteins is their selective extraction with phenol. This technique was first introduced by Steele and Busch[1353] and considerably refined by Allfrey and his co-workers.[1323,1324] Saline-washed nuclei or isolated chromatin from which the histones have been removed with acid are extracted by buffered phenol; the phenol extract is separated from the aqueous phase containing nucleic acids and the phosphoproteins are recovered by a sequence of dialyses against buffers containing urea, mercaptoethanol, and finally SDS. The soluble phosphoproteins resulting from this procedure are heterogeneous in polyacrylamide electrophoresis and show considerable heterogeneity and tissue specificity (Figure 63).[791,792,1323,1360] Partial fractionation of nuclear phosphoproteins can be achieved by manipulating the pH of buffered phenol used for their extraction (Table 18).[1360] Nuclear phosphoproteins are not limited to the higher animals since they were isolated from the slime mold *Physarum polycephalum*[1366] and from sea urchin embryos.[1367] In both experimental systems differentiation was accompanied by dramatic changes in qualitative and quantitative distribution of the individual protein bands representing the phosphoprotein fraction. Additionally, in rat liver, developing sea urchin embryos, and most likely in other animal species, the cytoplasmic ribosomes and the cytosol fraction both contain considerable quantities of phenol-soluble proteins which are qualitatively (electrophoresis in polyacrylamide gels) different from a similar fraction found in chromatin.

All the discussed techniques for obtaining nuclear nonhistone proteins, perhaps with the exception of nuclear phosphoproteins, do not differentiate between the functionally or structurally specific protein groups. An entirely new concept for the isolation of nonhistone proteins which exhibit affinity for homologous DNA was conceived by Alberts et al.[1368,1369] Since in the bacteria, certain proteins with genetic regulatory functions interact selectively with the DNA (RNA polymerase subunits, various repressor factors, etc.) these authors prepared columns of cellulose containing firmly bound DNA. By slowly passing DNA-free *E. coli* extracts through such an *E. coli* DNA-cellulose column, they isolated specific DNA-binding proteins from these bacteria. Kleinsmith et al.[1356] applied this new technique to eukaryotic systems and demonstrated the existence of nonhistone proteins in rat liver chromatin which associated selectively with rat liver DNA. These proteins, most likely part of the nuclear phosphoprotein fraction, required the presence of urea to associate with the DNA selectively and could be dissociated from their complexes with DNA by elution with 0.6 *M* NaCl. The existence of proteins in eukaryotic chromatin which bind selectively to the DNA of the same species was confirmed by Teng et al.[792] who reconstituted isolated phosphoproteins to DNA in the presence of 2 *M* Nacl and 5.0 *M* urea in Tris-HCl buffer pH 8.0. The specific complexes were isolated by gradually decreasing the salt and urea concentration followed by ultracentrifugation of the DNA-protein mixtures on sucrose gradients. Nuclear phosphoproteins were found to bind only to the DNA of the same species. Similar specificity was observed by Wakabayashi and Hnilica[1370] who used agarose-bound DNA for their studies on the specificity of chromatin proteins.

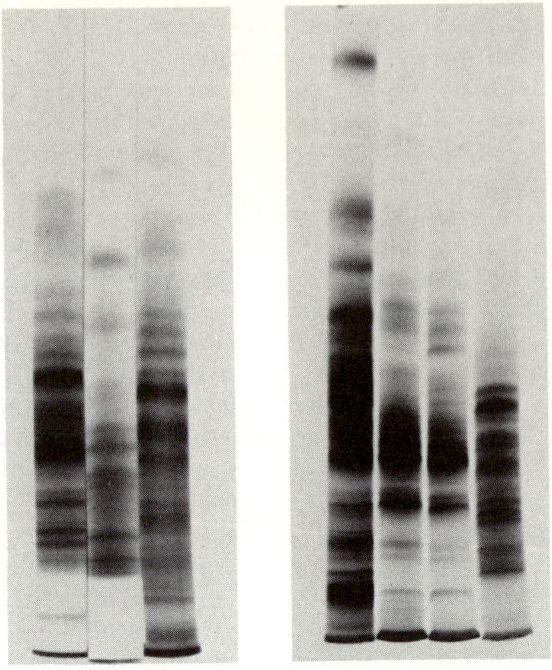

FIGURE 63. A. Phenol-soluble proteins (phosphoproteins) from rat liver (RL), calf thymus (CT) and Novikoff hepatoma (NH) chromatin. The extraction was performed as described by Teng et al.[792] Polyacrylamide electrophoresis in the presence of SDS. Coomassie Brilliant Blue staining. (Courtesy of Carlleen M. Groves.) B. Phenol-soluble proteins from rat liver chromatin (CH), rat liver mitochondrial (MI) and ribosomal (RI) fractions and from 105,000 x g supernatant-cytosol (CS). Conditions of the extraction and electrophoresis were the same as in A. (Courtesy of Judith Thomson.)

If the selective regulation of DNA genes is really determined by specific nonhistone proteins, it can be assumed that these proteins must be tissue specific. On the other hand, there is no obvious need for species specificity of regulatory proteins, especially not in higher animals where homologous tissues are both structurally and functionally alike. It is possible that each parti-

TABLE 18

The Effect of pH on the Phenol Extraction of Nuclear Nonhistone Proteins

	Phenol pH 8.4			Phenol pH 9.4		
Amino acid	Liver	RBC	IRBC	Liver	RBC	IRBC
Lysine	7.0	6.0	6.6	7.6	6.3	6.3
Histidine	2.6	2.3	3.4	2.2	2.2	2.3
Arginine	6.1	6.0	4.9	6.5	7.5	5.9
Aspartic acid	**9.3**	**8.2**	**8.7**	**10.0**	8.7	**8.8**
Threonine	5.0	5.5	5.3	4.7	5.0	4.7
Serine	5.8	6.0	5.8	5.5	5.8	6.1
Glutamic acid	**11.8**	**11.4**	**10.9**	**13.7**	**15.3**	**14.7**
Proline	5.3	5.3	5.0	5.0	4.4	4.5
Glycine	8.2	7.6	7.5	7.5	6.2	6.8
Alanine	8.5	8.9	9.5	8.0	9.1	9.3
Half cystine	1.3	0.9	1.5	1.2	0.7	0.9
Valine	6.4	6.6	6.9	6.0	5.9	6.8
Methionine	1.6	1.7	1.6	1.7	1.4	1.8
Isoleucine	4.6	4.6	4.2	4.3	4.0	4.0
Leucine	**9.5**	**10.7**	**10.9**	**9.4**	**11.4**	**10.7**
Tyrosine	3.2	2.9	2.6	3.1	2.5	2.6
Phenylalanine	3.8	4.1	4.8	3.6	3.5	3.9
Acidic/basic	1.3	1.4	1.3	1.5	1.5	1.6

All values are percent of total moles of amino acids recovered. The proteins were obtained by extracting acid-washed chromatin from goose liver or red blood cells suspended in a mixture of Tris buffer pH 8.4 containing 0.01 *M* EDTA, 1% 2-mercaptoethanol with phenol, saturated with the same buffer. After extraction, the aqueous phase and interface were titrated to pH 9.4 with concentrated NaOH and the phenol extraction was repeated. Details of this procedure are in Reference 1360.

Liver – chicken liver
RBC – mature erythrocytes
IRBC – immature, regenerating erythroid cells

In addition to being different in their amino acid composition, the nonhistone proteins extracted with phenol at pH 8.4 differed electrophoretically from those extracted at pH 9.4.[1360]

Features distinguishing each protein mixture are printed in a boldface type.

cular type of cell possesses a set of factors determining its differentiated state and that such factors changed very little during the evolution of higher animals. Indeed, when compared, the electrophoretic patterns of chromosomal nonhistone proteins showed much larger differences between individual tissues in the same animal than between homologous tissues of different species.[103] As evidenced by nuclear transplantation experiments, substances exist in the cytoplasm, most likely proteins, which are capable of completely reprogramming the restrictional specificity of the transplanted nucleus.

Using DNA-cellulose columns, Salas and Green[1371] recently detected a differential synthesis of proteins with an affinity for DNA during the cell cycle of mouse embryonic fibroblasts 3T3 and 3T6. The biosynthesis of one specific protein fraction appears associated with the control of DNA synthesis in these cells. Although this protein may have originated in the cytoplasm, it nevertheless shows the value of the

DNA-column technique for the isolation of cellular proteins with specific regulatory functions. Another successful application of the selective binding of regulatory proteins by homologous DNA was the isolation of r-RNA synthesis inhibitor from maturing *Xenopus* oocytes.[1372] The efficiency of the chromatography on DNA columns can be further improved by binding the DNA covalently to agarose, or by trapping it into acrylamide polymer.[1373,1374] Such agarose-bound DNA was found to bind the DNA polymerase enzyme from HeLa cells about 50 times better than a corresponding DNA-cellulose matrix.

Heterogeneity and Tissue Specificity of Nuclear Nonhistone Proteins

The heterogeneity of nuclear nonhistone proteins was first realized from their NH_2-terminal amino acid analyses,[109,1353] but only the advent of polyacrylamide gel electrophoresis revealed the full scope of their heterogeneity. Using gels containing urea, SDS, and special buffer solutions, several investigators reported numerous electrophoretic bands in nonhistone protein preparations from various tissues.[103,104,790-792,1044,1323,1350,1351,1354-1362,1375-1377] As estimated from their electrophoretic migration in polyacrylamide gels containing SDS, the molecular weights of chromatin nonhistone proteins range between about 10,000 and 200,000 daltons.[103,1362] Marushige et al.[1359] isolated a nonhistone protein fraction from rat liver chromatin whose molecular weight in 0.1% SDS was 14,300 daltons. However, partial removal of SDS caused this protein to form aggregates with molecular weights of more than 100,000 daltons. Therefore, caution should be exerted if patterns of nonhistone proteins are compared which were electrophoresed in the absence of detergent.

As opposed to the histones, the amino acid composition of the nuclear nonhistone proteins shows an excess of dicarboxylic amino acids over the basic amino acids (Table 19). For this reason, the nuclear nonhistone proteins are frequently called nuclear acidic proteins. The tissue specificity of nuclear nonhistone proteins was shown mostly by comparing their electrophoretic patterns in polyacrylamide gels.[103,323,790-792,1350,1351,1355-1357,1360,1361,1366,1367,1375,1376] Generally, the differences between tissues from the same animal are limited to the presence of additional bands or to quantitative variations of corresponding protein bands in various tissues. Interestingly, there are more differences in the nonhistone proteins between the individual tissues of the same animal than between homologous tissues in different species.[103] This seems to suggest that the differentiated state of cells in various animal species is maintained by similar, if not identical factors (proteins). These proteins are tissue specific but most likely not species specific. Qualitative differences among nonhistone proteins from several tissues are illustrated in Figure 64.

The tissue specificity of nonhistone proteins detected by electrophoretic analysis presents a refreshing change from the monotonous similarity of histone patterns in various tissues and animals. However, caution should be exerted in attaching much importance to such findings. Only several years ago reports of tissue and species specificity of histones appeared in the literature which were later explained by either variable aggregation or

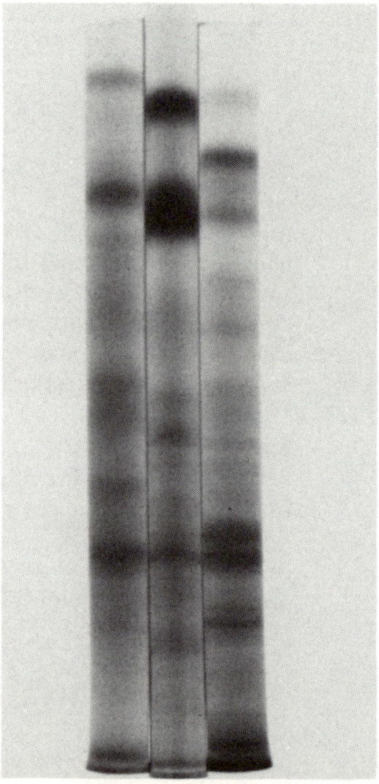

RK CT RL

FIGURE 64. Nonhistone proteins from rat liver (RL), calf thymus (CT), and rat kidney (RK) obtained from acid-extracted chromatin by solubilization in SDS and ultracentrifugation (supernatant) for 36 hr. Polyacrylamide gels containing SDS. Coomassie Brilliant Blue staining. (Courtesy of Carlleen M. Groves.)

TABLE 19

Amino Acid Composition of Various Nuclear Nonhistone Proteins

Amino acid	a	b	c	d	e	f	g
Lysine	9.2	3.4	4.2	6.9	6.0	6.7	7.6
Histidine	0.8	2.5	1.7	1.4	2.5	2.4	2.8
Arginine	3.4	5.6	5.3	3.2	5.9	7.1	5.9
Aspartic acid	**11.8**	**16.9**	**12.1**	**10.1**	**10.3**	**9.5**	**10.5**
Threonine	7.6	7.6	5.5	4.9	3.8	4.9	5.6
Serine	8.4	7.0	**12.5**	**16.9**	4.4	5.3	6.9
Glutamic acid	**10.1**	**11.0**	**14.4**	**17.6**	**12.2**	**12.3**	**11.5**
Proline	4.2	6.0	**10.0**	4.3	6.5	5.1	5.2
Glycine	9.2	6.6	ND	**12.5**	**15.3**	**9.6**	8.6
Alanine	9.2	7.0	9.7	7.7	6.7	7.5	8.3
Half cystine	3.4	ND	ND	ND	ND	ND	ND
Valine	6.7	6.5	5.3	3.6	6.1	6.3	6.4
Methionine	2.5	1.7	1.8	0.8	1.9	2.0	1.4
Isoleucine	3.4	4.1	2.8	2.4	4.1	5.2	4.9
Leucine	4.2	7.6	6.7	4.0	6.8	8.9	8.3
Tyrosine	3.4	2.4	3.2	0.8	3.3	3.2	2.6
Phenylalanine	2.5	4.4	3.2	1.7	4.0	3.6	3.5
Acidic/basic	1.63	2.42	2.36	2.40	1.56	1.34	1.34

All values represent percent of total moles of amino acids recovered. ND – not determined

a c-RNA binding protein from Novikoff hepatoma (integral composition).[1411]
b Alkali-soluble mixture of nonhistone proteins from nuclei of *Strongylocentrotus purpuratus* blastula. The aspartic acid-rich proteins disappear during later development.[311]
c DNA-associated protein material extractable with hot perchloric acid from Ehrlich ascites dehistonized nuclei.[960]
d Phosphoprotein fraction isolated from Novikoff hepatoma chromatin.
e Phenol-soluble proteins (phosphoproteins) from rat liver.[792]
f Phenol-soluble proteins (phosphoproteins) from rat kidney.[792]
g Phenol-soluble proteins (phosphoproteins) from slime mold *Physarum polycephalum.*[1366]

Features distinguishing each protein mixture are printed in a boldface type.

limited proteolysis of these proteins. Johns and Forrester[32] observed that during its isolation, chromatin can bind considerable quantities of acidic proteins, most likely of cytoplasmic origin. While such proteins remain firmly attached in the presence of 0.14 *M* NaCl, increasing the concentration of NaCl to 0.35 *M* removed most of these acidic proteins. Since the washing of chromatin with 0.35 *M* NaCl does not alter its transcriptional properties or its thermostability,[47,182,1357] this treatment is recommended for comparative studies on nonhistone proteins to avoid isolation artifacts. Aware of the possibility of chromatin contamination by cytoplasmic proteins, Smart (quoted in Reference 103) analyzed proteins removed from isolated chromatin by extraction with 0.3 *M* NaCl. In rat liver, this treatment removed approximately 10 to 15% of the chromatin nonhistone content. However, the extracted protein contained the complete electrophoretic protein pattern found for rat liver, slightly enriched in the high molecular weight protein bands. Similar results were obtained by Wilhelm et al.[1377] This indicates that if chromatin is isolated according to the procedure developed by Bonner et al.,[23] or if isolated nuclei are used for chromatin preparation,[1377] the resulting material is not appreciably contaminated by cytoplasmic proteins.

Perhaps the most convincing evidence of tissue specificity of the chromatin nonhistone proteins was recently presented by Chytil and Spelsberg.[1378] Confirming an early report of Henning et al.,[1379] these authors demonstrated considerable immunochemical specificity of the dehistonized or intact chromatin samples from various tissues of the chick. Marked changes of the antigenic specificity of nonhistone protein-DNA complexes (dehistonized chromatin) occurred during the differentiation of chick oviduct,[1380] embryonal rat liver,[1381] and neoplasia.[1382] Since the specific antisera were shown to be directed only against the nonhistone proteins and not the DNA, this indicates that tissue differentiation is accompanied by specific changes in the nonhistone proteins, resulting in changed immunochemical specificity of the differentiated chromatin. Interestingly, the immunochemical specificity of chromatin does not parallel its tissue-specific transcription of RNA. Removal of the chromatin phosphoproteins which completely abolishes the transcriptional specificity of DNA restriction did not affect the immunospecificity of such chromatin.[1382]

Biosynthesis of Nuclear Nonhistone Proteins

Early studies using the incorporation of labeled amino acids into acid insoluble chromatin proteins (nonhistone proteins) as the measure of their rates of biosynthesis showed that in nondividing tissues these proteins are labeled much more rapidly than the nuclear histones (for review see Spelsberg et al.,[1348] McClure and Hnilica,[1383] and Baserga and Stein[1349]). The differences in labeling between histones and nonhistone proteins in chromatin may become quite large in specialized tissues. In chicken erythrocytes, Sadgopal and Kabat[543] found ^{14}C-leucine to be incorporated into the nuclear nonhistone proteins about 35 times more extensively than into the histones. A similar situation exists in maturing rat spermatozoa.

An interesting correlation can be seen between the biosynthetic rates of nuclear nonhistone proteins and gene activation followed by cellular proliferation. Liver stimulated by hormones,[730,1323,1348] phenobarbital,[1385] or partial hepatectomy[168,535] increases its nuclear nonhistone protein and RNA synthesis shortly after the stimulus. Similarly the effect of steroid hormones on their target tissues,[1106,1121,1196,1272,1318-1320,1322,1330,1331,1348,1349,1386] of isoproterenol on mouse salivary glands,[566,1349] and stimulation of various cultured cells[546,549,1349,1387,1388] all result in an early increase in the incorporation of labeled amino acids into the nuclear nonhistone proteins. This early increase precedes the proliferative period characterized by DNA and histone synthesis. Administration of puromycin inhibits this early biosynthetic activity and the affected cells do not proliferate.

As opposed to the histones, whose biosynthesis is tightly coupled to DNA replication during the S phase under normal circumstances, most nonhistone proteins incorporate labeled amino acids during all parts of the cell cycle, including mitosis.[547,567,1389-1392] Indeed, as was pointed out by McClure and Hnilica[547,1389] and by Stein and Borun,[1391,1392] the labeling pattern of the nonhistone proteins in chromatin is approximately inverse of that reported for histones, i.e., their specific activities are proportionally lowest in the mid S period of the cell cycle. In their recent experiments Stein and Borun[1391] showed that the biosynthesis of chromatin nonhistone proteins is insensitive to the inhibitors of DNA synthesis and that electrophoretically different proteins are biosynthesized during the G_1, S, and G_2 phases of the cell cycle. Similarly the turnover rates of pulse-labeled nonhistone proteins separated by polyacrylamide gel electrophoresis in the presence of SDS differed during the individual periods of the cell cycle.[1392] Although it can be anticipated that individual nuclear or chromatin nonhistone proteins will be biosynthesized and utilized at different time points of the cell cycle, the discussed findings represent a good introduction to the selection of protein fractions with specific gene regulatory properties.

In addition to the increased biosynthetic rates preceding cellular proliferation, stimulation of the phosphorylation of nuclear nonhistone proteins was reported in lymphocytes stimulated with phytohemagglutinin,[761] in testosterone stimulated rat ventral prostate nuclei,[1393,1394] and in other systems.[792,1395] As was already mentioned, the nuclear phosphoproteins are tissue specific,[791,792,1323] exhibit sequence-specific interactions with homologous DNA,[792,1356] and show considerable changes, both in their quantitative or qualitative proportions and biosynthetic rates during cellular differentiation[1366] and embryonic development.[1367,1396] The group of nuclear phosphoproteins may indeed prove to be the fraction of nuclear nonhistone proteins containing genetic activator and repressor proteins similar to those found in bacteria.

Teng et al.[792] pointed out the similarities between nuclear phosphoproteins in eukaryotic cells and the σ factors of bacterial RNA polymerases. The selective transcription of DNA genes in bacteria appears to be regulated through the recognition of individual gene initiation (promoter) sites by the σ factors of RNA polymerase molecules.[1213,1396] Interaction between these promoter sites and the σ factors leads to the initiation of transcription of the selected structural genes. After initiation, the σ factor is released from the transcribing enzyme and can interact with another molecule of RNA polymerase. As was already mentioned, Martelo et al.[796] found that the *E. coli* polymerase σ factor can be phosphorylated in vitro by animal phosphokinases. This process was stimulated by cyclic AMP and the phosphorylated enzyme was more active in transcribing DNA than the controls. Based on the resemblance of the σ factor phosphorylation to the phosphorylation of chromosomal proteins in eukaryotes, Teng et al.[792] concluded: "Further developments in this area promise to reveal protein phosphorylation as a general mechanism for influencing the interaction between DNA and proteins which regulate transcription in prokaryotic and eukaryotic cells."

The sites of nuclear nonhistone protein synthesis appear to be largely in the cytoplasm. Stein and Baserga[1398] labeled exponentially growing HeLa-S_3 cells with short pulses of ^{3}H-leucine. The pulses were followed by chases with unlabeled leucine for various time intervals or by transfer into media containing cycloheximide to inhibit the protein synthesis. The labeling kinetics of nuclear and cytoplasmic fractions showed that the bulk of the nonhistone proteins is made in the cytoplasm and rapidly transferred into the nucleus. Essentially identical conclusions were reached by Wu and Warner,[1399] who also used HeLa-S_3 cells and leucine (^{14}C) for their experiments. Since histones (for references see Chapter 4) and ribosomal proteins[1400] also are made in the cytoplasm, it appears that the cell nucleus does not biosynthesize the majority of its proteins.

The nature of nuclear nonhistone protein synthesis briefly discussed here points to a different function of these proteins as compared with nuclear histones. While histone biosynthesis is tightly coupled to DNA replication, the nonhistone proteins are most actively labeled outside the S period of the cell cycle. Physiological stimulation of selected cells or tissues results first in the activation of RNA synthesis followed later by a proliferative phase characterized by a wave of DNA replication. The biosynthesis of nuclear nonhistone proteins increases considerably in the early portion of this schedule, during the RNA synthesis. The histones, on the other hand, are biosynthesized most actively later, when the DNA replicates. If the nonhistone proteins of chromatin play a role in maintaining the restricted state of DNA in differentiated cells, they most likely function as variable repressors or activators as contrasted with the permanently restricting histones.

The recent explosion of interest in nuclear nonhistone proteins brings about a continuously increasing stream of publications dealing with various aspects of these interesting macromolecules. It should be realized, however, that much erroneous information was accumulated concerning the much simpler and less heterogeneous histones. The formidable complexity of the nonhistone protein fractions in the cell nucleus together with the difficulties associated with their isolation and fractionation cautions against premature conclusions concerning the tissue specificity, heterogeneity, and possible functions of these proteins.

Chapter 10

CONCLUSIONS AND PERSPECTIVES

Despite the more than two decades of intensive research which resulted in the publication of several thousand pertinent scientific papers, the question of the biological function of histones cannot yet be answered with certainty. Although experiments with reconstituted DNA-histone complexes, as well as with isolated chromatin, point to the histones as repressors of DNA genes, their simple structure, lack of tissue specificity, and their inability to interact selectively with the DNA indicate the need of additional factors modulating the repressive action of histones. The recent elucidation of the amino acid sequences of four of the five main histone fractions present in eukaryotic cells made possible meaningful investigations on how the histones and DNA affect each other during their interactions. As was indicated by the irregular distribution of basic amino acids in histones, only certain parts of the histone molecules associate with the DNA, leaving the neutral and hydrophobic portions free to interact with other components of chromatin.[447,475] As dictated by the structural differences between individual histone fractions and by the specific modifications of their amino acids (acetylation, phosphorylation, methylation, thiolation), each kind of histone differs in its modifying effect on the conformation of DNA[918,919,923,1048,1052] and on the thermodynamic stability of the DNA-histone complexes.[903]

It was suggested by Monod et al.[1401] that the inducer-repressor molecules in bacteria are most likely allosteric proteins which should exhibit the following properties:

1. Extreme stereospecificity of the inducer-repressor-operator interactions.[1402]
2. The interaction must be virtually immediate and reversible.[1403]
3. A single gene and, presumably, a single specific macromolecular constituent, the repressor, are responsible for the specificity of the interaction.[1404]
4. Single mutations of this gene abolish the repressor-effector interaction while conserving the repressor-operator activity.[1405]

An allosteric protein possessing two sites, one which binds the operator and the other which binds the (positive or negative) effector, conforms to these requirements. Recent discoveries of specific bacterial repressor proteins and of the mechanisms by which these proteins regulate the transcription of specific genes are in excellent agreement with the postulated properties of repressor-activator molecules. As follows from the chemical nature of histones and from their principally electrostatic interactions with the DNA, histones alone cannot function as repressors of specific genes in eukaryotic cells. More likely they associate with larger segments of the chromatin DNA, perhaps determined by cytodifferentiation to remain transcriptionally inactive. In prokaryotes and primitive eukaryotes in which cells do not differentiate into tissues, typical histones were not demonstrated.

The biological function of histones must be rather general since, with the exception of the Fl (I) histones, these proteins remained virtually unchanged during the evolution of many animal and plant species. It is difficult to escape the impression that the amino acid sequence of each histone fraction is evolutionary fixed and essential for its biological function. Any mutation deviating from the evolutionary stereotype of histones proved lethal to the affected individual. The necessity for the proper histones to be present at the time of DNA replication is manifested by the temporal coincidence of the biosynthesis and turnover of both the DNA and histones in normal cells. Apparently, once assembled, the DNA-histone complex in chromatin remains virtually intact representing a structural and functional unity.[562,570] However, the coupling of DNA and histone synthesis is not absolute, since in certain cellular systems the inhibition of DNA synthesis does not affect histone biosynthesis for several hours. On the other hand, the turnover of residues modifying the already assembled histone molecules (acetylation, phosphorylation, thiolation, and methylation) is extensive and does not (in most cells) correlate directly to the DNA replication.[673,1406]

Experiments using artificial DNA-histone complexes or partially dehistonized chromatin as templates for the in vitro RNA synthesis established that histones effectively inhibit this reaction, perhaps by preventing the RNA polymerase from moving along the DNA backbone.[1244,1407] The transcriptionally inactive heterochromatin contains essentially the same amount of histones as euchromatin which has been shown to be the site of active RNA synthesis.[705,834,904,985] To explain this phenomenon, Frenster[1408] suggested that in heterochromatin, the histones are bound directly to the DNA, while in euchromatin they associate with other macromolecules (RNA, nonhistone proteins, etc.), leaving the DNA available for transcription. Since the histones increase the thermal stability of the DNA in chromatin, the nature of the histone binding in the two types of chromatin should be detectable by analysis of the thermal denaturation profiles of these two chromatin types.

As indicated by DNA-RNA hybridization studies, the tissue specificity of hybridizable RNA biosynthesized in vitro on normal and reconstituted chromatin templates depends on the origin of the nonhistone protein fraction.[47,1341,1344] Although the conditions employed in these experiments did not allow the hybridization of nonrepetitive RNA species, it is quite likely that the tissue-specific transcription of unique RNA is also regulated by the nonhistone proteins or perhaps by their complexes with special RNA. Nevertheless, the histones are essential for the transcription of selected genes, since without them blocking the bulk of the DNA in chromatin, the transcription is nonspecific. It appears as if the operative genes in differentiated cells were protected from associating with the histones. This protection is most likely mediated by the nonhistone proteins. Dehistonized chromatin templates for nonspecific, DNA-like RNA. Restoration of the histones to such chromatin retains only the transcription of selected RNA species specific for the tissue from which the chromatin was isolated. As was shown by Liau et al.,[664] this specificity of protection applies also to the biosynthesis of r-RNA. Isolated nucleoli contain sufficient amounts of endogenous RNA polymerase to biosynthesize r-RNA in vitro. Removal of the histones by trypsin does not inactivate the polymerase enzyme and trypsinized nucleoli biosynthesize RNA which resembles DNA in its base composition. The addition of histones inhibits transcription of the nonspecific DNA and leaves only the r-RNA cistrons available to the polymerase. The RNA biosynthesized by such "reconstituted" nucleoli is again the GC-rich, r-RNA. In addition to demonstrating that in nucleoli the r-RNA cistrons are somehow shielded from the histones, this experiment also shows that the nucleolar (endogenous) RNA polymerase can effectively transcribe any DNA sequences and is not permanently affixed to the r-RNA cistrons.

Additional support indicating that the unique feature of the chromatin nonhistone proteins is in conferring the differentiated state to eukaryotic cells comes from studies on the binding of the hormone-receptor protein complexes to the chromatin of the target tissues. Chromatin from the target tissue of a specific hormone, e.g., estradiol and chicken oviduct, is programmed to receive the complex of the hormone with its cytoplasmic receptor protein. Chromatin preparations from nontarget tissues are unable to bind this complex efficiently.[1201-1203] This binding specificity appears to be determined by a fraction of the chromatin nonhistone proteins, since it can be transferred to the chromatin of nontarget tissues by reconstituting this nonhistone protein fraction to such chromatin preparations.[1380]

Although it is presently unknown how the tissue-specific restriction of DNA genes in eukaryotic cells operates, it can be concluded that the cooperation of at least three components of chromatin is essential for cellular differentiation. The histones are necessary for the restriction of large blocks of DNA which are probably permanently inactivated. The nonhistone proteins, on the other hand, most likely regulate the expression of those specific genes whose transcription is determined during differentiation and who respond to the immediate physiological circumstances of the cell.

Another macromolecular species said to direct the tissue specific transcription of genes in chromatin is chromosomal RNA. Several years ago Huang and Bonner[1409] reported the presence in chromatin of a new RNA species which they named chromosomal- or c-RNA. Although these authors first concluded that the c-RNA is covalently bound to the histones, their later investigations showed that the chromatin nonhistone proteins are the binding site for this DNA.[739,1339,1342,1343,1410,1411] The c-RNA-nonhistone protein complex associates with the histones, presumably by hydrogen bonds, which can be dissociated by guanidine hydrochloride.[23,1410] In chromatin, the c-RNA is resistant to RNase, perhaps because it exists in the form of a hybrid with complementary DNA.

The c-RNA is relatively small, about 40 to 60 nucleotides (approximate sedimentation coefficient 4s) and contains, in addition to the common bases, 5 to 10% of dihydropyrimidine.[1412] It hybridizes to native DNA, it is complementary (by hybridization studies) to the repetitive DNA sequences, and it is species and tissue-specific.[23,739,1339,1343,1413-1416] The c-RNA appears to be unrelated to the t-RNA and other well-defined RNA species. According to recent reports, it may be derived from the rapidly labeled,. heterogeneous high molecular weight RNA (Hn-RNA).[1417] During the compensatory hypertrophy (regeneration) of liver the biosynthesis of new RNA species (rapidly labeled nuclear RNA) can be detected in the remaining liver tissue within 2 hr after hepatectomy.[722,776,1418] According to Mayfield and Bonner,[1419] the appearance of new species of rapidly labeled nuclear RNA (Hn-RNA) is almost immediately followed by an abrupt increase in the nuclear c-RNA (about 4 hr after hepatectomy). The ability of isolated liver chromatin to template for DNA dependent RNA synthesis (exogenous *E. coli* polymerase) also increased, reaching a maximum several hours after the peak of c-RNA, i.e., 8-12 hr after surgery. The peak of DNA and histone synthesis occurs in regenerating rat liver between 18 and 22 hr.[267,535,742] Many properties described for the c-RNA from various tissues suggest that it may play an important role in gene activation. Indeed, the existence of a special class of RNA molecules responsible for activating genes was predicted by Sypherd and Strauss[1420] and by Britten and Davidson.[1325] Since the existence of c-RNA was questioned by several investigators,[1421-1424] confirmatory work from other laboratories is most desirable. It is also necessary to determine whether the chromatin nonhistone proteins which interact with homologous DNA are ribonucleoproteins containing RNA species similar to the c-RNA described by Bonner and his associates. It is noteworthy that a treatment of chromatin which obliterates most if not all of its RNA content also deletes its capability to template for tissue-specific species of RNA.[739,1339] As was already mentioned, possibly the whole complex on nonhistone proteins with covalently bound RNA is necessary to maintain the tissue-specific restriction of chromatin DNA.

Although the function of nonhistone proteins in the nucleus together with chromosomal RNA is quite compatible with the predictions of several models describing gene regulation in eukaryotic cells, the experimental evidence is only circumstantial and open to some criticisms. Among others, the tissue-specific transcription of unique RNA species must be studied, testing the possible regulatory roles of individual ions[1425-1428] and various species of RNA polymerase found in eukaryotic cells.[1429-1431] Undoubtedly, many of the recent discoveries concerning gene regulation in prokaryotes can be applied to the much more complex situation in eukaryotic cells. It should be remembered, however, that the survival of fully differentiated cells in higher organisms frequently depends on precise biochemical coordinations within the individual, limited proliferation, high stability and a relative irreversibility of the differentiated state, etc. These features, which help the cells in higher organisms to live in harmony, are considerable handicaps in the constant struggle for survival of prokaryotes. It may reasonably be expected that the regulatory principles operating in prokaryotes underwent considerable modifications or were abandoned during further evolution. Perhaps the prokaryotes and eukaryotes evolved independently from a common primitive ancestor. Nevertheless, the simple beauty of the operon model substantially influenced the contemporary thoughts of most investigators attempting to construct similar models for gene regulation in the more complex higher organisms.

A highly heterogeneous fraction of high molecular weight RNA (between 10 and 100s) was found in the cell nucleus.[1432,1433] It turns over very rapidly and its base composition (42 to 48% G + C) is unlike ribosomal precursor RNA (60 to 65% G + C).[1432,1434-1437] This heterogeneous Hn-RNA is confined to the cell nucleus.[1433,1438,1439] It may be the precursor of m-RNA or c-RNA and some of its properties are those to be expected for specific regulatory RNA.[1417] The Hn-RNA contains tissue-specific nucleotide sequences and hybridizes mostly to the reiterated DNA.[1439-1444] Assuming that cells do not produce large quantities of useless RNA, the Hn-RNA must serve some useful purpose within the cell nucleus. The hypothesis of gene regulation in higher organisms suggested by Britten and

Davidson[1325] requires the existence of special activator RNAs which are confined to the nucleus. Perhaps at least a part of the HnRNA functions in this capacity.

According to the Britten-Davidson model, five basic elements operate the transcription of selected genes in eukaryotic cells. Genes regulated in this fashion code for specific products (e.g., enzymes, structural proteins, etc.). Individual metabolic sequences frequently require the simultaneous activation of more than one gene. Genes transcribing RNA coding for proteins are termed *producer genes* (structural genes in bacteria). A set of producer genes whose products carry out a closely related set of functions forms a *battery of genes*. In living cells the genome is repressed except for those genes which are selectively activated. The gene activation occurs by transfer and interactions of diffusable regulatory molecules. Hence, the individual genes comprising a battery do not need to be adjacent and can be even located on different chromosomes. The sequence is initiated by the interaction of a selected molecule (*activator*) with specific DNA site, the *sensor gene*. The function of this sensor gene is to recognize a specific inducing molecule (e.g., a metabolite, hormone, etc.) and through its binding to initiate the transcription of producer (structural) genes. Since most of the inducing agents are simple molecules unable to bind specifically to the DNA, the inducing agent most likely complexes first with another molecule, probably a protein, and the complex then binds to the sensor gene DNA in a sequence-specific way. An example of such an initiation complex is the interaction of cytoplasmic receptor proteins with steroid hormones and the subsequent stereospecific binding of such complexes to the chromatin of the target tissues.

The sensor gene is adjacent to an *integrator gene*. The formation of a complex between the activator and sensor gene activates the transcription of the integrator gene which codes for a specific RNA, the *activator RNA*. This RNA carries the information to a *receptor gene* which in turn is adjacent to a *producer* (*structural*) *gene* whose transcription it activates. In other words, the activator RNA binds in a sequence-specific manner (e.g., hybridization between the single stranded activator RNA and the double stranded receptor gene DNA) to the receptor DNA thereby initiating the transcription of the producer (structural) gene. The producer gene transcribes m-RNA coding for specific proteins. Essentially, the sequence consists of two transcriptional units, whose activation results in the biosynthesis of two kinds of RNA, the *activator* and *messenger RNA*. The advantage of this two-step procedure is its greater versatility in the simultaneous activation of several transcriptional complexes (batteries) which contain identical integrator or receptor genes. This schedule is schematically illustrated in Figures 65A and B and, as can be seen, producer genes not sharing a given receptor sequence can be activated simultaneously. In their recent publications, Britten and Davidson extended this model to accumulate evolutionary and developmental changes.[1445,1446]

A similar model was independently conceived by Georgiev.[1447] Although it accommodates the existence of large RNA molecules, special nuclear RNA, alteration of transcription after the integration of viral genome, etc., it offers less versatility as compared with the Britten-Davidson scheme. In this model (Figure 66) the operon in higher organisms consists of a noninformative zone (promoter-proximal acceptor zone) and an informative zone (promoter-distal structural zone). The transcription is initiated by activation of acceptor loci adjacent to the promoter site in the noninformative zone. The activation mechanism is similar to the activator-sensor gene interaction postulated by Britten and Davidson and requires stereospecific recognition by the activator (protein) of certain base sequences on the DNA. There can be more than one acceptor locus in each operon and different operons can contain identical acceptor loci. Hence, one activator (protein) can initiate transcription of several operons, each transcribing different m-RNA species. After transcribing the noninformative zone (the acceptor sites) the polymerase continues to transcribe the informative zone containing a collection of structural (producer) genes. The whole operon is transcribed which results in a very large RNA molecule (D-RNA). After its release, this giant RNA is degraded to smaller fragments, the informative zone producing m-RNA species which are subsequently transferred into the cytoplasm.

As shown by visible and electron microscopy, physicochemical measurements, etc., the structure of chromatin is complex and is determined by the interactions of its components, primarily DNA and histones. Undoubtedly, the structural features of

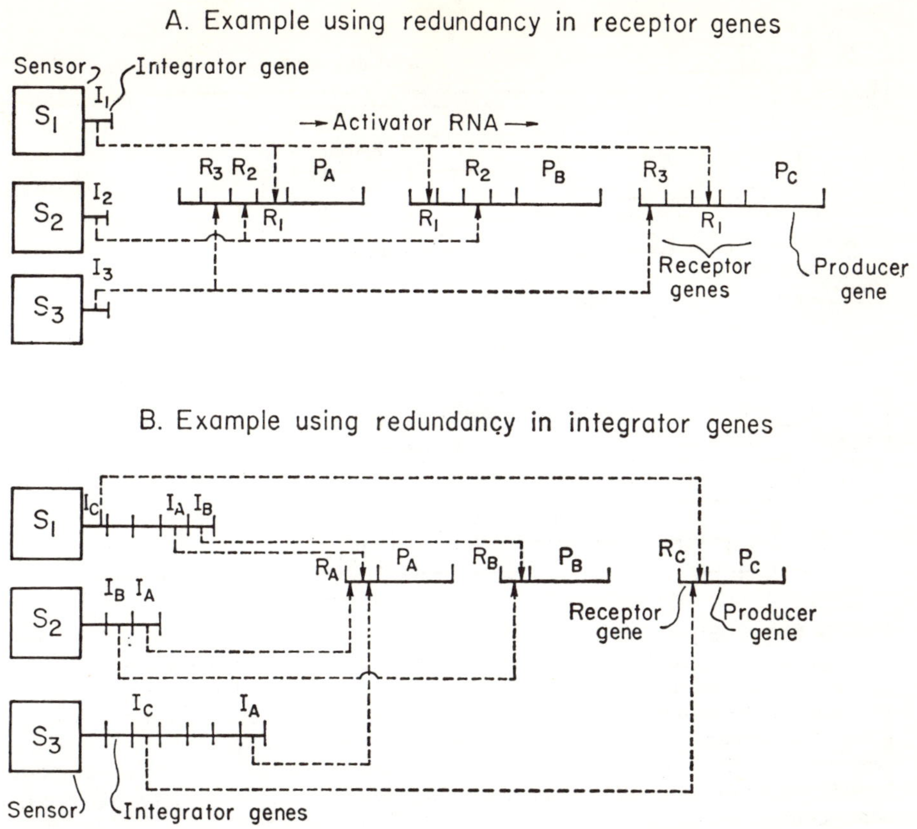

FIGURE 65A. Britten-Davidson model for gene regulation in higher organisms. Types of integrative system within the model. (A) Integrative system depending on redundancy among the regulator genes. (B) Integrative system depending on redundancy among the integrator genes. These diagrams schematize the events that occur after the three sensor genes have initiated transcription of their integrator genes. Activator RNAs diffuse (symbolized by dotted line) from their sites of synthesis, the integrator genes, to receptor genes. The formation of a complex between them leads to active transcription of the producer genes P_A, P_B, and P_C.[1325]

chromatin reflect its function and must be considered in constructing models of gene regulation. It was suggested by Frenster[985,1083,1084,1408] that in transcriptionally active chromatin, histones are displaced from the DNA by nuclear polyanions such as nonhistone proteins. In such displaced areas DNA strands separate and are stabilized by hybridization with de-repressor RNA. The sequence-specific de-repressor RNA hybridizes only with one DNA strand leaving the other free to template for m-RNA synthesis. Although oversimplified, this concept considers the transcriptionally active (euchromatin) and inactive (heterochromatin) forms of the chromosomes in higher organisms, which while structurally different, are quite similar in their chemical composition. A more complicated model involving histone-DNA, histone-histone and histone-nonhistone protein interactions was presented recently by Tsanev and Sendov.[1448]

Another model of DNA organization in the chromatin of higher organisms was recently presented by Crick.[1449] According to this model, the DNA in chromatin is either globular or fibrous. The globular DNA contains control regions in which the strands are unpaired and the fibrous DNA codes for m-RNA. In the globular DNA, a DNA segment (loop) becomes unpaired and available for a regulatory function (promoter sites for RNA polymerase, an operator sequence, initiator sequence for DNA replication, etc.), due to the untwisting effect produced by the interactions of histones with the double helical double helix of the DNA forming the globule. Although according to its author,[1449] this model is speculative, not fully detailed, and raises at least as many

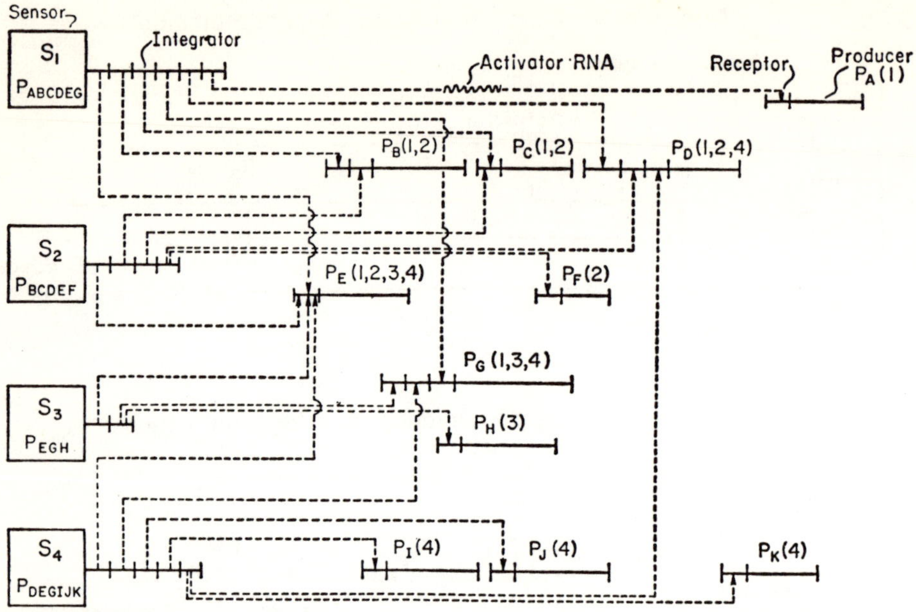

FIGURE 65B. Britten-Davidson model for gene regulation in higher organisms. This diagram is intended to suggest the existence of overlapping batteries of genes and to show how, according to the model, control of their transcription might occur. The dotted lines symbolize the diffusion of activator RNA from its sites of synthesis, the integrator genes, to the receptor genes. The numbers in parentheses show which sensor genes control the transcription of the producer genes. At each sensor the battery of producer genes activated by that sensor is listed. In reality many batteries will be much larger than those shown and some genes will be part of hundreds of batteries.[1325]

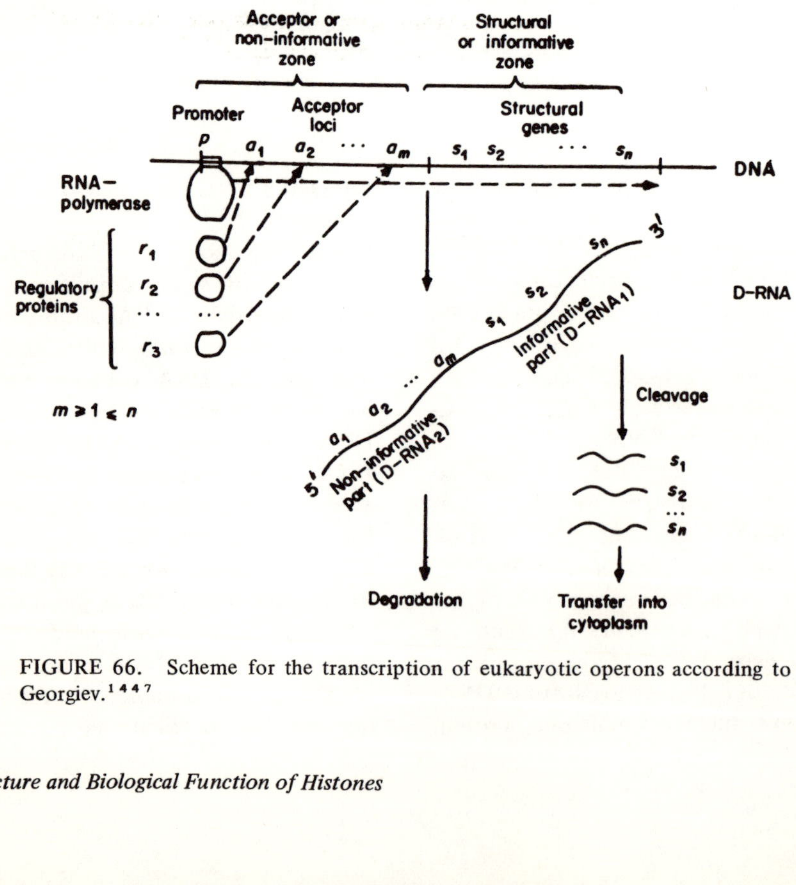

FIGURE 66. Scheme for the transcription of eukaryotic operons according to Georgiev.[1447]

questions as it attempts to answer, it offers a much needed stimulus to interpretations of the operon theory fitted to the greater complexity of higher organisms. The nucleus and cytoplasm form a functional unit in which they dependently influence each other to assure cellular survival and reproduction. As demonstrated by nuclear transplantation experiments,[1450] the specificity of gene restriction in chromatin determined during differentiation is reversible. All the factors necessary for the cell to differentiate are contained in the cytoplasm since the transplanted nucleus assumes biosynthetic activities characteristic of the cell donating the cytoplasm. It may be rewarding to seek macromolecules regulating gene expression in the cytoplasm in addition to the experiments with isolated nuclei and chromatin.

A molecular biologist who is used to working with bacterial systems and who is exposed to the speculations and experiments concerning the biological functions of histones and other nuclear proteins in higher organisms is undoubtedly reminded of gazing into a crystal ball. The nebulous picture is beginning to clear, however, and there is little doubt that when unraveled, the gene regulation in higher organisms will prove to be at least as exciting as that of the bacterial chromosome.

Acknowledgments

The author wishes to express his sincere appreciation for critical reading of the manuscript and numerous suggestions to Dr. Michael McClure, Department of Biochemistry, The University of Texas at Houston, M. D. Anderson Hospital and Tumor Institute. I am also most grateful to Dr. Bruno Jirgensons, Department of Biochemistry and to Dr. A. T. Ansevin, Department of Physics, The University of Texas at Houston, M. D. Anderson Hospital and Tumor Institute, for their comments and reading the chapters concerning the conformation of histones and chromatin, to Dr. Thomas C. Spelsberg, Department of Obstetrics and Gynecology, Vanderbilt University, to Dr. John A. Wilhelm, Department of Biologie Animale, University of Geneva, Switzerland, and to Dr. Alan W. Johnson, Department of Biology, Northern Michigan University for their most helpful discussions and criticism. The invaluable assistance of Miss Susan Getz in the preparation of this manuscript is gratefully acknowledged. I am especially thankful to all my colleagues who allowed me to read their unpublished manuscripts or to reproduce parts of their published research, furnishing in many instances the original photographs for reproduction. Original studies by the author discussed in this book were aided by grants from the United States Public Health Service (CA 07746 and HD 05803), The Robert A. Welch Foundation (G 138), and the American Cancer Society (E 388).

REFERENCES

1. **Luck, J. M.,** Histone chemistry: The pioneers, in *The Nucleohistones,* Bonner, J. and Ts'o, P. O. P., Eds., Holden-Day, San Francisco, 1964, 3.
2. **Miescher, F.,** Die Spermatozoen einiger Wirbelthieren, in *Die Histochemischen und Physiologischen Arbeiten von F. Miescher,* Vogel Vlg., Leipzig, 1897, 359.
3. **Kossel, A.,** *The Protamines and Histones,* Longmans, Green and Co., London, 1928.
4. **Edlbacher, S.,** Albrecht Kossel zum Gedächtnis, *H. S. Z. Physiol. Chem.,* 177, 1, 1928.
5. **Huiskamp, W.,** Über die Eiweisskörper der Thymusdrüse, *H. S. Z. Physiol. Chem.,* 32, 145, 1901.
6. **Huiskamp, W.,** Über die Elektrolyse der Salze des Nucleohistons und Histons, *H. S. Z. Physiol. Chem.,* 34, 32, 1901.
7. **Huiskamp, W.,** Beiträge zur Kenntnis des Thymusnucleohistons, *H. S. Z. Physiol. Chem.,* 39, 55, 1903.
8. **Lilienfeld, L.,** Über Leukozyten und Blutgerinnung, *DuBois-Reymon's Arch. Physiol.,* 4, 167, 1892.
9. **Lilienfeld, L.,** Zur Chemie der Leukozyten, *H. S. Z. Physiol. Chem.,* 18, 473, 1894.
10. **Kossel, A. and Dakin, H. D.,** Über Salmin und Clupein, *H. S. Z. Physiol. Chem.,* 41, 407, 1904.
11. **Kossel, A. and Dakin, H. D.,** Beiträge zum System der einfachsten Eiweisskörper, *H. S. Z. Physiol. Chem.,* 40, 565, 1904.
12. **Kossel, A. and Kutscher, F.,** Bieträge zur Kenntnis der Eiweisskörper, *H. S. Z. Physiol. Chem.,* 31, 165, 1900.
13. **Kossel, A. and Edlbacher, S.,** Beiträge zur chemischem Kenntnis der Echinodermen, *H. S. Z. Physiol. Chem.,* 94, 264, 1915.
14. **Kossel, A. and Staudt, W.,** Zur Kenntnis der basischen Proteine, *H. S. Z. Physiol. Chem.,* 159, 172, 1926.
15. **Felix, K.,** Protamines, *Adv. Protein Chem.,* 15, 1, 1960.
16. **Felix, K., Fischer, H., and Krekels, A.,** Protamines and nucleoprotamines, *Progr. Biophys. Biophys. Chem.,* 6, 1, 1956.
17. **Hogeboom, G. H.,** Fractionation of cell components of animal tissues, in *Methods in Enzymology,* Vol. 1, Colowick, S. P. and Kaplan, N. O., Eds., Academic Press, New York, 1955, 16.
18. **Dounce, A. L.,** The isolation of nuclei from tumor cells, *Exp. Cell Res.,* Suppl. 9, 126, 1963.
19. **Siebert, G.,** Isolation of nuclei, in *Methods in Cancer Research,* Vol. 2, Busch, H., Ed., Academic Press, New York, 1967, 287.
20. **Busch, H., Hodnett, J. L., Morris, H. P., Neogy, R., and Unuma, T.,** Isolation of nuclei and nucleoli of Morris hepatoma cells, in *Methods in Cancer Research,* Vol. 4, Busch, H., Ed., Academic Press, New York, 1968, 179.
21. **Wang, T. Y.,** The isolation and purification of mammalian cell nuclei, in *Methods in Enzymology,* Vol. 12a, Grossman, L. and Moldave, K., Eds., Academic Press, New York, 1967, 417.
22. **Busch, H.,** Isolation and purification of nuclei, in *Methods in Enzymology,* Vol. 12a, Grossman, L. and Moldave, K., Eds., Academic Press, New York, 1967, 421.
23. **Bonner, J., Chalkley, G. R., Dahmus, M., Fambrough, D., Fujimura, F., Huang, R. C. C., Huberman, J., Jensen, R., Marushige, K., Ohlenbusch, H., Olivera, B. M., and Widholm, J.,** Isolation and characterization of chromosomal nucleoproteins, in *Methods in Enzymology,* Vol. 12b, Grossman, L. and Moldave, K., Eds., Academic Press, New York, 1968, 3.
24. **Davison, P. F. and Butler, J. A. V.,** The fractionation and composition of histones from thymus nucleoprotein, *Biochim. Biophys. Acta,* 15, 439, 1954.
25. **Daly, M. M. and Mirsky, A. E.,** Histones with high lysine content, *J. Gen. Physiol.,* 38, 405, 1955.
26. **Ginzburg-Teitz, Y., Kaufmann, E., and Traub, A.,** A one step method for the isolation of nuclei from Krebs-2 tumor cells, *Biochim. Biophys. Acta,* 134, 211, 1967.
27. **Johnson, A. W. and Hnilica, L. S.,** *In vitro* RNA synthesis and nuclear proteins of isolated sea urchin embryo nuclei, *Biochim. Biophys. Acta,* 224, 518, 1970.
28. **Butler, J. A. V., Davison, P. F., James, D. W. F., and Shooter, K. V.,** The histones of calf thymus deoxyribonucleoprotein. I. Preparation and homogeneity, *Biochim. Biophys. Acta,* 13, 224, 1954.
29. **Hnilica, L. S., Edwards, L. J., and Hey, A. E.,** Studies on nuclear proteins. II. Quantitative distribution of histone fractions in various tissues, *Biochim. Biophys. Acta,* 124, 109, 1966.
30. **Allfrey, V. G.,** Nuclear ribsomes, messenger-RNA and protein synthesis, *Exp. Cell Res.,* Suppl. 9, 183, 1963.
31. **Widholm, J. W. and Bonner, J.,** Sedimentaion and Actinomycin D binding studies of partially denatured crab dAT, *Biochemistry,* 5, 1753, 1966.
32. **Johns, E. W. and Forrester, S.,** Studies on nuclear proteins. The binding of extra acidic proteins to DNP during the preparation of nuclear proteins, *Eur. J. Biochem.,* 8, 547, 1969.
33. **Wilhelm, J. A., McClure, M. E., and Hnilica, L. S.,** Studies on the structure and composition of rat chromatin, in preparation.
34. **Johns, E. W., Phillips, D. M. P., Simson, P., and Butler, J. A. V.,** Improved fractionations of arginine-rich histones from calf thymus, *Biochem. J.,* 77, 631, 1960.
35. **Hnilica, L. S.,** The extractibility of calf thymus histone factions, *Experientia,* 15, 139, 1959.
36. **Johns, E. W. and Butler, J. A. V.,** Further fractionations of histones from calf thymus, *Biochem. J.,* 82, 15, 1962.

37. **Dounce, A. L., Seaman, F., and MacKay, M.,** Effects of the pH of isolation on the quantities and properties of histones extractable from calf thymus and rat liver nuclei, *Arch. Biochem. Biophys.,* 117, 550, 1966.
38. **Murray, K.,** The acid extraction of histones from calf thymus deoxyribonucleoprotein, *J. Mol. Biol.,* 15, 409, 1966.
39. **Phillips, D. M. P. and Johns, E. W.,** A study of the proteinase content and the chromatography of thymus histones, *Biochem. J.,* 72, 538, 1959.
40. **Dounce, A. L. and Umaña, R.,** Proteases of isolated cell nuclei, *Biochemistry,* 1, 811, 1962.
41. **Furlan, M. and Jericijo, M.,** Protein catabolism in thymus nuclei. I. Hydrolysis of nucleoproteins by proteases present in calf thymus nuclei, *Biochim. Biophys. Acta,* 147, 135, 1967.
42. **Furlan, M. and Jericijo, M.,** Protein catabolism in thymus nuclei. II. Binding of histone-splitting nuclear proteases to deoxyribonucleic acid, *Biochim. Biophys. Acta,* 147, 145, 1967.
43. **Stellwagen, R. H., Reid, B. R., and Cole, R. D.,** Degradation of histones during the manipulation of isolated nuclei and deoxyribonucleoprotein, *Biochim. Biophys. Acta,* 155, 581, 1968.
44. **Panyim, S., Jensen, R. H., and Chalkley, R.,** Proteolytic contamination of calf thymus nucleohistone and its inhibition, *Biochim. Biophys. Acta,* 160, 252, 1968.
45. **Bartley, J. and Chalkley, R.,** Further studies of a thymus nucleohistone-associated protease, *J. Biol. Chem.,* 245, 4286, 1970.
46. **Panyim, S., Bilek, D., and Chalkley, R.,** An electrophoretic comparison of vertebrate histones, *J. Biol. Chem.,* 246, 4206, 1971.
47. **Spelsberg, T. C., Hnilica, L. S., and Ansevin, A. T.,** Proteins of chromatin in template restriction. III. The macromolecules in specific restriction of the chromatin DNA, *Biochim. Biophys. Acta,* 228, 550, 1971.
48. **Zubay, G. and Doty, P.,** The isolation and properties of deoxyribonucleoprotein particles containing single nucleic acid molecules, *J. Mol. Biol.,* 1, 1, 1959.
49. **Commerford, S. L., Hunter, M. J., and Oncley, J. L.,** The preparation and properties of calf liver deoxyribonucleoprotein, *J. Biol. Chem.,* 238, 2123, 1963.
50. **Hnilica, L. S. and Busch, H.,** Fractionation of the histones of Walker 256 carcinosarcoma by combined chemical and chromatographic techniques, *J. Biol. Chem.,* 238, 918, 1963.
51. **Hirschbein, L. and Rozencwajg, R.,** Action du diisopropylphosphofluorure et du pH dans l'association et la dissociation des histones du placenta humain, *C. R. Acad. Sci. Paris,* 251, 1309, 1960.
52. **Shapiro, R., Cohen, B. I., and Servis, R. E.,** Specific deamination of RNA by sodium bisulphite, *Nature,* 227, 1047, 1970.
53. **Hayatsu, H., Wataya, Y., and Kai, K.,** The addition of sodium bisulfite to uracil and to cytosine, *J. Am. Chem. Soc.,* 92, 724, 1970.
54. **Smillie, L. B., Marko, A. M., and Butler, G. C.,** The separation of calf thymus nucleic acid and histone by alcohol-salt precipitation, *Can. J. Biochem. Physiol.,* 33, 263, 1955.
55. **Bakay, B., Kirschner, L. B., and Toennies, G.,** Dialyzability of histones, *Biochim. Biophys. Acta,* 24, 329, 1957.
56. **Luck, J. B., Rasmussen, P. S., Satake, K., and Tsvetikov, A. N.,** Further studies on the fractionation of calf thymus histone, *J. Biol. Chem.,* 233, 1407, 1958.
57. **Crampton, C. F., Stein. W. H., and Moore, S.,** Comparative studies on chromatographically purified histones, *J. Biol. Chem.,* 225, 363, 1957.
58. **Phillips, D. M. P.,** The histones, *Progr. Biophys. Biophys. Chem.,* 12, 211, 1961.
59. **Hnilica, L. S., Johns, E. W., and Butler, J. A. V.,** Observations on the species and tissue specificity of histones, *Biochem. J.,* 82, 123, 1962.
60. **Lindh, N. O. and Brantmark, B. L.,** Preparation of histones by the use of Reinecke salt, *Anal. Biochem.,* 10, 415, 1955.
61. **Lindh, N. O. and Brantmark, B. L.,** Preparation and analysis of basic proteins, *Methods Biochem. Anal.,* 14, 79, 1966.
62. **Stedman, E. and Stedman, E.,** The basic proteins of cell nuclei, *Phil. Trans. Royal Soc.,* B 235, 565, 1957.
63. **Ui, N.,** Preparation, fractionation and properties of calf thymus histone, *Biochim. Biophys. Acta,* 25, 493, 1957.
64. **Bijvoet, P.,** Ethanol precipitation analysis of thymus histone, *Biochim. Biophys. Acta,* 25, 502, 1957.
65. **Tsumita, T. and Chargaff, E.,** Studies on nucleoproteins VI. The deoxyribonucleoprotein and the deoxyribonucleic acid of bovine tubercle bacilli (BCG), *Biochim. Biophys. Acta,* 29, 568, 1958.
66. **Zubay, G. and Watson, M. R.,** The absence of histone in the bacterium *Escherichia coli.* I. Preparation and analysis of nucleoprotein extract, *J. Biophys. Biochem. Cytol.,* 5, 51, 1959.
67. **Wilkins, M. H. F. and Zubay, G.,** The absence of histone in the bacterium *Escherichia coli.* II. X-ray diffraction of nucleoprotein extract, *J. Biophys. Biochem. Cytol.,* 5, 55, 1959.
68. **Bhagavan, N. V. and Atchley, W. A.,** Properties of a deoxyribonucleoprotein complex derived from *Bacillus subtilis, Biochemistry,* 4, 234, 1965.
69. **Leaver, J. L. and Cruft, H. J.,** Investigation into the distribution and properties of histones and other basic proteins in bacteria, *Biochem. J.,* 101, 665, 1966.
70. **Raaf, J. and Bonner, J.,** Properties of *Escherichia coli* deoxyribonucleoprotein as compared with the deoxyribonucleoprotein of higher organisms, *Arch. Biochem. Biophys.,* 125, 567, 1968.

71. **Takeuchi, M. and Tsugita, A.,** Isolation and characterization of nucleoprotein "nuclear fraction" from *Bacillus subtilis, J. Biochem.*, 67, 237, 1970.
72. **Makino, F. and Tsuzuki, J.,** Absence of histone in the blue-green alga *Anabaena cylindrica, Nature,* 231, 446, 1971.
73. **Dwivedi, R. S., Dutta, K. S., and Bloch, D. P.,** Isolation and characterization of chromatin from *Neurospora crassa, J. Cell Biol.*, 43, 51, 1969.
74. **Leighton, T. J., Dill, B. C., Stock, J. J., and Phillips, C.,** Absence of histones from the chromosomal proteins of fungi, *Proc. Natl. Acad. Sci. U. S.*, 68, 677, 1971.
75. **Mohberg, J. and Rusch, H. P.,** Isolation of the nuclear histones from myxomycete *Physarum polycephalum, Arch. Biochem. Biophys.*, 134, 577, 1969.
76. **Black, J. A. and Dixon, G. H.,** Evolution of protamine: a further example of partial gene duplication, *Nature,* 216, 152, 1967.
77. **Phillips, D. M. P.,** The primary structure of histones and protamines, in *Histones and Nucleohistones,* Phillips, D. M. P., Ed., Plenum Press, London, 1971, 47.
78. **Butler, J. A. V., Cohn, P., and Simson, P.,** The presence of basic proteins in microsomes, *Biochim. Biophys. Acta,* 38, 386, 1960.
79. **Lindsay, D. T.,** Electrophoretically identical histones from ribosomes and chromosomes of chicken liver, *Arch. Biochem. Biophys.*, 113, 687, 1966.
80. **Traut, R. R., Moore, P. B., Delius, H., Noller, H., and Tissieres, A.,** Ribosomal proteins of *Escherichia coli. I.* Demonstration of different primary structures, *Proc. Natl. Acad. Sci. U. S.*, 57, 1294, 1967.
81. **Traub, P., Hosokawa, K., Craven, G. R., and Nomura, M.,** Structure and function of *E. coli* ribosomes. IV. Isolation and characterization of functionally active ribosomal proteins, *Proc. Natl. Acad. Sci. U. S.*, 58, 2430, 1967.
82. **Fogel, S. and Sypherd, P. S.,** Chemical basis for heterogeneity of ribosomal proteins, *Proc. Natl. Acad. Sci. U. S.*, 59, 1329, 1968.
83. **Ansley, S. B., Campbell, L. L., and Sypherd, P. S.,** Isolation and amino acid composition of ribosomal proteins from *Bacillus stearothermophylus, J. Bacteriol.*, 98, 568, 1969.
84. **Kaltschmidt, E., Dzionara, M., and Wittmann, H. G.,** Ribosomal proteins. XV. Amino acid compositions of isolated ribosomal proteins from 30s and 50s subunits of *Escherichia coli, Mol. Gen. Genet.*, 109, 292, 1970.
85. **Gancevici, G., Stoian, I., and Marx, A.,** Cationic proteins in liver mitochondria, *Naturwissenschaften,* 52, 40, 1965.
86. **Nakao, A., Davis, W. J., and Einstein, E. R.,** Basic proteins from the acidic extract of bovine spinal cord. I. Isolation and characterization, *Biochim. Biophys. Acta,* 130, 163, 1966.
87. **Thomasi, L. G. and Kornguth, S. E.,** Purification and partial characterization of basic protein from pig brain, *J. Biol. Chem.*, 242, 4933, 1967.
88. **Eylar, E. H.,** Amino acid sequence of the basic protein of the myelin membrane, *Proc. Natl. Acad. Sci. U. S.*, 67, 1425, 1970.
89. **Uyemura, K., Tobari, C., and Hirano, S.,** Purifications and properties of basic proteins in pig spinal cord and peripheral nerve, *Biochim. Biophys. Acta,* 214, 190, 1970.
90. **Cruft, H. J., Hindley, J., Mauritzen, C. M., and Stedman, E.,** Amino acid composition of the six histones of calf thymocytes, *Nature,* 180, 1107, 1957.
91. **Davison, P. F. and Shooter, K. V.,** Sedimentation, electrophoretic and chromatographic studies of whole and fractionated calf thymus histones, *Bull. Soc. Chim. Belg.*, 65, 85, 1956.
92. **Crampton, C. F.,** Studies on the dissociation of histones from the nucleohistone of calf thymus, *J. Biol. Chem.*, 227, 495, 1957.
93. **Giannoni, G. and Peacocke, A. R.,** Thymus deoxyribonucleoprotein. III. Sedimentation behaviour, *Biochim. Biophys. Acta,* 68, 157, 1963.
94. **Stedman, E. and Stedman, E.,** Cell specificity of histones, *Nature,* 166, 780, 1950.
95. **Neelin, J. M. and Neelin, E. M.,** Zone electrophoresis of calf thymus histone in starch gel, *Can. J. Biochem. Physiol.*, 38, 355, 1960.
96. **Johns, E. W., Phillips, D. M. P., Simson, P., and Butler, J. A. V.,** The electrophoresis of histones and histone fractions on starch gels, *Biochem. J.*, 80, 189, 1961.
97. **MacPherson, A. and Murray, K.,** On the sub-fractionation of histones by zone electrophoresis in polyacrylamide gels, *Biochim. Biophys. Acta,* 104, 574, 1965.
98. **Reisfeld, R. A., Lewis, U. J., and Williams, D. E.,** Disc electrophoresis of basic proteins and peptides on polyacrylamide gels, *Nature,* 195, 281, 1962.
99. **Cruft, H. J.,** Electrophoresis and gel filtration of histones, in *The Nucleohistones,* Bonner, J. and Ts'o, P. O. P., Eds., Holden-Day, San Francisco, 1964, 72.
100. **Fambrough, D. M., Fujimura, F., and Bonner, J.,** Quantitative distribution of histone components in the pea plant, *Biochemistry,* 7, 575, 1968.
101. **Panyim, S. and Chalkley, R.,** High resolution acrylamide gel electrophoresis of histones, *Arch. Biochem. Biophys.*, 130, 337, 1969.
102. **Panyim, S. and Chalkley, R.,** The heterogeneity of histones. I. A quantitative analysis of calf histones in very long polyacrylamide gels, *Biochemistry,* 8, 3972, 1969.
103. **Elgin, S. C. R. and Bonner, J.,** Limited heterogeneity of the major nonhistone chromosomal proteins, *Biochemistry,* 9, 4440, 1970.

104. **Shirey, T. and Huang, R. C. C.,** Use of SDS, alone, to separate chromatin proteins from DNP of *Arbacia punctulata* sperm chromatin, *Biochemistry,* 8, 4138, 1969.
105. **Busch, H. and Mauritzen, C. M.,** Nuclear proteins, in *Methods in Cancer Research,* Vol. 3, Busch, H., Ed., Academic Press, New York, 1967, 391.
106. **Crampton, C. F., Moore, S., and Stein, W. H.,** Chromatographic fractionation of calf thymus histone, *J. Biol. Chem.,* 215, 787, 1955.
107. **Neelin, J. M., Callahan, P. X., Lamb, D. C., and Murray, K.,** The histones of chicken erythrocyte nuclei, *Can. J. Biochem.,* 42, 1743, 1964.
108. **Davis, J. R. and Busch, H.,** Chromatographic analysis of cationic nuclear proteins of number of neoplastic tissues, *Cancer Res.,* 20, 1208, 1960.
109. **Busch, H.,** *Histones and Other Nuclear Proteins,* Academic Press, New York, 1965.
110. **Murray, K.,** Histone nomenclature, in *The Nucleohistones,* Bonner, J. and Ts'o, P. O. P., Eds., Holden-Day, San Francisco, 1964, 15.
111. **Butler, J. A. V., Johns, E. W., and Phillips, D. M. P.,** Recent investigations on histones and their functions, *Progr. Biophys. Mol. Biol.,* 18, 209, 1968.
112. **Lowry, O. H., Rosebrough, N. J., Farr, A. L., and Randall, R. J.,** Protein measurement with the Folin phenol reagent, *J. Biol. Chem.,* 193, 265, 1951.
113. **Mirsky, A. E. and Pollister, A. W.,** Chromosin, a desoxyribose nucleoprotein complex of the cell nucleus, *J. Gen. Physiol.,* 30, 117, 1946.
114. **Sheth, K. and Cohen, L. H.,** A fluorometric protein assay suitable for histones based on lysine content, *Anal. Biochem.,* 37, 142, 1970.
115. **Shepherd, G. R. and Noland, B. J.,** A fluorescence assay suitable for histone solutions, *Anal. Biochem.,* 11, 443, 1965.
116. **Laurence, D. J. R.,** Interactions of calf thymus histone fractions in aqueous solutions with 8-anilinonaphthalene 1-sulphonic acid, *Biochem. J.,* 99, 419, 1966.
117. **Alfert, M. and Geschwindt, I. I.,** A selective staining method for the basic proteins of cell nuclei, *Proc. Natl. Acad. Sci. U.S.,* 39, 991, 1953.
118. **Alfert, M.,** La cytochimie de l'ADN et des nucléohistones, *Arch. J. Klaus-Stiftung für Vererbungsforschung, Socialanthropologie und Rassenhygiene,* 32, 553, 1957.
119. **Alfert, M.,** Variations in cytochemical properties of cell nuclei, *Exp. Cell Res.,* Suppl. 6, 227, 1958.
120. **Bloch, D. P. and Godman, G. C.,** Evidence of differences in the desoxyribonucleoprotein complex of rapidly proliferating and non-dividing cells, *J. Biophys. Biochem. Cytol.,* 1, 531, 1955.
121. **Bloch, D. P. and Godman, G. C.,** A microphotometric study of the syntheses of deoxyribonucleic acid and nuclear histone, *J. Biophys. Biochem. Cytol.,* 1, 17, 1955.
122. **Bloch, D. P. and Hew, H. Y. C.,** Changes in nuclear histones during fertilization and early embryonic development in the pulmonate snail, *Helix aspersa, J. Biochem. Biophys. Cytol.,* 8, 69, 1960.
123. **Ringertz, N. R. and Zetterberg, A.,** Cytochemical demonstration of histones and protamines. Mechanism and specificity of the alkaline bromphenol blue binding reaction, *Exp. Cell Res.,* 42, 243, 1966.
124. **Noeske, K.,** Stöchiometrische Probleme der quantitativen Fastgreen-Zytophotometrie bei der Histonbestimmung, *Histochemie,* 27, 243, 1971.
125. **Bloch, D. P.,** Histochemistry of Histones, in *Protoplasmologia,* Vol. 5, Fasc. 3, Springer Verlag, Vienna, 1966.
126. **Swift, J. H.,** The histones of polytene chromosomes, in *The Nucleohistones,* Bonner, J. and Ts'o, P. O. P., Eds., Holden-Day, San Francisco, 1964, 169.
127. **Berlowitz, L., Pallotta, D., and Pawlowski, P.,** Isolated histone fractions and the alkaline fast green reaction, *J. Histochem. Cytochem.,* 18, 334, 1970.
128. **Magun, B. E. and Kelly, J. W.,** A new fluorescent method with phenanthrene-quinone for the histochemical demonstration of arginine residues in tissues, *J. Histochem. Cytochem.,* 17, 821, 1969.
129. **Black, M. M., Speer, F. D., and Lillick, L. C.,** Acid extractable proteins in cancer cells. I. Staining with ammoniacal silver, *J. Natl. Cancer Inst.,* 25, 967, 1960.
130. **Rigler, R., Killander, D., Bolund, L., and Ringertz, N. R.,** Cytochemical characterization of deoxyribonucleoprotein in individual cell nuclei, *Exp. Cell Res.,* 55, 215, 1969.
131. **Love, R. and Liles, R. H.,** Differentiation of nucleoproteins by inactivation of protein-bound amino groups and staining with toluidine blue and ammonium molybdate, *J. Histochem. Cytochem.,* 7, 164, 1959.
132. **Love, R. and Walsh, R. J.,** Studies on the cytochemistry of nucleoproteins. II. Improved staining methods with toluidine blue and ammonium molybdate, *J. Histochem. Cytochem.,* 11, 188, 1963.
133. **Smetana, K. and Busch, H.,** Studies on the staining and localization of acidic nuclear proteins in the Walker 256 carcinosarcoma, *Cancer Res.,* 26, 331, 1966.
134. **Smetana, K.,** Basic histochemical and cytochemical methods, in *Methods in Cancer Research,* Vol. 2, Busch, H., Ed., Academic Press, New York, 1967, 361.
135. **Davison, P. F., James, D. W. F., Shooter, K. V., and Butler, J. A. V.,** The histones of calf thymus deoxyribonucleoprotein. II. Electrophoretic and sedimentation behaviour and a partial fractionation, *Biochim. Biophys. Acta,* 15, 415, 1954.

136. **Ui, N.**, Further studies on the preparation of calf thymus histone, *Bull. Chem. Soc. Jap.*, 30, 822, 1957.
137. **Smillie, L. B., Butler, G. C., and Smith, D. B.**, Properties of calf thymus histone isolated by three different methods, *Can. J. Biochem. Physiol.*, 36, 1, 1958.
138. **Gregoire, J. and Limozin, M.**, Recherches sur l'hétérogénéité et la composition des préparations de thymohistone, *Bull. Soc. Chim. Biol.*, 36, 15, 1954.
139. **Bakay, B., Kolb, J. J., and Toennies, G.**, On the component proteins of calf thymus nucleoprotein, *Arch. Biochem. Biophys.*, 58, 144, 1955.
140. **Ahlström, L.**, Nukleoproteide in normalen und cancerösen Zellen. V. Histon aus Thymuszellkernen, *Arkiv Kemi Mineral. Geol.*, 24A(31), 1947.
141. **Johns, E. W.**, Studies on histones 8. A degradation product of lysine-rich histone, *Biochem. J.*, 93, 161, 1964.
142. **Luck, J. M., Cook, H. A., Eldredge, N. T., Haley, M. I., Kupke, D. W., and Rasmussen, P. S.**, On the fractionation of thymus histones, *Arch. Biochem. Biophys.*, 65, 449, 1956.
143. **Neelin, J. M. and Butler, G. C.**, The fractionation of the histones of calf thymus and chicken erythrocytes by cation exchange chromatography with sodium salts, *Can. J. Biochem. Physiol.*, 37, 843, 1959.
144. **Stellwagen, R. H. and Cole, R. D.**, Danger of contamination in chromatographically prepared arginine-rich histones, *J. Biol. Chem.*, 243, 4452, 1968.
145. **Kinkade, J. M. and Cole, R. D.**, The resolution of four lysine-rich histones derived from calf thymus, *J. Biol. Chem.*, 241, 5790, 1966.
146. **Kinkade, J. M. and Cole, R. D.**, A structural comparison of different lysine-rich histones of calf thymus, *J. Biol. Chem.*, 241, 5798, 1966.
147. **Bustin, M. and Cole, R. D.**, Species and organ specificity in very lysine-rich histones, *J. Biol. Chem.*, 243, 4500, 1968.
148. **Bustin, M. and Cole, R. D.**, A study of the multiplicity of lysine-rich histones, *J. Biol. Chem.*, 244, 5286, 1969.
149. **Peterson, E. A. and Sober, H. A.**, Chromatography of proteins. I. Cellulose ion-exchange adsorbents, *J. Am. Chem. Soc.*, 78, 751, 1956.
150. **Davison, P. F.**, Histones from normal and malignant cells, *Biochem. J.*, 66, 703, 1957.
151. **Davison, P. F.**, Chromatography of histones, *Biochem. J.*, 66, 708, 1957.
152. **Davis, J. R. and Busch, H.**, Chromatographic analysis of radioactive cationic proteins of tissues of tumor-bearing rats, *Cancer Res.*, 19, 1157, 1959.
153. **Johnson, L. D., Driedger, A. and Marko, A. M.**, Chromatographic behavior of histones on Sephadex and carboxymethylcellulose, *Can. J. Biochem.*, 42, 795, 1964.
154. **Klyszejko, L., De Mende, M. S., and Khouvine, Y.**, Désoxyribonucléoprotéides du pancréas de boeuf. III. Fractionnement des histones, *Bull. Soc. Chim. Biol.*, 46, 111, 1964.
155. **Kobayashi, Y. and Iwai, K.**, Carboxymethylcellulose chromatography in 6 M urea and zone electrophoresis of calf thymus histone, *J. Biochem.*, 67, 465, 1970.
156. **Oh, Y. H.**, Spectroscopic studies of five purified histones from calf thymus, *J. Biol. Chem.*, 245, 6404, 1970.
157. **Iwai, K., Mita, T., Senshu, T., Shiomi-Yabuki, H., Ochiai, H., Ishikawa, K., and Hayashi, H.**, Chemical aspects of histones from calf thymus and from synchronously growing Tetrahymena, *Abstracts of the 7th Int. Congress of Biochem., Tokyo,* Vol. 3, 1967, 511.
158. **Senshu, T. and Iwai, K.**, Fractionation of calf thymus histone by a new method of CM-cellulose chromatography, *J. Biochem.*, 67, 473, 1970.
159. **Hnilica, L. S.**, Proteins of the cell nucleus, *Progr. Nucl. Acid Res. Mol. Biol.*, 7, 25, 1967.
160. **Hnilica, L. S., Kappler, H. A., and Jordan, J. J.**, Asymetry in the distribution of basic amino acid residues in the moderately lysine-rich histone F2b from calf thymus, *Experientia*, 26, 353, 1970.
161. **Cruft, H. R.**, The fractionation of histones on Sephadex G75, *Biochim. Biophys. Acta*, 54, 611, 1961.
162. **Hnilica, L. S. and Bess, L. G.**, Fractionation of calf thymus histone fractions 2a and 3 on Sephadex, *Anal. Biochem.*, 8, 521, 1964.
163. **Hnilica, L. S.**, The fractionation of arginine-rich histones from calf thymus, *Experientia*, 21, 124, 1965.
164. **Hnilica, L. S. and Bess, L. G.**, The heterogeneity of arginine-rich histones, *Anal. Biochem.*, 12, 421, 1965.
165. **Sautière, P., Dautrevaux, M., and Biserte, G.**, Fractionnement des histones riches en arginine du thymus de veau par chromatographie de gel-filtration, *Bull. Soc. Chim. Biol.*, 47, 821, 1965.
166. **Mauritzen, C. M., Starbuck, W. C., Saroja, I. S., Taylor, C. W., and Busch, H.**, The fractionation of arginine-rich histones from fetal calf thymus by exclusion chromatography, *J. Biol. Chem.*, 242, 2240, 1967.
167. **Bellair, J. T. and Mauritzen, C. M.**, The fractionation of β-histone from chicken erythrocyte nuclei, *Aust. J. Biol. Sci.*, 18, 160, 1965.
168. **Hnilica, L. S., Kappler, H. A., and Hnilica, V. S.**, Biosynthesis of histones and acidic nuclear proteins under different conditions of growth, *Science*, 150, 1470, 1965.
169. **Bellair, J. T. and Mauritzen, C. M.**, The fractionation of α-histones from chicken erythrocyte nuclei. II. Fractions obtained by exclusion chromatography, *Aust. J. Biol. Sci.*, 17, 1001, 1964.
170. **Edwards, P. A. and Shooter, K. V.**, A study of calf thymus histone fraction F2(b) by gel filtration, *Biochem. J.*, 120, 61, 1970.

171. **Bustos-Valdez, S. E. and Dounce, A. L.,** Comparison of rat liver and calf thymus histone fractions by molecular exclusion chromatography and ultracentrifugation, *Arch. Biochem. Biophys.*, 136, 134, 1970.
172. **Phillips, D. M. P. and Clarke, M.,** Behaviour of histones in exclusion chromatography and gel electrophoresis in relation to their molecular weight, *J. Chromatogr.*, 46, 321, 1970.
173. **De Nooij, E. H. and Westenbrink, H. G. K.,** Isolation of a homogeneous lysine-rich histone from calf thymus, *Biochim. Biophys. Acta*, 62, 608, 1962.
174. **Wilhelm, F. X., Wilhelm, M. L., and Champagne, M. H.,** Preparation de deux fractions d'histones par extraction saline, *Eur. J. Biochem.*, 16, 103, 1970.
175. **Georgiev, G., Ananieva, L., and Kozlov, J.,** Stepwise removal of protein from deoxyribonucleoprotein complex and de-repression of the genome, *J. Mol. Biol.*, 22, 365, 1966.
176. **Ohlenbusch, H. H., Olivera, B. M., Tuan, D., and Davidson, N.,** Selective dissociation of histones from calf thymus nucleoprotein, *J. Mol. Biol.*, 25, 299, 1967.
177. **Ohba, Y.,** Structure of nucleohistone. I. Hydrodynamic behaviour, *Biochim. Biophys. Acta*, 123, 76, 1966.
178. **Fambrough, D. and Bonner, J.,** Selective dissociation of pea bud nucleohistone, *Biochim. Biophys. Acta*, 154, 601, 1968.
179. **Wilhelm, X. and Champagne, M.,** Dissociation de la nucléoprotéine d'érythrocytes de poulets par les sels, *Eur. J. Biochem.*, 10, 102, 1969.
180. **Murray, K., Vidali, G., and Neelin, J. M.,** The stepwise removal of histones from chicken erythrocyte nucleoprotein, *Biochem. J.*, 107, 207, 1968.
181. **Johns, E. W. and Forrester, S.,** Interactions between the lysine-rich histone Fl and deoxyribonucleic acid, *Biochem. J.*, 111, 371, 1969.
182. **Spelsberg, T. C. and Hnilica, L. S.,** Proteins of chromatin in template restriction. I. RNA synthesis *in vitro, Biochim. Biophys. Acta*, 228, 202, 1971.
183. **Handler, S. C., Smart, J. E., and Bonner, J.,** Studies on the mechanism of removal of histones from chromatin by deoxycholate, *Biochim. Biophys. Acta*, 236, 253, 1971.
184. **Smart, J. E. and Bonner, J.,** Selective dissociation of histones from chromatin by sodium deoxycholate, *J. Mol. Biol.*, 58, 651, 1971,
185. **Seligy, V. L. and Neelin, J. M.,** Transcription properties of stepwise acid extracted chicken erythrocyte chromatin, *Biochim. Biophys. Acta*, 213, 380, 1970.
186. **Murray, K.,** Stepwise removal of histones from native deoxyribonucleoprotein by titration with acid at low temperature and some properties of the resulting partial nucleoproteins, *J. Mol. Biol.* 39, 125, 1969.
187. **Kleinman, L.,** Dissociation of histones from calf thymus chromatin and reassociation to the DNA in the presence of salt and urea, *Fed. Proc.*, 30, 1084a, 1971.
188. **Senshu, T.,** Selective extraction of histones with sodium chloride and urea, *Biochim. Biophys. Acta*, 236, 349, 1971.
189. **Johns, E. W.,** A method for selective extraction of histone fractions f2(a)1 and f2(a)2 from calf thymus deoxyribonucleoprotein at pH 7, *Biochem. J.*, 105, 611, 1967.
190. **Bartkowiak, J., Klyszejko-Stefanowicz, L., Wierzbicki, R., and Zablocka, E.,** Selective extraction of histones from normal and irradiated calf lymph nodes, *Int. J. Radiat. Biol.*, 20, 173, 1971.
191. **Cruft, H. J., Mauritzen, C. M., and Stedman, E.,** The nature and physicochemical properties of histones, *Trans. R. Soc.*, B 241, 93, 1957.
192. **Cruft, H. J., Mauritzen, C. M., and Stedman, E.,** The isolation of β-histone from calf thymocytes and the factors affecting its aggregation, *Proc. R. Soc.*, B 149, 21, 1958.
193. **Cruft, H. J., Mauritzen, C. M., and Stedman, E.,** The isolation and properties of 1.6 sδ -histone from calf thymocytes, *Proc. R. Soc.*, B 149, 36, 1958.
194. **Bellair, J. T. and Mauritzen, C. M.,** The fractionation of α-histones from chicken erythrocyte nuclei. I. The effect of ethanol concentration and pH on the fractional precipitation of α-histones, *Aust. J. Biol. Sci.*, 17, 990, 1964.
195. **Yokotsuka, K., Yoshida, M., and Shimura, K.,** Fractionation of histones by counter-current distribution, *J. Biochem.*, 60, 595, 1966.
196. **Butler, J. A. V., Power, D. F., and Palau, J.,** Countercurrent-distribution studies on histones, *Biochem. J.*, 102, 539, 1967.
197. **Chanda, S. K. and Dounce, A. L.,** Separation of histones from normal rat liver nuclei into easily extractable and residual histone and characterization of the fractions, *Arch. Biochem. Biophys.*, 146, 441, 1971.
198. **Van der Westhuyzen, D. R. and Von Holt, C.,** A new procedure for the isolation and fractionation of histones, *FEBS Lett.*, 14, 333, 1971.
199. **Johns, E. W.,** A method for the differentiation of histone fractions F2(a)1 and F2(a)2 by starch gel electrophoresis, *J. Chromatogr.*, 34, 429, 1968.
200. **Sung, M. T., Dixon, G. H., and Smithies, O.,** Phosphorylation and synthesis of histones in regenerating rat liver, *J. Biol. Chem.*, 246, 1358, 1971.
201. **Sung, M. and Dixon, G. H.,** Modification of histones during spermiogenesis in trout: A molecular mechanism for altering histone binding to DNA, *Proc. Natl. Acad. Sci. U.S.*, 67, 1616, 1970.
201a. **Dixon, G. H.,** The basic proteins of trout testis chromatin: Aspects of their synthesis, post-synthetic modifications and binding to DNA, 5th Karolinska Symposium on *Gene Transcription in Reproductive Tissues* in press.

202. **Paoletti, R. A. and Huang, R. C. C.,** Characterization of sea urchin sperm chromatin and its basic proteins, *Biochemistry*, 8, 1615, 1969.
203. **Fambrough, D. M. and Bonner, J.,** Limited molecular heterogeneity of plant histones, *Biochim. Biophys. Acta*, 175, 113, 1969.
204. **Johnson, A. W., Ward, D. N., Wilhelm, J. A., and Hnilica, L. S.,** Fractionation and analysis of histones from sea urchin sperm and embryos, in preparation.
205. **Machicao, F. and Sonnenbichler, J.,** Cellogel electrophoresis of histones, *Biochim. Biophys. Acta,* 236, 360, 1971.
206. **Johns, E. W.,** Studies on histones 7. Preparative methods for histone fractions from calf thymus, *Biochem. J.*, 92, 55, 1964.
207. **Bustin, M. and Cole, R. D.,** The applicability of extraction by trichloroacetic acid to the preparation of very lysine-rich histones from mammary gland, *Arch. Biochem. Biophys.*, 127, 457, 1968.
208. **Maskos, C. and Mejbaum-Katzenellenbogen, W.,** Lysine-rich histone in sulphosalicylic acid extract from muscles, *Acta Biochim. Polonica,* 18, 39, 1971.
209. **Murray, K., Bradbury, E. M., Crane-Robinson, C., Stephens, R. M., Haydon, A. J., and Peacocke, A. R.,** The dissociation of chicken erythrocyte deoxyribonucleoprotein and some properties of its partial nucleoproteins, *Biochem. J.*, 120, 859, 1970.
210. **Teller, D. C., Kinkade, J. M., and Cole, R. D.,** The molecular weight of lysine-rich histone, *Biochem. Biophys. Res. Commun.*, 20, 739, 1965.
211. **Haydon, A. J. and Peacocke, A. R.,** Sedimentation equilibrium and other physicochemical studies on the lysine-rich fraction of calf thymus histones, *Biochem. J.*, 110, 243, 1968.
212. **Bustin, M. and Cole, R. D.,** Bisection of a lysine-rich histone by N-bromosuccinimide, *J. Biol. Chem.*, 244, 5291, 1969.
213. **Reid, B. R. and Cole, R. D.,** Biosynthesis of a lysine-rich histone in isolated calf thymus nuclei, *Proc. Natl. Acad. Sci. U. S.*, 51, 1044, 1964.
214. **Murray, K.,** The basic proteins of cell nuclei, *Annu. Rev. Biochem.*, 34, 209, 1965.
215. **Murray, K.,** The heterogeneity of histones, in *The Nucleohistones,* Bonner, J. and Ts'o, P. O. P., Eds., Holden-Day, San Francisco, 1964, 21.
216. **Stellwagen, R. H. and Cole, R. D.,** Comparison of histones obtained from mammary gland at different stages of development and lactation, *J. Biol. Chem.*, 243, 4456, 1968.
217. **Kinkade, J. M.,** Qualitative species differences and quantitative tissue differences in the distribution of lysine-rich histones, *J. Biol. Chem.*, 244, 3375, 1969.
218. **Bustin, M., Rall, S. C., Stellwagen, R. H., and Cole, R. D.,** Histone structure: Asymmetric distribution of lysine-rich residues in lysine-rich histone, *Science,* 163, 391, 1969.
219. **Rall, S. C. and Cole, R. D.,** Amino acid sequence and sequence variability of the amino terminal regions of lysine-rich histones, *J. Biol. Chem.*, 246, 7175, 1971.
220. **Bustin, M. and Cole, R. D.,** Regions of high and low cationic charge in a lysine-rich histone, *J. Biol. Chem.*, 245, 1458, 1970.
221. **Fambrough, D. M.,** Nuclear protein fractions, in *Handbook of Molecular Cytology,* Lima-de-Faria, A., Ed., North Holland Publ. Co., Amsterdam, 1969, 437.
222. **Phillips, D. M. P.,** The presence of acetyl groups in histones, *Biochem. J.*, 87, 258, 1963.
223. **Jones, G. M. T. and Cole, R. D.,** personal communication.
224. **Langan, T. A.,** Phosphorylation of proteins of the cell nucleus, in *Some Regulatory Mechanisms for Protein Synthesis in Mammalian Cells,* San Pietro, A., Lamborg, M., and Kenney, F. T., Eds., Academic Press, New York, 1968, 101.
225. **Langan, T.,** Histone phosphorylation: stimulation by adenosine 3',5'-monophosphate, *Science,* 162, 579, 1968.
226. **Langan, T.,** Phosphorylation of liver histone following the administration of glucagon and insulin, *Proc. Natl. Acad. Sci. U. S.*, 64, 1276, 1969.
227. **Ord, M. G. and Stocken, L. A.,** Metabolic properties of histones from rat liver and thymus gland, *Biochem. J.*, 98, 888, 1966.
228. **Kleinsmith, L. J., Allfrey, V. G., and Mirsky, A. E.,** Phosphoprotein metabolism in isolated lymphocyte nuclei, *Proc. Natl. Acad. Sci. U. S.*, 55, 1182, 1966.
229. **Evans, K., Hohmann, P., and Cole, R. D.,** Chromatographic resolution of lysine-rich histones unaffected by phosphatase or ribonuclease treatment, *Biochim. Biophys. Acta,* 221, 128, 1970.
230. **Sherod, D., Johnson, G., and Chalkley, R.,** Phosphorylation of mouse ascites tumor cell lysine-rich histone, *Biochemistry,* 9, 4611, 1970.
231. **Langan, T. A.,** Action of adenosine 3', 5'-monophosphate-dependent histone kinase *in vivo, J. Biol. Chem.*, 244, 5763, 1969.
232. **Dixon, G. H., Marushige, K., Ling, V., Sung, M., and Wigle, D. T.,** Phosphorylation of histones in maturing trout testis, *Fed. Proc.*, 28, 599, 1969.
233. **Tonino, G. J. M. and Rozijn, T. H.,** The occurrence of histones in yeast, *Biochim. Biophys. Acta,* 124, 427, 1966.
234. **Edwards, L. J. and Hnilica, L. S.,** The specificity of histones in nucleated erythrocytes, *Experientia,* 24, 228, 1968.

235. **Backström, S.**, Basic proteins in the sea urchin embryo *(Paracentrotus lividus), Acta Embryol. Morphol. Exp.*, 8, 20, 1965.
236. **Ord, M. G. and Stocken, L. A.**, Variations in the phosphate content and the thiol/disulphide ratio of histones during the cell cycle. Studies with regenerating rat liver and sea urchins, *Biochem. J.*, 107, 403, 1968.
237. **Asao, T.**, Behaviour of histones and cytoplasmic basic proteins during embryogenesis of the Japanese newt, *Triturus pyrrhogaster, Exp. Cell Res.*, 58, 243, 1969.
238. **Asao, T.**, Regional histone changes in embryos of the newt *Triturus pyrrhogaster* during early development, *Exp. Cell. Res.*, 61, 255, 1970.
239. **Orengo, A. and Hnilica, L. S.**, *In vivo* incorporation of labeled amino acids into nuclear proteins of the sea urchin embryos, *Exp. Cell Res.*, 62, 331, 1970.
240. **Panyim, S., Chalkley, R., Spiker, S., and Oliver, D.**, Constant electrophoretic mobility of the cysteine-containing histone in plants and animals, *Biochim. Biophys. Acta*, 214, 216, 1970.
241. **Cohn, N. S. and Quatrano, R. S.**, Extraction and analysis of histone protein from the roots of three plant species, *Experientia*, 21, 237, 1965.
242. **Iwai, K.**, Histones in rice embryos and in *Chlorella*, in *The Nucleohistones*, Bonner, J. and Ts'o, P. O. P., Eds., Holden-Day, San Francisco, 1964, 59.
243. **Iwai, K., Shiomi, H., Ando, T., and Mita, T.**, The isolation of histone from *Tetrahymena, J. Biochem.*, 58, 312, 1965.
244. **Lee, Y. C. and Scherbaum, D. H.**, Nucleohistone composition in stationary and division synchronized *Tetrahymena* cultures, *Biochemistry*, 5, 2067, 1966.
245. **Johns, E. W. and Butler, J. A. V.**, Studies on histones. 4. The histones of wheat germ, *Biochem. J.*, 84, 436, 1962.
246. **Fambrough, D. M. and Bonner, J.**, On the similarity of plant and animal histones, *Biochemistry*, 5, 2563, 1966.
247. **Dick, C.**, Isolation of nuclei and histones from roots of *Vicia faba, Arch. Biochem. Biophys.*, 124, 431, 1968.
248. **Pallotta, D. and Berlowitz, L.**, Histones of two insecet species, *Biochim. Biophys. Acta*, 200, 538, 1970.
249. **Yoshida, M., Yokotsuka, K., and Shimura, K.**, Amino acid composition and heterogeneity of histones from posterior silkgland, *J. Biochem.*, 60, 586, 1966.
250. **Yokotsuka, K., Yoshida, M., and Shimura, K.**, Fractionation of histones from calf thymus and posterior silkgland by countercurrent distribution, *J. Biochem.*, 64, 129, 1968.
251. **Cohen, L. H. and Gotchel, B. V.**, Histones of polytene and nonpolytene nuclei of *Drosophila melanogaster, J. Biol. Chem.*, 246, 1841, 1971.
252. **Cozcolluela, C. and Subirana, J. A.**, Somatic histones from *Mytilus edulis, Biochim. Biophys. Acta*, 154, 242, 1968.
253. **Subirana, J. A., Palau, J., Cozcolluela, C., and Ruiz-Carillo, A.**, Very lysine-rich histone of echinoderms and molluscs, *Nature*, 228, 992, 1970.
254. **Phelan, J. J., Subirana, J. A., and Cole, R. D.**, An unusual group of lysine-rich histones from gonads of a sea cucumber, *Holothuria tubulosa*, in preparation.
255. **Iwai, K., Hamana, K., and Yabuki, H.**, Histone fractions of protozoa: their correlation with mammalian histones, *J. Biochem.*, 68, 597, 1970.
256. **Kinkade, J. M.**, Differences in the quantitative distribution of lysine-rich histones in neoplastic and normal tissues, *Proc. Soc. Exp. Biol. Med.*, 137, 1131, 1971.
257. **DeLange, R. J., Fambrough, D. M., Smith, E. L., and Bonner, J.**, Calf and pea histone IV. III. Complete amino acid sequence of pea seedling histone. IV. Comparson with the homologous calf thymus histone, *J. Biol. Chem.*, 244, 5669, 1969.
258. **Combard, A. and Vendrely, R.**, Analytical study of the degradation of nucleohistone during calf thymus chromatin autolysis, *Biochem. J.*, 118, 875, 1970.
259. **Hnilica, L. S.**, Studies on nuclear proteins. I. Observation on the tissue and species specificity of the moderately lysine-rich histone fraction 2b, *Biochim. Biophys. Acta*, 117, 163, 1966.
260. **Iwai, K., Ishikawa, K., and Hayashi, H.**, Amino acid sequence of slightly lysine-rich histone, *Nature*, 226, 1056, 1970.
261. **Hnilica, L. S., Taylor, C. W., and Busch, H.**, Analysis of peptides of the moderately lysine-rich histone fraction F2b of the Walker tumor and other tissues, *Exp. Cell Res.*, Suppl. 9, 367, 1963.
262. **Palau, J. and Butler, J. A. V.**, Trout-liver histones, *Biochem. J.*, 100, 779, 1966.
263. **Hnilica, L. S., Taylor, C. W., and Busch, H.**, Peptides of histone fraction 2b, in *The Nucleohistones*, Bonner, J. and Ts'o, P. O. P., Eds., Holden-Day, San Francisco, 1964, 84.
264. **Laurence, D. J. R., Phillips, D. M. P., and Butler, J. A. V.**, A comparison of histone fractions from an "osteogenic" rat tumour and calf thymus, *Arch. Biochem. Biophys.*, 113, 338, 1966.
265. **Johns, E. W.**, The fractionation of the histones of group F2(b) from calf thymus, *Eur. J. Biochem.*, 4, 437, 1968.
266. **Gutierrez, R. M. and Hnilica, L. S.**, Tissue specificity of histone phosphorylation, *Science*, 157, 1324, 1967.
267. **Gutierrez-Cernosek, R. M. and Hnilica, L. S.**, Histone synthesis and phosphorylation in regenerating rat liver, *Biochim. Biophys. Acta*, 247, 348, 1971.
268. **Shepherd, G. R., Noland, B. J., and Hardin, J. M.**, Histone phosphorylation in synchronized mammalian cell cultures, *Arch. Biochem. Biophys.*, 142, 299, 1971.

269. **Phillips, D. M. P. and Johns, E. W.,** A fractionation of the histones of group F2(a) from calf thymus, *Biochem. J.,* 94, 127, 1965.
270. **Bellair, J. T. and Mauritzen, C. M.,** Comparison of β-histones isolated from chicken erythrocyte, liver and spleen, and ox thymus, *Biochim. Biophys. Acta,* 133, 263, 1967.
271. **Starbuck, W. C., Mauritzen, C. M., Taylor, C. W., Saroja, I. S., and Busch, H.,** A large scale procedure for isolation of the glycine-rich, arginine-rich histone and the arginine-rich, lysine-rich histone in a highly purified form, *J. Biol. Chem.,* 243, 2038, 1968.
272. **Phillips, D. M. P.,** N-terminal acetyl-peptides from two calf thymus histones, *Biochem. J.,* 107, 135, 1968.
273. **Leclerc, J., Martinage, A., Dautrevaux, M., and Biserte, G.,** Identification of a contaminating protein in pH 7 prepared F2(a) histone, *Biochim. Biophys. Acta,* 194, 342, 1969.
273a. **Sautière, M. P., Tyrou, D., Laine, B., Mizon, J., Lambelin-Breynaert, M. D., Ruggin, P., and Biserte, G.,** Structure primaire de l'histone riche en arginine et en lysine du thymus de Veau, *C. R. Acad. Sci. Paris,* 274, 1422, 1972.
273b. **Hayashi, H. and Iwai, K.,** Calf thymus alanine-rich, leucine-rich histone: Sequences of the tryptic peptides and characteristic distributions of the basic and other residues in the molecule, *J. Biochem.,* 70, 543, 1971.
274. **DeLange, R. J. and Smith, E. L.,** Histones: structure and function, *Annu. Rev. Biochem.,* 40, 279, 1971.
275. **Leclerc, J., Martinage, A., Moschetto, Y, and Biserte, G.,** A comparative study of histone fractions F2(a)1 and F2(a)2, *Eur. J. Biochem.,* 11, 261, 1969.
276. **Blazsek, V. A. and Bukaresti, L.,** The study of the cysteine content of the chicken erythrocyte histone by polarography, *Experienta,* 20, 369, 1964.
277. **Phillips, D. M. P.,** Cysteine in calf-thymus histones, *Biochem. J.,* 97, 669, 1965.
278. **Ord, M. G. and Stocken, L. A.,** Phosphate and thiol groups in histone f3 from rat liver and thymus nuclei, *Biochem. J.,* 102, 631, 1967.
279. **DeLange, R. J., Fambrough, D. M., Smith, E. L., and Bonner, J.,** Calf and pea histone IV. I. Amino acid compositions and the identical COOH-terminal 19-residue sequence, *J. Biol. Chem.,* 243, 5906, 1968.
280. **Sautière, P., Starbuck, W. C., Roth, C., and Busch, H.,** Structural analysis of the glycine-rich, arginine-rich histone. I. Sequence of the carboxyl-terminal portion, *J. Biol. Chem.,* 243, 5899, 1968.
281. **DeLange, R. J., Fambrough, D. M., Smith, E. L., and Bonner, J.,** Calf and pea histone IV. II. The complete amino acid sequence of calf thymus histone IV; presence of ε-N-acetyllysine, *J. Biol. Chem.,* 244, 319, 1969.
282. **Quagliarotti, G., Ogawa, Y., Taylor, C. W., Sautière, P., Jordan, J. J., Starbuck, W. C., and Busch, H.,** Structural analysis of the glycine-rich, arginine-rich histone. II. Sequence of the half of the molecule containing the aromatic amino acids, *J. Biol. Chem.,* 244, 1796, 1969.
283. **Ogawa, Y., Quagliarotti, G., Jordan, J., Taylor, C. W., Starbuck, W. C., and Busch, H.,** Structural analysis of the glycine-rich, arginine-rich histone. III. Sequence of the amino terminal half of the molecule containing the modified lysine residues and the total sequence, *J. Biol. Chem.,* 244, 4387, 1969.
284. **Gershey, E. L., Vidali, G., and Allfrey, V. G.,** Chemical studies of histone acetylation. The occurrence of ε-N-acetyllysine in the f2al histone, *J. Biol. Chem.,* 243, 5018, 1968.
285. **Murray, K.,** The occurrence of epsilon-N-methyl lysine in histones, *Biochemistry,* 3, 10, 1964.
286. **Paik, W. K. and Kim, S.,** ε-N-dimethyllysine in histones, *Biochem. Biophys. Res. Commun.,* 27, 479, 1967.
287. **Subirana, J. A.,** Specific aggregation products of histone fractions (presence of cysteine in F2al from echinoderms), *FEBS Lett.,* 16, 133, 1971.
288. **Sautière, P., Breynaert, M. D., Moschetto, Y., and Biserte, G.,** Sèquence complète des acides aminés de l'histone riche en arginine du thymus de porc, *C. R. Acad. Sci. Paris,* 271, 364, 1970.
289. **Wilson, R. K., Starbuck, W. C., Taylor, C. W., Jordan, J., and Busch, H.,** Structure of the glycine-rich, arginine-rich histone of the Novikoff hepatoma, *Cancer Res.,* 30, 2942, 1970.
290. **Sautière, P., Tyrou, D., Moschetto, Y., and Biserte, G.,** Structure primaire de l'histone riche en glycine et en arginine isolée de la tumeur de chloroleucèmie du rat, *Biochimie,* 53, 479, 1971.
291. **Desai, L., Ogawa, Y., Mauritzen, C. M., Taylor, C. W., and Starbuck, W. C.,** Carboxyl-terminal sequence of the glycine-arginine-rich histone from bovine lymphosarcoma, Novikoff hepatoma and fetal calf thymus, *Biochim. Biophys. Acta,* 181, 146, 1969.
292. **Sautière, P., Tyrou, D., Leclerq, M., and Biserte, G.,** Sèquence C-terminale de l'histone riche en glycine et en arginine isolée de la tumeur de chloroleucèmie du rat, *C. R. Acad. Sci. Paris,* 271, 1131, 1970.
293. **Desai, L. S. and Foley, G. E.,** Homologies in amino acid composition and structure of histone F2al isolated from human leukemic cells, *Biochem. J.,* 119, 165, 1970.
294. **Desai, L. S. and Foley, G. E.,** Studies on the nucleic acids of human lymphocytic cells: acetylation of histones, *Arch. Biochem. Biophys.,* 141, 552, 1970.
295. **Dayhoff, M. O.,** *Atlas of Protein Sequence and Structure,* 4:39, Natl. Biomed. Research Found., Silver Springs, 1969, 362.
296. **Kedes, L. H. and Birnstiel, M. L.,** Reiteration and clustering of DNA sequences complementary to histone messenger RNA, *Nature New Biol.,* 230, 165, 1971.
297. **Brandt, W. F. and Von Holt, C.,** Purification and characterization of the cysteine-containing F3 histone from chicken erythrocytes, *FEBS Lett.,* 14, 338, 1971.

298. **Fambrough, D. M. and Bonner, J.**, Sequence homology and role of cysteine in plant and animal arginine-rich histones, *J. Biol. Chem.*, 243, 4434, 1968.
299. **Yokotsuka, K. and Shimura, K.**, Fractionation of histones by counter-current distribution and structure of arginine-rich histone from calf thymus, in *Histones and Gene Function*, Iwai, K., Ed., The Research Group on Histones and Gene Function in Japan, 1969, 20.
300. **Yokotsuka, K., Kikuchi, A., and Shimura, K.**, in preparation.
301. **DeLange, R. J., Smith, E. L., and Bonner, J.**, Calf thymus histone. III. Sequence of the amino- and carboxyl-terminal regions and of the regions containing lysyl residues modified by acetylation and methylation, *Biochem. Biophys. Res. Commun.*, 40, 989, 1970.
302. **Phillips, D. M. P. and Simson, P.**, Identification of some peptides from an arginine-rich histone and their bearing on the structure of deoxyribonucleohistone, *Biochem. J.*, 82, 236, 1962.
303. **Panyim, S., Sommer, K. R., and Chalkley, R.**, Oxidation of the cysteine-containing histone F3. Detection of an evolutionary mutation in a conservative histone, *Biochemistry*, 10, 3911, 1971.
304. **Bauer, K.**, Homology of a partial sequence of calf thymus histone IV with several nonhistone proteins, *Int. J. Prot. Res.*, 3, 165, 1971.
305. **Mauritzen, C. M. and Stedman, E.**, Cell specificity of β-histones in domestic fowl, *Proc. R. Soc.*, B 150, 299, 1959.
306. **Mauritzen, C. M. and Stedman, E.**, Cell specificity of β-histones from ox, *Proc. R. Soc.*, B 153, 80, 1960.
307. **Laurence, D. J. R., Simson, P., and Butler, J. A. V.**, Studies on histones. 5. The histones of the Crocker sarcoma and spontaneous mammary tumours of mice, *Biochem. J.*, 87, 200, 1963.
308. **Busch, H. Hnilica, L. S., Chien, S. C., Davis, J. R., and Taylor, C. W.**, Isolation and purification of RP2-L, a nuclear protein fraction of the Walker 256 carcinosarcoma, *Cancer Res.*, 22, 637, 1962.
309. **Hnilica, L. S.**, The specificity of histones in chicken erythrocytes, *Experientia*, 20, 13, 1964.
310. **Senshu, T.**, Fractionation of histones from chicken erythrocyte and sea urchin sperm by CM-cellulose chromatography with ethanolic formic acid, *Biochim. Biophys. Acta*, 243, 323, 1971.
311. **Hnilica, L. S. and Johnson, A. W.**, Fractionation and analysis of nuclear proteins in sea urchin embryos, *Exp. Cell Res.*, 63, 261, 1970.
312. **Autran, J. C. and Bourdet, A.**, Histones de germes et de farine de blé, *C. R. Acad. Sci. Paris*, 271, 2050, 1970.
313. **Autran, J. C. and Bourdet, A.**, Constitution électrophorétique des histones de germes de diffèrent types de blés, *C. R. Acad. Sci. Paris*, 272, 2732, 1971.
314. **Towill, L. E. and Nooden, L. D.**, Comparison of histones from different organs of maize seedlings, *Plant Physiol.*, 47, 19, 1971.
315. **Spiker, S. and Chalkley, R.**, Electrophoretic analysis of histones from Gibberellic acid-treated dwarf peas, *Plant Physiol.*, 47, 342, 1971.
316. **Gorovsky, M. A.**, Studies on nuclear structure and function in *Tetrahymena pyriformis*, *J. Cell. Biol.*, 47, 631, 1970.
317. **Hamana, K. and Iwai, K.**, Fractionation and characterization of *Tetrahymena* histone in comparison with mammalian histone, *J. Biochem.*, 69, 1097, 1971.
318. **Duerre, J. A. and Gaitonde, M. K.**, Isolation and electrophoretic identification of histones of rat brain and liver, *J. Neurochem.*, 18, 1921, 1971.
319. **Netrawali, M. S.**, On the presence of histones in *Euglena gracilis var. Bacillaris*, *Exp. Cell Res.*, 63, 422, 1970.
320. **Stenroos, O. O. and Reichard, S. M.**, Metamorphic alterations in nuclear RNA and histones, *Am. Zool.*, 10, 321, 1970.
321. **Neelin, J. M. and Butler, G. C.**, A composition of histones from chicken tissues by zone electrophoresis in starch gel, *Can. J. Biochem. Physiol.*, 39, 485, 1961.
322. **Vendrely, R., Genty, N., and Coirault, Y.**, Étude comparée d'histones d'èrythrocytes chez diverses espèces animales, *Bull. Soc. Chim. Biol.*, 47, 2233, 1965.
323. **Spelsberg, T. C. and Sarkissian, I. V.**, Isolation and electrophoresis of nuclear proteins of bean, *Phytochem.*, 7, 2083, 1968.
324. **Lindsay, D. T.**, Histones from developing tissues of the chicken: heterogeneity, *Science*, 144, 420, 1964.
325. **Driedger, A., Johnson, L. D., and Marko, A. M.**, Electrophoresis of histones on polyacrylamide gels, *Can. J. Biochem. Physiol.*, 41, 2507, 1963.
326. **Beeson, J. L. and Triplett, E. L.**, Localization and characterization of rat and chicken histones, *Exptl. Cell Res.*, 48, 61, 1967.
327. **Neidle, A. and Waelsch, H.**, Histones: species and tissue specificity, *Science*, 145, 1059, 1964.
328. **Kischer, C. W., Gurley, L. R., and Shepherd, G. R.**, Nuclear histones and early embryogenesis of the chick, *Nature*, 212, 304, 1966.
329. **Kischer, C. W. and Hnilica, L. S.**, Analysis of histones during organogenesis, *Exp. Cell Res.*, 48, 424, 1967.
330. **Shepherd, G. R. and Gurley, L. R.**, High resolution disc electrophoresis of histones. I. An improved method, *Anal. Biochem.*, 14, 356, 1966.
331. **Gurley, L. R. and Shepherd, G. R.**, High-resolution disc electrophoresis of histones. II. Application to whole and fractionated preparations from various laboratories, *Anal. Biochem.*, 14, 364, 1966.
332. **Billen, D. and Hnilica, L. S.**, Inhibition of DNA synthesis by histones, in *The Nucleohistones*, Bonner, J. and Ts'o, P. O. P., Eds., Holden-Day, San Francisco, 1964, 289.

333. **Sandberg, A. L., Liss, M., and Stollar, B. D.,** Rabbit antibodies induced by calf thymus histone- serum albumin complexes, *J. Immunol.*, 98, 1182, 1967.
334. **Stollar, B. D. and Ward, M.,** Rabbit antibodies to histone fractions as specific reagents for preparative and comparative studies, *J. Biol. Chem.*, 245, 1261, 1970.
335. **Schmidt, F., Schwenke, K. D., and Gassmann, B.,** Zur Immunologie der Histone aus Rattenhepatom und normaler Rattenleber, *Naturwissenschaften*, 54, 567, 1967.
336. **Sluyser, M., Rümke, P., and Hekman, A.,** Antigenicity of histones: Comparative studies on histones with very high lysine content from various sources, *Immunochemistry*, 6, 494, 1969.
337. **Fukazawa, C. and Shimura, K.,** The demonstration of species specificity of histones by immunological techniques, *Biochim. Biophys. Acta*, 154, 619, 1968.
338. **Black, M. M. and Ansley, H. R.,** Histone staining with ammoniacal silver, *Science*, 143, 693, 1964.
339. **Black, M. M., Ansley, H., and Mandl, R.,** On cell specificity of histones, *Arch. Pathol.*, 78, 350, 1964.
340. **Black, M. M. and Ansley, H. R.,** Histone specificity revealed by ammoniacal silver staining, *J. Histochem. Cytochem.*, 14, 177, 1966.
341. **Neelin, J. M.,** Histones from chicken erythrocyte nuclei, in *The Nucleohistones*, Bonner, J. and Ts'o, P. O. P., Eds., Holden-Day, San Francisco, 1964, 66.
342. **Johns, E. W. and Diggle, J. H.,** A method for the large scale preparation of the avian erythrocyte specific histone F2c, *Eur. J. Biochem.*, 11, 495, 1969.
343. **Champagne, M., Mazen, A., and Wilhelm, X.,** Histones d'erythrocytes de poulets. I. Fractionnement des histones totales et isolement d'une histone specifique, *Bull. Soc. Chim. Biol.*, 50, 1261, 1968.
344. **Vidali, G. and Neelin, J. M.,** A comprehensive fractionation procedure for avian erythrocyte histones, *Eur. J. Biochem.*, 5, 330, 1968.
345. **Vidàli, G. and Neelin, J. M.,** The stability of acid soluble chromosomal protein from avian erythrocytes, *Can. J. Biochem.*, 46, 781, 1968.
346. **Neelin, J. M. and Vidali, G.,** Ribosomal proteins from goose reticulocytes are not histones, *Can. J. Biochem.*, 46, 1507, 1968.
347. **Adams, G. H. M. and Neelin, J. M.,** A modified fractionation procedure for avian erythrocyte histones, *Can. J. Biochem.*, 47, 1121, 1969.
348. **Greenaway, P. J. and Murray, K.,** Heterogeneity and polymorphism in chicken erythrocyte histone fraction V, *Nature New Biol.*, 229, 233, 1971.
349. **Greenaway, P. J.,** Studies on the primary structure of chicken erythrocyte histone fraction V, *Biochem. J.*, 124, 319, 1971.
350. **Purkayastha, R. and Neelin, J. M.,** Comparison of histones from avian erythroid tissues by zone electrophoresis, *Biochim. Biophys. Acta*, 117, 468, 1966.
351. **Neelin, J. M.,** Variability in cell-specific and common histones of avian erythrocytes, *Can. J. Biochem.*, 46, 241, 1968.
352. **Dick, C. and Johns, E. W.,** A quantitative comparison of histones from immature and mature erythroid cells of the duck, *Biochim. Biophys. Acta,* 175, 414, 1969.
353. **Champagne, M., Pouyet, J., Ouellet, L., and Garel, A.,** Histones d'érythrocytes de poulets. II. Étude physicochimique de l'histone spécifique, *Bull. Soc. Chim. Biol.*, 52, 377, 1970.
354. **Diggle, J. H. and Peacocke, A. R.,** The molecular weights and association of histones of chicken erythrocytes, *FEBS Lett.*, 18, 138, 1971.
355. **Tsai, Y. H. and Hnilica, L. S.,** Histones of turtle erythrocytes, in preparation.
356. **Palau, J., Ruiz-Carrillo, A., and Subirana, J. A.,** Histones from sperm of the sea urchin *Arbacia lixula, Eur. J. Biochem.*, 7, 209, 1969.
357. **Vendrely, R. and Picaud, M.,** Etude comparée des histones d'érythrocytes de differentes especes animales, *Exp. Cell. Res.*, 49, 13, 1968.
358. **Nelson, R. D. and Yunis, J. J.,** Species and tissue specificity of very lysine-rich and serine-rich histones, *Exp. Cell Res.*, 57, 311, 1969.
359. **Hnilica, L. S. and Edwards, L. J.,** The evolution of histones, *Abstracts of the 7th. Int. Congress Biochem.*, Tokyo, Vol. 3, 1967, 513.
360. **Marushige, K. and Dixon, G. H.,** Developmental changes in chromosomal composition and template activity during spermatogenesis in trout testis, *Dev. Biol.*, 19, 397, 1969.
361. **Johnson, A. W. and Rothstein, H.,** Amphibian lens histones and their relation to the cell cycle, *J. Gen. Physiol.*, 55, 688, 1970.
362. **Panyim, S. and Chalkley, R.,** A new histone found only in mammalian tissues with little cell division, *Biochem. Biophys. Res. Commun.*, 37, 1042, 1969.
363. **Daly, M. M., Mirsky, A. E., and Ris, H.,** The amino acid composition and some properties of histones, *J. Gen. Physiol.*, 34, 439, 1951.
364. **Fischer, H. and Kreuzer, L., Gallin,** *H. S. Z. Physiol. Chem.*, 293, 176, 1953.
365. **Ando, T. and Sawada, F.,** Studies on protamines. IX. Fractionation of clupeine, *J. Biochem.*, 49, 252, 1961.

366. **Ingles, C. J. and Dixon, G. H.,** Phosphorylation of protamine during spermatogenesis in trout testis, *Proc. Natl. Acad. Sci. U.S.,* 58, 1011, 1967.

367. **Ando, T. and Watanabe, S.,** New method for fractionation of protamines and the amino acid sequences of salmine and three components of iridine, *Int. J. Prot. Res.,* 1, 221, 1969.

368. **Ando, T., Iwai, K., Ishii, S. I., Azegami, M., and Nakahara, C.,** Chemical structure of one component of clupeine, *Biochim. Biophys. Acta,* 56, 628, 1962.

369. **Ando, T. and Suzuki, K.,** The amino acid sequence of the second component of clupeine, *Biochim. Biophys. Acta,* 121, 427, 1966.

370. **Ando, T. and Suzuki, K.,** Amino acid sequence of the third component of clupeine, *Biochim. Biophys. Acta,* 140, 375, 1967.

371. **Iwai, K., Nakahara, C., and Ando, T.,** Studies on protamines. XV. The complete amino acid sequence of the Z component of clupeine. Application of N→O acyl rearrangement and selective hydrolysis in sequence determination, *J. Biochem.,* 69, 493, 1971.

372. **Smithies, O., Connel, G. E., and Dixon, G. H.,** Chromosomal rearrangements and the evolution of haptoglobin genes, *Nature,* 196, 232, 1962.

373. **Black, J. A. and Dixon, G. H.,** Evolution of protamine: a further example of partial gene duplication, *Nature,* 216, 152, 1967.

374. **Felix, K. and Hashimoto, C.,** Ernuete Untersuchungen über die chemische Struktur von Protaminen, *H. S. Z. Physiol. Chem.,* 330, 205, 1963.

375. **Ando, T. and Hashimoto, C.,** Studies on protamines. IV. The basic proteins of the testis cell nuclei of adult and immature rainbow-trouts in breeding season, *J. Biochem.,* 45, 453, 1958.

376. **Waldschmidt-Leitz, E. and Gutermann, H.,** Über die Struktur der Protamine. VII. Über den Einfluss des Alters von Fischen auf die Zusammensetzung ihrer Protamine, *H. S. Z. Physiol. Chem.,* 344, 50, 1966.

377. **Ando, T. and Sawada, F.,** Studies on protamines. VIII. Chromatographic studies on the heterogeneity of iridine, *J. Biochem.,* 48, 886, 1960.

378. **Ando, T. and Sawada, F.,** Studies on protamines. X. Heterogeneity of clupeine from single herring, *J. Biochem.,* 51, 416, 1962.

379. **Wilkins, M. H. F.,** Molecular structure of deoxyribonucleoproteins, in *Nucleoproteins,* Stoops, R., Ed., Interscience, New York, 1959, 45.

380. **Wilkins, M. H. F.,** Physical studies on the molecular structure of deoxyribose nucleic acid and nucleoprotein, *Cold Spring Harbor Symp. Quant. Biol.,* 21, 75, 1956.

381. **Akinrimisi, E. O., Bonner, J., and Ts'o, P. O. P.,** Binding of basic proteins to DNA, *J. Mol. Biol.,* 11, 128, 1965.

382. **Bloch, D. P.,** A catalog of sperm histones, *Genetics,* 61, Suppl. 1, 93, 1969.

383. **Hultin, T.,** Some physiological effects of basic sperm protein, *Arkiv Kemi. Min. Geol.,* 24B,(12), 1947.

384. **Hultin, T. and Herne, R.,** Amino acid analysis of a basic protein fraction from sperm nuclei of some different invertebrates, *Arkiv Kemi. Min. Geol.,* A 26,(20), 1948.

385. **Hamer, D.,** The composition of the basic proteins of echinoderm sperm, *Biol. Bull.,* 108, 35, 1955.

386. **Subirana, J. A. and Palau, J.,** Histone-like proteins from the sperm of echinoderms, *Exp. Cell Res.,* 53, 471, 1968.

387. **Messineo, L.,** Amino acid analysis of glycine-extracted deoxyribonucleoproteins, *Arch. Biochem. Biophys.,* 133, 92, 1969.

388. **Marushige, K. and Ozaki, H.,** Properties of isolated chromatin from sea urchin embryo, *Dev. Biol.,* 16, 474, 1967.

389. **Subirana, J. A.,** Nuclear proteins from a somatic and germinal tissue of the echinoderm *Holothuria tubulosa, Exp. Cell Res.,* 63, 253, 1970.

390. **Subirana, J. A., Palau, J., Pladellorens, M., Cozcolluela, C., and Ruiz-Carillo, A.,** Proteines associades a l'ADN en l'esperma i els sens canvis en l'espermiogènesi, *Treballs de la Societat Catalana de Biologia,* 30, 31, 1971.

391. **Das, N. K., Micou-Eastwood, J., and Alfert, M.,** Cytochemical and biochemical properties of basic proteins of *Urechis* acrosomes, *J. Cell Biol.,* 35, 455, 1967.

392. **Hnilica, L. S., McClure, M. E., and Spelsberg, T. C.,** Histone biosynthesis and the cell cycle, in *The Histones and Nucleohistones,* Phillips, D. M. P., Ed., Plenum Press, London, 1971, 187.

393. **Marushige, K. and Dixon, G. H.,** Transformation of trout testis chromatin, *J. Biol. Chem.,* 246, 5799, 1971.

394. **Wigle, D. T. and Dixon, G. H.,** A new histone from trout testis, *J. Biol. Chem.,* 246, 5636, 1971.

395. **Bloch, D. P. and Hew, H. Y. C.,** Schedule of spermatogenesis in the pulmonate snail *Helix aspersa,* with special reference to histone transition, *J. Biophys. Biochem. Cytol.,* 7, 515, 1960.

396. **Bloch, D. P.,** Symposium: Synthetic processes in the cell nucleus. I. Histones synthesis in non-replicating chromosomes, *J. Histochem. Cytochem.,* 10, 137, 1962.

397. **Das, C. C., Kaufmann, B. P., and Gay, H.,** Autoradiographic evidence of synthesis of an arginine-rich histone during spermiogenesis in *Drosophila melanogaster, Nature,* 204, 1008, 1964.

398. **Das, C. C., Kaufmann, B. P., and Gay, H.,** Histone-protein transition in *Drosophila melanogaster.* I. Changes during spermatogenesis, *Exp. Cell Res.,* 35, 507, 1964.

399. **Bloch, D. P. and Brack, S. D.,** Evidence for the cytoplasmic synthesis of nuclear histone during spermiogenesis in the grasshopper *Chortophaga viridifasciata (de Geer), J. Cell Biol.,* 22, 327, 1964.

400. **Monesi, V.**, Autoradiographic evidence of a nuclear histone synthesis during mouse spermiogenesis in the absence of detectable quantities of nuclear ribonucleic acid, *Exp. Cell Res.*, 36, 683, 1964.
401. **Monesi, V.**, Synthetic activities during spermatogenesis in the mouse, *Exp. Cell Res.*, 39, 197, 1965.
402. **Lison, L.**, Variation de la basophilie pendant la maturation du spermatozoide chez le rat et sa signification histochimique, *Acta Histochem.*, 2, 47, 1955.
403. **Vaughn, J. C.**, The relationship of the "sphere chromatophile" to the fate of displaced histones following histone transition in rat spermiogenesis, *J. Cell Biol.*, 31, 257, 1966.
404. **Gledhill, B. L., Gledhill, M. P., Rigler, R., and Ringertz, N. R.**, Changes in deoxyribonucleoprotein during spermiogenesis in the bull, *Exp. Cell Res.*, 41, 652, 1966.
405. **Zirkin, B. R.**, The protein composition of nuclei during spermiogenesis in the leopard frog, *Rana pipiens, Chromosoma*, 31, 231, 1970.
406. **Deleon, R. P. and Vaughn, J. C.**, Histone metabolism during spermatogenesis in the toad *Bufo americanus, Am. Zool.*, 10, 317, 1970.
407. **Bril-Petersen, R. and Westenbrink, H. G. K.**, A structural basic protein as a counterpart of deoxyribonucleic acid in mammalian spermatozoa, *Biochim. Biophys. Acta*, 76, 152, 1963.
408. **Henricks, D. M. and Mayer, D. T.**, Characterization of the basic proteins associated with DNA in mammalian spermatozoa, *Proc. Soc. Exp. Biol. Med.*, 119, 769, 1965.
409. **Henricks, D. M. and Mayer, D. T.**, Isolation and characterization of a basic keratin-like protein from mammalian spermatozoa, *Exp. Cell Res.*, 40, 402, 1965.
410. **Coelingh, J. P., Rozijn, T. H., and Monfoort, C. H.**, Isolation and partial characterization of a basic protein from bovine sperm heads, *Biochim. Biophys. Acta*, 188, 353, 1969.
411. **Lam, D. M. K. and Bruce, W. R.**, The biosynthesis of protamine during spermatogenesis of the mouse: extraction, partial characterization, and site of synthesis, *J. Cell Physiol.*, 78, 13, 1971.
412. **Vaughn, J. C., Chaitoff, J., Deleon, R., Garland, C., and Thomson, L.**, Changing nuclear histone patterns during development. II. Isolation and partial characterization of "decapodine" from sperm cells of the crab *Emerita analoga, Exp. Cell Res.*, 54, 362, 1969.
413. **Vaughn, J. C. and Locy, R. D.**, Changing nuclear histone patterns during development. III. The deoxyribonucleic acid content of spermatogenic cells in the crab *Emerita analoga, J. Histochem. Cytochem.*, 17, 591, 1969.
414. **Langreth, S. G.**, Spermiogenesis in Cancer crabs, *J. Cell Biol.*, 43, 575, 1969.
415. **Chevaillier, P.**, Contribution à l'étude du complexe ADN-histone dans le spermatozoide du pagure *Eupagurus bernhardus* (Crustace decapode), *J. Microscop.*, 5, 739, 1966.
416. **Chevaillier, P.**, Étude cytochimique ultrastructurale des nucléoprotéines dans le spermatozoide du pagure *Eupagurus bernhardus* (Crustace decapode), *J. Microscop.*, 7, 107, 1968.
417. **Chevaillier, P.**, Mise en évidence et étude cytochimique d'une proteine basique extranucléaire dans les spermatozöides des crustacés décapodes, *J. Cell Biol.*, 32, 547, 1967.
418. **Vaughn, J. C. and Hinsch, G. W.**, Absorbance melting profile analysis of DNA and chromatin in somatic cells and sperm of the spider crab, *Libinia emarginata, Biol. Bull.*, 139, 441, 1970.
419. **Pladellorens, M. and Subirana, J. A.**, Ultrastructural study of chromatin changes during spermiogenesis in the echinoderms *Holothuria polli* and *Holothuria tubulosa, Seventh Int. Congr. Electron Microscop.*, Grenoble, Abstracts, 1970, 645.
420. **Sheridan, W. F. and Stern, H.**, Histones in meiosis, *Exp. Cell Res.*, 45, 323, 1967.
421. **Hotta, Y. and Stern, H.**, A DNA-binding protein in meiotic cells of *Lilium, Dev. Biol.*, 26, 87, 1971.
422. **Zirkin, B. R.**, The fine structure of nuclei in mature sperm. I. Application of the Langmuir trough-critical point method to histone-containing sperm nuclei, *J. Ultrastruct. Res.*, 36, 237, 1971.
423. **Isawa, M., Allfrey, V. G., and Mirsky, A. E.**, Composition of the nucleus and chromosomes in the lampbrush stage of the newt oocyte, *Proc. Natl. Acad. Sci. U.S.*, 50, 811, 1963.
424. **Davenport, R. and Davenport, J. C.**, A cytochemical study of cytoplasmic basic proteins in the ascidian oocyte, *J. Cell Biol.*, 25, 319, 1965.
425. **Davenport, R. and Davenport, J. C.**, Cytoplasmic basic proteins in the oocytes of three species of molluscs, *Exp. Cell Res.*, 39, 74, 1965.
426. **Davenport, R. and Davenport, J. C.**, A cytochemical study of cytoplasmic basic proteins in echinoderm oogenesis, *Exp. Cell Res.*, 42, 429, 1966.
427. **Bäckström, S.**, A complex between basic proteins and acid polysaccharides in sea urchin oocytes and eggs, *Acta Embryol. Morphol. Exp.*, 9, 37, 1966.
428. **Bäckström, S.**, Distribution of basic proteins in centrifuged sea urchin eggs, *Exp. Cell Res.*, 43, 578, 1966.
429. **Horn, E. C.**, Extranuclear histone in the amphibian oocyte, *Proc. Natl. Acad. Sci. U.S.*, 49, 257, 1962.
430. **Moore, B. C.**, Histones and differentiation, *Proc. Natl. Acad. Sci. U.S.*, 50, 1018, 1963.
431. **Cowden, R. C.**, Cytochemical studies on cytoplasmic RNA-associated basic proteins in oocytes, somatic cells, and ribosomes, *Histochemie*, 6, 226, 1966.
432. **Repsis, L. C.**, Acid soluble nuclear proteins in developmental stages of *Lytechinus variegatus, Exp. Cell Res.*, 48, 146, 1968.

433. **Thaler, M. M., Cox, M. C. L., and Villee, C. A.,** Histones in early embryogenesis. Developmental aspects of composition and synthesis, *J. Biol. Chem.*, 245, 1479, 1970.
434. **Benttinen, L. C. and Comb, D. G.,** Early and late histones during the sea urchin development, *J. Mol. Biol.*, 57, 355, 1971.
435. **Neelin, J. M. and Connel, G. E.,** Zone electrophoresis of chicken erythrocyte histone in starch gel, *Biochim. Biophys. Acta,* 31, 539, 1959.
436. **Biserte, G., Sautière, P., Houard, C., and Devys, M.,** Étude comparative du comportement en électrophorèse en gel d'amidon des histones de thymus de veau, *C. R. Séances Soc. Biol.*, 156, 1823, 1962.
437. **Sung, M. and Smithies, O.,** Differential elution of histones from geltrapped nuclei, *Biopolymers,* 7, 39, 1969.
438. **McAllister, H. C., Wan, Y. C., and Irvin, J. L.,** Electrophoresis of histones and histone fractions on polyacrylamide gels, *Anal. Biochem.*, 5, 321, 1963.
439. **Vande Woude, G. F. and Davis, F. F.,** The fractionation of histones by electrophoresis in polyacrylamide gel, *Anal. Biochem.*, 12, 444, 1965.
440. **Johns, E. W.,** The electrophoresis of histones in polyacrylamide gel and their quantitative determination, *Biochem. J.*, 104, 78, 1967.
441. **Johns, E. W. and Forrester, S.,** The identification of the five main histone fractions by comparative electrophoresis in polyacrylamide gel, *J. Chromatogr.*, 55, 429, 1971.
442. **Cruft, H. J., Mauritzen, C. M., and Stedman, E.,** Abnormal properties of histones from malignant cells, *Nature,* 174, 580, 1954.
443. **Toennies, G. and Bakay, B.,** Characterization and total recovery of the component proteins of a deoxyribonucleoprotein, *Nature,* 176, 696, 1955.
444. **Ui, N.,** Sedimentation and diffusion studies of calf thymus histones, *Bull. Chem. Soc. Japan,* 30, 801, 1957.
445. **Ui, N.,** The aggregation of calf thymus histone, *Bull. Chem. Soc. Japan,* 30, 806, 1957.
446. **Johns, E. W.,** The preparation and characterization of histones, in *The Histones and Nucleohistones,* Phillips, D. M. P., Ed., Plenum Press, London, 1971, 1.
447. **Bradbury, E. M. and Crane-Robinson, C.,** Physical and conformational studies of histones and nucleohistones, in *The Histones and Nucleohistones,* Phillips, D. M. P., Ed., Plenum Press, London, 1971, 85.
448. **Trautman, R. and Crampton, C. F.,** Application of the Archibald principle for the ultracentrifugal determination of the molecular weight in urea solutions of histone fractions from calf thymus, *J. Am. Chem. Soc.*, 81, 4036, 1959.
449. **Fish, W. W., Reynolds, J. A., and Tanford, C.,** Gel chromatography of proteins in denaturing solvents. Comparison between sodium dodecyl sulfate and guanidine hydrochloride as denaturants, *J. Biol. Chem.*, 245, 5166, 1970.
450. **Williams, J. G., and Gratzer, W. B.,** Limitations of the detergent-polyacrylamide gel electrophoresis method for molecular weight determination of proteins, *J. Chromatogr.*, 57, 121, 1971.
451. **Edwards, P. A. and Shooter, K. S.,** Ultracentrifuge studies of histone fractions from calf thymus deoxyribonucleoprotein, *Biochem. J.*, 114, 227, 1969.
452. **Bradbury, E. M. and Crane-Robinson, C.,** Physical studies of the molecular configurations of histone and of nucleohistone, in *The Nucleohistones,* Bonner, J. and Ts'o, P. O. P., Eds., Holden-Day, San Francisco, 1964, 117.
453. **Bradbury, E. M., Price, W. C., Wilkinson, G. R., and Zubay, G.,** Polarized infrared studies of nucleoproteins. II. Nucleohistones, *J. Mol. Biol.*, 4, 50, 1962.
454. **Elliot, A., Bradbury, E. M., and Zubay, G.,** Optical rotatory dispersion studies of solid films and solutions of histone, *J. Mol. Biol.*, 4, 61, 1962.
455. **Urnes, P. and Doty, P.,** Optical rotation and the conformation of polypeptides and proteins, *Adv. Protein Chem.*, 16, 401, 1961.
456. **Jirgensons, B. and Hnilica, L. S.,** The conformational changes of calf thymus histone fractions as determined by the optical rotatory dispersion, *Biochim. Biophys. Acta,* 109, 241, 1965.
457. **Bradbury, E. M., Crane-Robinson, C., Phillips, D. M. P., Johns, E. W., and Murray, K.,** Conformational investigation of histones, *Nature,* 205, 1315, 1965.
458. **Ramm, E. I., Bolotina, I. A., Volkenstein, M. V., and Vorobyev, V. I.,** Studies on the conformation of the histones by the method of optical rotatory dispersion, *Mol. Biol. USSR,* 2, 818, 1968.
459. **Ramm, E. I., Birstein, T. M., Bolotina, I. A., Volkenstein, M. V., Vorobyev, V. I., Dmitrenko, L. V., and Nekrasova, T. N.,** Investigation of histone structure, *Mol. Biol. USSR,* 4, 118, 1970.
460. **Jirgensons, B.,** *Optical Rotatory Dispersion of Proteins and Other Macromolecules,* Springer-Verlag, Berlin, 1969.
461. **Tamburro, A. M., Scatturin, A., and Vidali, G.,** On the conformation of histones: CD studies on the lysine- and serine-rich fractions from avian erythrocytes, *Int. J. Protein Chem.*, 2, 127, 1970.
462. **Jirgensons, B. and Hnilica, L. S.,** The ability of anionic detergents to fold disordered polypeptide chains of histones in the α-helical form, *J. Am. Chem. Soc.*, 88, 2431, 1966.
463. **Jirgensons, B.,** Effect of anionic detergents on the optical rotatory dispersion of proteins, *J. Biol. Chem.*, 241, 4855, 1966.
464. **Jirgensons, B.,** Effects of n-propyl alcohol and detergents on the optical rotatory dispersion of α-chymotrypsinogen, β-casein, histone fraction F1, and soybean trypsin inhibitor, *J. Biol. Chem.*, 242, 912, 1967.

464a. **Jirgensons, B., Hnilica, L. S., and Capetillo, S. C.,** Viscosity and conformation of calf thymus histones, *Macromol. Chem.*, 97, 216, 1966.

465. **Wagner, T. E.,** Circular dichroism study of the F2al histone in the presence of polyvinylphosphate and DNA, *Nature,* 227, 65, 1970.

466. **Tuan, D. Y. H. and Bonner, J.,** Optical absorbance and optical rotatory dispersion studies on calf thymus nucleohistone, *J. Mol. Biol.,* 45, 59, 1969.

467. **Boublik, M., Bradbury, E. M., Crane-Robinson, C., and Rattle, H. W. E.,** Optical rotatory dispersion and nuclear magnetic resonance studies of helix→coil transitions in poly-L-arginine, poly-L-lysine, and histones, *Eur. J. Biochem.,* 12, 258, 1970.

468. **Boublik, M., Bradbury, E. M., and Crane-Robinson, C.,** An investigation of the conformational changes of histones F1 and F2al by proton magnetic resonance spectroscopy, *Eur. J. Biochem.,* 14, 486, 1970.

469. **Boublik, M., Bradbury, E. M., Crane-Robinson, C., and Johns, E. W.,** An investigation of the conformational changes of histone F2b by high resolution nuclear magnetic resonance, *Eur. J. Biochem.,* 17, 151, 1970.

470. **Blout, E. R.,** The dependence of the conformation of polypeptides and proteins upon amino-acid composition, in *Polyamino Acids, Polypeptides, and Proteins,* Stahman, M., Ed., University of Wisconsin Press, Madison, 1962, 275.

471. **Davies, D. R.,** A correlation between amino acid composition and protein structure, *J. Mol. Biol.,* 9, 605, 1964.

472. **Guzzo, A. V.,** The influence of amino-acid sequence on protein structure, *Biophys. J.,* 5, 809, 1965.

473. **Prothero, J. W.,** Correlation between the distribution of amino acids and alpha helices, *Biophys. J.,* 6, 367, 1966.

474. **Schiffer, M. and Edmundson, A. B.,** Use of helical wheels to represent the structures of proteins and to identify segments with helical potential, *Biophys. J.,* 7, 121, 1967.

475. **Boublik, M., Bradbury, E. M., Crane-Robinson, C., and Rattle, H. W. E.,** Proton magnetic resonance studies of the interactions of histones F1 and F2b with DNA, *Nature New Biol.,* 229, 149, 1971.

476. **Bradbury, E. M., Cary, P. D., Crane-Robinson, C., Riches, P. L., and Johns, E. W.,** Proton magnetic resonance studies of conformational changes in cleaved halves of histone F2b, *Nature New Biol.,* 233, 265, 1971.

477. **Suzuki, K. and Ando, T.,** Studies on the conformation of clupeine, *J. Biochem.,* 63, 403, 1968.

478. **Stedman, E. and Stedman, E.,** The chemical nature and functions of the components of cell nuclei, *Cold Spring Harbor Symp. Quant. Biol.,* 12, 224, 1947.

479. **Allfrey, V. G.,** Biosynthetic reactions in the cell nucleus, in *Aspects of Protein Synthesis,* Anfinsen, C. B., Ed., Academic Press, New York, 1970, 247.

480. **Busch, H. and Smetana, K.,** Nucleolar structural proteins and enzymes, in *The Nucleolus,* Academic Press, New York, 1970, 319.

481. **Allfrey, V. G., Mirsky, A. E., and Osawa, S.,** Protein synthesis in isolated cell nuclei, *Nature,* 176, 1042, 1955.

482. **Allfrey, V. G., Mirsky, A. E., and Osawa, S.,** Protein synthesis in isolated cell nuclei, *J. Gen. Physiol.,* 40, 451, 1957.

483. **Allfrey, V. G., Hopkins, J. W., Frenster, J. H., and Mirsky, A. E.,** Reactions governing incorporation of amino acids into the proteins in isolated cell nuclei, *Ann. N. Y. Acad. Sci.,* 88, 722, 1960.

484. **Allfrey, V. G., Littau, V. C., and Mirsky, A. E.,** Methods for the purification of thymus nuclei and their application to studies of nuclear protein synthesis, *J. Cell Biol.,* 21, 213, 1964.

485. **Reid, B. R., Stellwagen, R. H., and Cole, R. D.,** Further studies on the biosynthesis of very lysine-rich histones in isolated nuclei, *Biochim. Biophys. Acta,* 155, 593, 1968.

486. **Schultze, B., Citoler, P., Hempel, K., Citoler, K., and Maurer, W.,** Cytoplasmic protein synthesis in cells of various types and its correlation to nuclear protein synthesis, *Symp. Int. Soc. Cell Biol.,* 4, 107, 1965.

487. **Mattingly, S. A.,** Nuclear protein synthesis in *Vicia faba, Exp. Cell Res.,* 29, 314, 1963.

488. **Sirlin, J. L.,** The nucleolus, *Progr. Biophys. Biophys. Chem.,* 12, 27, 1962.

489. **Birnstiel, M. L. and Flamm, W. G.,** Intranuclear site of histone synthesis, *Science,* 145, 1435, 1964.

490. **Birnstiel, M. L. and Hyde, B. B.,** Protein synthesis by isolated pea nucleoli, *J. Cell Biol.,* 18, 41, 1963.

491. **Hnilica, L. S., Liau, M. C., and Hurlbert, R. B.,** Biosynthesis and composition of histones in Novikoff hepatoma nuclei and nucleoli, *Science,* 152, 521, 1966.

492. **Sekeris, C. E., Schmid, W., Gallwitz, D., and Lukacz, I.,** Protein Synthese in Rattenleber-Chromatin und Chromatin-Fraktionen, *Angew. Chem.,* 78, 601, 1966.

493. **Ono, H. and Terayama, H.,** Amino acid incorporation into proteins in isolated rat liver nuclei, *Biochim. Biophys. Acta,* 166, 175, 1968.

494. **Zimmerman, E. F., Hackney, J., Nelson, P., and Arias, I. M.,** Protein synthesis in isolated nuclei and nucleoli of HeLa cells, *Biochemistry,* 8, 2636, 1969.

495. **Gallwitz, D. and Mueller, G. C.,** Protein synthesis in nuclei isolated from HeLa cells, *Eur. J. Biochem.,* 9, 431, 1969.

496. **Helmsing, P. J.,** Protein synthesis of polytene nuclei *in vitro, Biochim. Biophys. Acta,* 224, 579, 1970.

497. **Helmsing, P. J.,** Isolation and partial characterization of the labeled product from polytene nuclei incubated *in vitro* with amino acids, *Biochim. Biophys. Acta,* 232, 733, 1971.

498. **Burdman, J. A., Haglid, K., and Dravid, A. R.,** Protein synthesis in fractions from isolated brain cell nuclei, *J. Neurochem.,* 17, 669, 1970.

499. **Bloch, D. P.,** Genetic implications of histone behavior, *J. Cell. Comp. Physiol.,* 62, Part II, 87, 1963.

500. **Ling, V., Trevithick, J. R., and Dixon, G. H.,** The biosynthesis of protamine in trout testis. I. Intracellular site of synthesis, *Can. J. Biochem.,* 47, 51, 1969.

501. **Dixon, G. H. and Smith, M.,** Nucleic acids and protamine in salmon testes, *Progr. Nucleic Acid Res. Mol. Biol.,* 8, 9, 1968.

502. **Robbins, E. and Borun, T. W.,** The cytoplasmic synthesis of histones in HeLa cells and its temporal relationship to DNA replication, *Proc. Natl. Acad. Sci. U.S.,* 57, 409, 1967.
503. **Borun, T. W., Scharff, M., and Robbins, E.,** Rapidly labeled, polysome associated RNA having the properties of histone messenger, *Proc. Natl. Acad. Sci. U.S.,* 58, 1977, 1967.
504. **Gallwitz, D. and Mueller, G. C.,** Histone synthesis *in vitro* by cytoplasmic microsomes from HeLa cells, *Science,* 163, 1351, 1969.
505. **Gallwitz, D. and Mueller, G. C.,** Histone synthesis *in vitro* on HeLa cell microsomes. The nature of the coupling to deoxyribonucleic acid synthesis, *J. Biol. Chem.,* 244, 5947, 1969.
506. **Gallwitz, D. and Mueller, G. C.,** RNA from HeLa cell microsomes with properties of histone messenger, *FEBS Lett.,* 6, 83, 1970.
507. **Griffin, A. C., Ward, V. C., Wade, J., and Ward, D. N.,** Synthesis and release of basic proteins or peptides by tumor ribosomes, *Biochim. Biophys. Acta,* 72, 500, 1963.
508. **Nemer, M. and Lindsay, D.,** Evidence that the s-polysomes of early sea urchin embryos may be responsible for the synthesis of chromosomal histones, *Biochem. Biophys. Res. Commun.,* 35, 156, 1969.
509. **Kedes, L. H., Gross, P. R., Cognetti, G., and Hunter, A. L.,** Synthesis of nuclear and chromosomal proteins on light polyribosomes during cleavage in the sea urchin embryos, *J. Mol. Biol.,* 45, 337, 1969.
510. **Moav, B. and Nemer, M.,** Histone synthesis. Assignment to a special class of polyribosomes in sea urchin embryos, *Biochemistry,* 10, 881, 1971.
511. **Johnson, A. W. and Hnilica, L. S.,** Cytoplasmic and nuclear basic protein synthesis during early sea urchin development, *Biochim. Biophys. Acta,* 246, 141, 1971.
512. **Liew, C. C., Haslett, G. W., and Allfrey, V. G.,** N-acetyl-seryl-t-RNA and polypeptide chain initiation during histone biosynthesis, *Nature,* 226, 414, 1970.
513. **Pederson, T. and Robbins, E.,** Absence of translational control of histone synthesis during HeLa cell life cycle, *J. Cell Biol.,* 45, 509, 1970.
514. **Goldstein, L.,** Localization of nucleus-specific protein as shown by transplantation experiments in *Amoeba proteus, Exp. Cell Res.,* 15, 635, 1958.
515. **Byers, T. J., Platt, D. B., and Goldstein, L.,** The cytonucleoproteins of amoebae. I. Some chemical properties and intracellular distribution, *J. Cell Biol.,* 19, 453, 1963.
516. **Byers, T. J., Platt, D. B., and Goldstein, L.,** The cytonucleoproteins of amoebae. II. Some aspects of cytonucleoprotein behavior and synthesis, *J. Cell Biol.,* 19, 467, 1963.
517. **Goldstein, L. and Prescott, D. M.,** Proteins in nucleocytoplasmic interactions. I. The fundamental characteristics of the rapidly migrating proteins and the slow turnover proteins of *Amoeba proteus* nucleus, *J. Cell Biol.,* 33, 637, 1967.
518. **Goldstein, L. and Prescott, D. M.,** Proteins in nucleocytoplasmic interactions. II. Turnover and changes in nuclear protein distribution with time and growth, *J. Cell Biol.,* 33, 637, 1967.
519. **Prescott, D. M. and Goldstein, L.,** Proteins in nucleocytoplasmic interactions. III. Redistributions of nuclear proteins during and following mitosis in *Amoeba proteus, J. Cell Biol.,* 39, 404, 1968.
520. **Prescott, D. M.,** Turnover of chromosomal and nuclear proteins, in *The Nucleohistones,* Bonner, J. and Ts'o, P. O. P., Eds., Holden-Day, San Francisco, 1964, 193.
521. **Kroeger, H., Jacob, J., and Sirlin, J. L.,** The movement of nuclear protein from the cytoplasm to the nucleus of salivary cells, *Exp. Cell Res.,* 31, 416, 1963.
522. **Butler, J. A. V. and Laurence, D. J. R.,** Relative metabolic activities of histones in tumours and liver, *Br. J. Cancer,* 14, 758, 1960.
523. **Allfrey, V. G., Mirsky, A. E., and Stern, H.,** The chemistry of the cell nucleus, *Adv. Enzymol.,* 16, 411, 1955.
524. **Samarina, O. P.,** Incorporation of labelled amino acids into protein fractions of nuclei of the liver and of the Ehrlich ascites carcinoma cells, *Biokhimiya,* 26, 61, 1961.
525. **Bustos-Valdes, S. E., Deisseroth, A., and Dounce, A. L.,** Metabolic studies of histones and residual proteins of rat liver nuclei, *Arch. Biochem. Biophys.,* 126, 848, 1968.
526. **Stellwagen, R. H. and Cole, R. D.,** Comparison of histones obtained from mammary gland at different stages of development and lactation, *J. Biol. Chem.,* 243, 4456, 1968.
527. **MacGillivray, A. J. and Monjardino, J. P. P.,** An investigation of acid-soluble proteins of human leukocytes in relation to fraction RP2-L, a component of neoplastic cells, *Biochem. J.,* 107, 799, 1968.
528. **Laurence, D. J. R. and Butler, J. A. V.,** Metabolism of histones in malignant tissues and liver of the rat and mouse, *Biochem. J.,* 96, 53, 1965.
529. **Umaña, R., Updike, S., Randall, J., and Dounce, A. L.,** Histone metabolism, in *The Nucleohistones,* Bonner, J. and Ts'o, P. O. P., Eds., Holden-Day, San Francisco, 1964, 200.
530. **Dick, C. and Johns, E. W.,** The biosynthesis of the five main histone fractions of rat thymus, *Biochim. Biophys. Acta,* 174, 380, 1969.
531. **Bondy, S. C.,** The synthesis and decay of histone fractions and of a deoxyribonucleic acid in the developing avian brain, *Biochem. J.,* 123, 465, 1971.
532. **Gurley, L. R. and Hardin, J. M.,** The metabolism of histone fractions. III. Synthesis and turnover of histone F1, *Arch. Biochem. Biophys.,* 136, 392, 1970.

533. **Gurley, L. R. and Hardin, J. M.,** The metabolism of histone fractions. I. Synthesis of histone fractions during the life cycle of mammalian cells, *Arch. Biochem. Biophys.,* 128, 285, 1968.
534. **Holbrook, D. J., Irvin, J. L., Irvin, E. M., and Rotherham, J.,** Incorporation of glycine into protein and nucleic acid fractions of nuclei of liver and hepatoma, *Cancer Res.,* 20, 1329, 1960.
535. **Holbrook, D. J., Evans, J. H., and Irvin, J. L.,** Incorporation of labeled precursors into proteins and nucleic acids of nuclei of regenerating liver, *Exp. Cell Res.,* 28, 120, 1962.
536. **Evans, J. H., Holbrook, D. J., and Irvin, J. L.,** Changes in content of nuclear proteins and nucleic acids in regenerating liver, *Exp. Cell Res.,* 28, 126, 1962.
537. **Jungmann, R. A., Schweppe, J. S., and Lestina, F. A.,** Studies on adrenal histones. Characterization, biosynthesis, enzymatic phosphorylation, and acetylation of histones from a human adrenal carcinoma, *J. Biol. Chem.,* 245, 4321, 1970.
538. **Ohly, K. W., Mehta, N. G., Mourkides, G. A. and Alivisatos, S. G. A.,** Isolation and characterization of histones from Ehrlich ascites tumor after *in vitro* labeling with lysine ^{14}C, *Arch. Biochem. Biophys.,* 118, 631, 1967.
539. **Chakravorty, A. K. and Biswas, B. B.,** Biosynthesis of histones, *Ind. J. Biochem.,* 2, 62, 1965.
540. **Gronow, M.,** Imbalances in DNA and histone synthesis in the rat liver during neonatal carcinogenesis induced by dimethylnitrosamine, *Eur. J. Cancer,* 7, 341, 1971.
541. **Henderson, T. R., Dacus, D. M., and Henderson, R. F.,** Deuterium isotope effects on histone synthesis in avian erythrocytes, *Arch. Biochem. Biophys.,* 122, 599, 1967.
542. **Freedman, M. L., Honig, G. R., and Rabinowitz, M.,** The role of newly synthesized RNA on nuclear histones synthesis by chicken immature erythrocytes, *Exp. Cell Res.,* 44, 263, 1966.
543. **Sadgopal, A. and Kabat, D.,** Synthesis of chromosomal proteins during the maturation of chicken erythrocyte, *Biochim. Biophys. Acta,* 190, 486, 1969.
544. **Chalkley, G. R. and Maurer, H. R.,** Turnover of template-bound histone, *Proc. Natl. Acad. Sci. U.S.,* 54, 498, 1965.
545. **Stellwagen, R. H. and Cole, R. D.,** Chromosomal proteins, *Annu. Rev. Biochem.,* 38, 951, 1969.
546. **Marzluff, W. F., McCarty, K. S., and Turkington, R. W.,** Insulin-dependent synthesis of histones in relation to the mammary epithelial cell cycle, *Biochim. Biophys. Acta,* 190, 517, 1969.
547. **McClure, M. E. and Hnilica, L. S.,** Temporal patterns of chromatin biosynthesis and activity of synchronized Chinese hamster cell culture, *J. Cell Biol.,* 47, 133a, 1970.
548. **Holoubek, V.,** Metabolism of different histone fractions in Ehrlich ascites cells after environmental change, *Experientia,* 27, 140, 1971.
549. **Stellwagen, R. H. and Cole, R. D.,** Histone biosynthesis in the mammary gland during development and lactation, *J. Biol. Chem.,* 244, 4878, 1969.
550. **Hohmann, P. and Cole, R. D.,** Hormonal effects on amino acid incorporation into lysine-rich histones, *Nature,* 223, 1064, 1969.
551. **Hohmann, P. and Cole, R. D.,** Hormonal effects on amino acid incorporation into lysine-rich histones in the mouse mammary gland, *J. Mol. Biol.,* 58, 533, 1971.
552. **Hohmann, P., Cole, R. D., and Bern, H. A.,** Comparison of lysine-rich histones in various normal and neoplastic mouse tissues, *J. Natl. Cancer Inst.,* 47, 337, 1971.
553. **Gurley, L. R., Hardin, J. M., and Walters, R. A.,** The response of histone fractions to X-irradiation in cultured chinese hamster cells, *Biochem. Biophys. Res. Commun.,* 38, 290, 1970.
554. **Gurley, L. R. and Walters, R. A.,** Response of histone turnover and phosphorylation to X-irradiation, *Biochemistry,* 10, 1588, 1971.
555. **Prescott, D. M. and Bender, M. A.,** Synthesis and behavior of nuclear proteins during the cell life cycle, *J. Cell. Comp. Physiol.,* 62, Suppl. 1, 175, 1963.
556. **Prescott, D. M. and Bender, M. A.,** Synthesis of RNA and protein during mitosis in mammalian tissue culture cells, *Exp. Cell Res.,* 26, 260, 1962.
557. **Prensky, W. and Smith, H. H.,** Incorporation of ^{3}H-arginine in chromosomes of *Vicia faba, Exp. Cell Res.,* 34, 525, 1964.
558. **Cave, M. D.,** Incorporation of tritium-labeled thymidine and lysine into chromosomes of cultured human leukocytes, *J. Cell Biol.,* 29, 209, 1966.
559. **Hancock, R. and Ryser, H. J. P.,** Histones in prophase and their possible role in nuclear membrane breakdown, *Nature,* 213, 70, 1967.
560. **Piha, R. S., Cuénod, M., and Waelsch, H.,** Metabolism of histones of brain and liver, *J. Biol. Chem.,* 241, 2397, 1966.
561. **Piha, R. S. and Rajamäki, A.,** The rate of turnover of mouse liver DNA compared to that of histones, *Suomen Kemistilehti,* B 39, 109, 1966.
562. **Byvoet, P.,** Metabolic integrity of deoxyribonucleohistones, *J. Mol. Biol.,* 17, 311, 1966.
563. **Byvoet, P.,** Metabolic integrity of deoxyribonucleohistones during enzyme induction, *Mol. Pharmacol.,* 3, 303, 1967.
564. **Murthy, L. D., Pradhan, D. S., and Sreenivasan, A.,** Effects of hydrocortisone upon metabolism of histones in rat liver, *Biochim. Biophys. Acta,* 199, 500, 1970.

565. **Allfrey, V. G., Daly, M. M., and Mirsky, A. E.,** Some observations on protein metabolism in chromosomes of non-dividing cells, *J. Gen. Physiol.,* 38, 415, 1954.
566. **Stein, G. and Baserga, R.,** The synthesis of acidic nuclear proteins in the prereplicative phase of the isoproterenol-stimulated salivary gland, *J. Biol. Chem.,* 245, 6097, 1970.
567. **Stein, G. and Baserga, R.,** Continued synthesis of nonhistone chromosomal proteins during mitosis, *Biochem. Biophys. Res. Commun.,* 41, 715, 1970.
568. **Spalding, J., Kajiwara, K., and Mueller, G. C.,** The metabolism of basic proteins in HeLa cell nuclei, *Proc. Natl. Acad. Sci. U. S.,* 56, 1535, 1966.
569. **Bloch, D. P., MacQuigg, R. A., Brack, S. D., and Wu, J. R. ,** The synthesis of deoxyribonucleic acid and histone in the onion root meristem, *J. Cell Biol.,* 33, 451, 1967.
570. **Hancock, R.,** Conservation of histones in chromatin during growth and mitosis *in vitro, J. Mol. Biol.,* 40, 457, 1969.
571. **Gurley, L. R. and Hardin, J.,** The metabolism of histone fractions. II. Conservation and turnover of histone fractions in mammalian cells, *Arch. Biochem. Biophys.,* 130, 1, 1969.
572. **Flamm, W. G. and Birnstiel, M. L.,** Inhibition of DNA replication and its effect on histone synthesis, *Exp. Cell Res.,* 33, 616, 1964.
573. **Ontko, J. A. and Moorehead, W. R.,** Histone synthesis after inhibition of deoxyribonucleic acid replication, *Biochim. Biophys. Acta,* 91, 658, 1964.
574. **Bases, R. and Mendez, F.,** Delayed inhibition of histone synthesis despite prompt inhibition of DNA synthesis in X-irradiated HeLa cells, *Radiat. Res.,* 47, 258, 1971.
575. **Yarbro, J. W., Niehaus, W. G., and Barnum, C. P.,** Effect of hydroxyurea on regenerating rat liver, *Biochem. Biophys. Res. Commun.,* 19, 592, 1965.
576. **Yarbro, J. W.,** Relationship of histone synthesis to DNA synthesis in mouse ascites tumor 6C3 HED, *Biochim. Biophys. Acta,* 145, 351, 1967.
577. **Zampetti-Bosseler, F., Malpoix, P., and Fievez, M.,** Differential inhibition of protein synthesis in the nucleus and cytoplasm of HeLa cells in the presence of Actinomycin, Puromycin and hydroxyurea, *Eur. J. Biochem.,* 9, 21, 1969.
578. **Chae, C. B., Williams, A., Krasny, H., Irvin, J. L., and Piantadosi, C.,** Inhibition of thymidine phosphorylation and DNA and histone synthesis in Ehrlich ascites carcinoma, *Cancer Res.,* 30, 2652, 1970.
579. **Jones, R. and Irvin, J. L.,** Effect of hydrocortisone on histone and DNA synthesis in regenerating rat liver, *Gastroenterology,* 58, 964, 1970.
580. **Stone, G. E.,** Change in histone/DNA ratio during population growth in *Tetrahymena, J. Cell Biol.,* 42, 837, 1969.
581. **Hnilica, L. S. and Trojanova, J.,** Inhibition of nuclear protein biosynthesis by colchicine and Actinomycin D, *7th Int. Congr. Biochem., Tokyo,* Abstracts Vol. 5, 857, 1967.
582. **Holoubek, V. and Rueckert, R. R.,** Studies on nuclear protein metabolism after infection of Ehrlich ascites cells with Maus-Eberfeld (ME) virus, *Biochem. Biophys. Res. Commun.,* 15, 166, 1964.
583. **Holoubek, V. and Crocker, T. T.,** Histones and RNA metabolism in Ehrlich ascites cells infected with Maus-Eberfeld virus, *Virology,* 43, 527, 1971.
584. **Munro, G. F., Dounce, A. L., and Lerman, S.,** Nuclear protein composition and metabolism of HeLa cells after infection with *herpes simplex* virus, *Cancer Res.,* 30, 379, 1970.
585. **Shimono, H. and Kaplan, A. S.,** Correlation between the synthesis of DNA and histones in polyoma virus-infected mouse embryo cells, *Virology,* 37, 690, 1969.
586. **Winocour, E. and Robbins, E.,** Histone synthesis in polyoma- and SV40-infected cells, *Virology,* 40, 307, 1970.
587. **Nilausen, K. and Green, H.,** Reversible arrest of growth in Gl of an established fibroblast line (3T3), *Exp. Cell Res.,* 40, 166, 1965.
588. **Suskind, R. G.,** Autoradiographic and cytochemical evidence for synthesis of a lysine- containing ribonucleoprotein in nucleoli inhibited by Actinomycin D, *J. Cell Biol.,* 24, 309, 1965.
589. **Marchis-Mouren, G. and Cozzone, A.,** Inhibition by Actinomycin D of valine incorporation into specific proteins of rat pancreas *in vivo, Biochemistry,* 5, 3684, 1966.
590. **Ackermann, W. W., Cox, D. C., and Dinka, S.,** Control of histone and DNA synthesis with canavanine, puromycin and poliovirus, *Biochem. Biophys. Res., Commun.,* 19, 745, 1965.
591. **Mirsky, A. E. and Ris, H.,** Variable and constant components of chromosomes, *Nature,* 163, 666, 1949.
592. **Vendrely, R. and Vendrely, C.,** La teneur du noyau cellulaire en acide désoxyribonucléique à travers les organes, les individus et les espèces animales, *Experientia,* 4, 434, 1948.
593. **Alfert, M.,** Quantitative cytochemical studies on patterns of nuclear growth, in *Fine Structure of Cells,* Int. Union of Biol. Sci. Ser. B, Interscience, New York, 1955, 157.
594. **Gall, J. G.,** Macromolecular duplication in the ciliated protozoan *Euplotes, J. Biophys. Biochem. Cytol.,* 5, 295, 1959.
595. **Prescott, D. M.,** Symposium: Synthetic processes in the cell nucleus. II. Nucleic acid and protein metabolism in the macronucleus of two ciliated protozoa, *J. Histochem. Cytochem.,* 10, 145, 1962.
596. **Prescott, D. M.,** The syntheses of total macronuclear protein, histone, and DNA during the cell cycle in *Euplotes eurystomus, J. Cell Biol.,* 31, 1, 1966.
597. **McLeish, J.,** Comparative microphotometric studies of DNA and arginine in plant nuclei, *Chromosoma,* 10, 686, 1959.

598. **Seed, J.,** The synthesis of DNA, RNA, and nuclear protein in normal and tumor strain cells. I. Fresh embryo human cells, *J. Cell Biol.*, 28, 233, 1966.
599. **Seed, J.,** The synthesis of DNA, RNA, and nuclear protein in normal and tumor strain cells. II. Fresh embryo mouse cells, *J. Cell Biol.*, 28, 249, 1966.
600. **Bloch, D. P. and Teng, C.,** The synthesis of deoxyribonucleic acid and nuclear histone of the X-chromosome of the *Rhenia spinosus* spermatocyte, *J. Cell Sci.*, 5, 321, 1969.
601. **Shapiro, I. M. and Levina, L. Y.,** Autoradiographic study on the time of nuclear protein synthesis in human leukocyte blood culture, *Exp. Cell Res.*, 47, 75, 1967.
602. **Shapiro, I. M. and Polikarpova, S. T.,** Autoradiographic study on the synthesis of nuclear proteins, *Chromosoma*, 28, 188, 1969.
603. **Shapiro, I. M. and Levina, L. Y.,** Incorporation of proteins labeled with ^{3}H lysine into chromosomes of human leukocytes cultured *in vitro, Chromosoma*, 25, 30, 1968.
604. **Das, N. K. and Alfert, M.,** Cytochemical studies on the concurrent synthesis of DNA and histone in primary spermatocytes of *Urechis caupo, Exp. Cell Res.*, 49, 51, 1968.
605. **Irvin, J. L., Holbrook, D. J., Evans, J. H., McAllister, H. C., and Stiles, E. P.,** Possible role of histones in regulation of nucleic acid synthesis, *Exp. Cell Res.*, Suppl. 9, 359, 1963.
606. **Orlova, L. V., Nedelyaeva, M. I., and Rodionov, V. M.,** Histone content in the cell nuclei of the regenerating rat liver, *Biokhimyia*, 34, 729, 1969.
607. **Orlova, L. V. and Rodionov, V. M.,** The time of histone synthesis in regenerating rat liver, *Exp. Cell Res.*, 59, 329, 1970.
608. **Orlova, L. V.,** Synthesis of histone proteins of the chromatin in the regenerating rat liver, *Biokhimiya*, 36, 423, 1971.
609. **Takai, S., Borun, T. W., Muchmore, J., and Lieberman, I.,** Concurrent synthesis of histone and deoxyribonucleic acid in liver after partial hepatectomy, *Nature*, 219, 860, 1968.
610. **Butler, J. A. V. and Cohn, P.,** Studies on histones. 6. Observations on the biosynthesis of histones and other proteins in regenerating liver, *Biochem. J.*, 87, 330, 1963.
611. **Tidwell, T., Allfrey, V. G., and Mirsky, A. E.,** The methylation of histones during regeneration of the liver, *J. Biol. Chem.*, 243, 707, 1968.
612. **Scherbaum, O. H. and Zeuthen, E.,** Induction of synchronous cell division in mass cultures of *Tetrahymena pyriformis, Exp. Cell Res.*, 6, 221, 1954.
613. **Hardin, J. A., Einem, G. E., and Lindsay, D. T.,** Simultaneous synthesis of histones and DNA in synchronously dividing *Tetrahymena pyriformis, J. Cell Biol.*, 32, 709, 1967.
614. **Jeffery, W. R., Stuart, K. D., and Frankel, J.,** The relationship between deoxyribonucleic acid replication and cell division in heat synchronized *Tetrahymena, J. Cell Biol.*, 46, 533, 1970.
615. **Zeuthen, E.,** Independent synchronization of DNA synthesis and of cell division in same culture of *Tetrahymena* cells, *Exp. Cell Res.*, 61, 311, 1970.
616. **Rusch, H. P.,** Some biochemical events in the growth cycles of *Physarum polycephalum, Fed. Proc.*, 28, 1761, 1969.
617. **Mohberg, J. and Rusch, H. P.,** Nuclear histones in *Physarum polycephalum* during growth and differentiation, *Arch. Biochem. Biophys.*, 138, 418, 1970.
618. **Petersen, D. F., Tobey, R. A., and Anderson, E. C.,** Synchronously dividing mammalian cells, *Fed. Proc.*, 28, 1771, 1969.
619. **Mueller, G. C.,** Biochemical events in the animal cell cycle, *Fed. Proc.*, 28, 1780, 1969.
620. **Tobey, R. A. and Ley, K. D.,** Isoleucine regulated regulation of genome replication in various mammalian cell lines, *Cancer Res.*, 31, 46, 1971.
621. **Stubblefield, E. and Klevecz, R.,** Synchronization of Chinese hamster cells by reversal of colcemid inhibition, *Exp. Cell Res.*, 40, 660, 1965.
622. **Sadgopal, A. and Bonner, J.,** The relationship between histone and DNA synthesis in HeLa cells, *Biochim. Biophys. Acta*, 186, 349, 1969.
623. **Littlefield, J. W. and Jacobs, P. S.,** The relation between DNA and protein synthesis in mouse fibroblasts, *Biochim. Biophys. Acta*, 108, 652, 1965.
624. **Studzinski, G. P. and Lambert, W. C.,** Thymidine as a synchronizing agent. I. Nucleic acid and protein formation in synchronous HeLa cultures treated with thymidine, *J. Cell. Physiol.*, 73, 109, 1969.
625. **Lambert, W. C. and Studzinski, G. P.,** Thymidine as a synchronizing agent. II. Partial recovery of HeLa cells from unbalanced growth, *J. Cell. Physiol.*, 73, 261, 1969.
626. **Meyn, R. E., Hewitt, R. R., and Humphrey, R. M.,** Evaluation of the cell cycle positions obtained with the blocking-synchronizing agents, thymidine and hydroxyurea, submitted for publication.
627. **Galavazi, G. and Bootsma, D.,** Synchronization of mammalian cells *in vitro* by inhibition of the DNA synthesis. II. Population dynamics, *Exp. Cell Res.*, 41, 438, 1966.
628. **Swift, H.,** The constancy of desoxyribose nucleic acid in plant nuclei, *Proc. Natl. Acad. Sci. U. S.*, 36, 643, 1950.
629. **Hotta, Y., Ito, M., and Stern, H.,** Synthesis of DNA during meiosis, *Proc. Natl. Acad. Sci. U. S.*, 56, 1184, 1966.

630. **Bogdanov, Y. F., Liapunova, N. A., Sherudilo, A. I., and Antropova, N. E.,** Uncoupling of DNA and histone synthesis prior to phase I of meiosis in the cricket *Grillus (acheta) domesticus, L., Exp. Cell Res.*, 52, 59, 1968.
631. **Henderson, S. A.,** RNA synthesis during male meiosis and spermiogenesis, *Chromosoma*, 15, 345, 1964.
632. **Muckenthaler, F. A.,** Autoradiographic study of nucleic acid synthesis during spermatogenesis in the grasshopper *Melanoplus differentialis, Exp. Cell Res.*, 35, 531, 1964.
633. **Taylor, H.,** Autoradiographic detection of incorporation of P^{32} into chromosomes during meiosis and mitosis, *Exp. Cell Res.*, 4, 164, 1953.
634. **Stern, H. and Hotta, Y.,** Biochemistry of meiosis, in *Handbook of Molecular Cytology,* Lima-de-Faria, A., Ed., North Holland Publ. Co., Amsterdam, 1969, 520.
635. **Antropova, E. N. and Bogdanov, Y. F.,** Cytophotometry of DNA and histone in meiosis of *Pyrrhocoris apterus, Exp. Cell Res.*, 60, 40, 1970.
636. **Felix, K., Goppold-Krekels, A., and Lehmann, H.,** Die Bildung der Nukleoprotamine wahrend der Spermiogenese, *H. S. Z. Physiol. Chem.*, 312, 57, 1958.
637. **Alfert, M.,** Chemical differentiation of nuclear proteins during spermatogenesis in the salmon, *J. Biophys. Biochem. Cytol.*, 2, 109, 1956.
638. **Ingles, C. J., Trevithick, J. R., Smith, M., and Dixon, G. H.,** Biosynthesis of protamine during spermatogenesis in salmonid fish, *Biochem. Biophys. Res. Commun.*, 22, 627, 1966.
639. **Ling, V. and Dixon, G. H.,** The biosynthesis of protamine in trout testis. II. Polysome patterns and protein synthetic activities during testis maturation, *J. Biol. Chem.*, 245, 3035, 1970.
640. **Wigle, D. T. and Dixon, G. H.,** Transient incorporation of methionine at the N-terminus of protamine newly synthetized in trout testis cells, *Nature*, 227, 676, 1970.
641. **Ling, V., Jergil, B., and Dixon, G. H.,** The biosynthesis of protamine in trout testis. III. Characterization of protamine components and their synthesis during testis development, *J. Biol. Chem.*, 246, 1168, 1971.
641a. **Evans, K., Konigsberg, P., and Cole, R. D.,** Displacemment of histones from deoxyribonucleoprotein by protamine, *Arch. Biochem. Biophys.*, 141, 389, 1970.
642. **Marushige, K. and Dixon, G. H.,** Transformation of trout testis chromatin, *J. Biol. Chem.*, 246, 5799, 1971.
642a. **Mirsky, A. E. and Ris, H.,** The composition and structure of isolated chromosomes, *J. Gen. Physiol.*, 34, 475, 1951.
643. **Izquierdo, L. and Marticorena, P.,** Basic proteins of mouse ova and blastocytes, *Experientia*, 26, 646, 1970.
644. **Immers, J., Markman, B., and Runnström, J.,** Nuclear changes in the course of development of the sea urchin studied by means of Hale staining, *Exp. Cell Res.*, 47, 425, 1967.
645. **Allfrey, V. G., Littau, V. C., and Mirsky, A. E.,** On the role of histones in regulating ribonucleic acid synthesis in the cell nucleus, *Proc. Natl. Acad. Sci. U. S.*, 49, 414, 1963.
646. **Runnström, J.,** Structural changes of the chromatin in early development of sea urchins, *Exp. Cell Res.*, 48, 691, 1967.
647. **Agrell, I. P. S. and Christensson, E. G.,** Changes of histone composition in the developing chick embryo, *Nature*, 207, 638, 1965.
648. **Orengo, A. and Hnilica, L. S.,** Histones of the sea urchin *Strongylocentrotus purpuratus* during early embryonic development, *J. Cell Biol.*, 43, 99A, 1969.
649. **Silver, D. J. and Comb, D. G.,** Acetic acid-soluble proteins in the developing sea urchin, *Dev. Biol.*, 16, 107, 1967.
650. **Johnson, A. W., Wilhelm, J. A., and Hnilica, L. S.,** Nuclear basic protein acetylation during early sea urchin development, *Biochim. Biophys. Acta*, in press.
651. **Gineitis, A. A., Vinogradova, I. A., and Vorobyev, V. I.,** The changes of the histone composition at early stages of embryogenesis of sea urchin and loach, *Tsitologia*, 10, 581, 1968.
652. **Gineitis, A. A., Vinogradova, I. A., Nivinska, S., and Vorobyev, V. I.,** Histones from the sperm and embryos of the loach *Misgurnus fossilis, Tsitologia*, 12, 198, 1970.
653. **Vorobyev, V. I., Gineitis, A. A., and Vinogradova, I. A.,** Histones in embryogenesis, *Exp. Cell Res.*, 57, 1, 1969.
654. **Crane, C. M. and Villee, C. A.,** The synthesis of nuclear histones in early embryogenesis, *J. Biol. Chem.*, 246, 719, 1971.
655. **Gross, P. R.,** The control of protein synthesis in embryonic development and differentiation, *Curr. Top. Dev. Biol.*, 2, 1, 1967.
656. **Terman, S. A. and Gross, P. R.,** Translation-level control of protein synthesis during early development, *Biochem. Biophys. Res. Commun.*, 21, 595, 1965.
657. **Terman, S. A.,** Relative effect of transcription-level and translation-level control of protein synthesis during early development of sea urchin, *Proc. Natl. Acad. Sci. U. S.*, 65, 985, 1970.
658. **Stavy, L. and Gross, P. R.,** Availability of mRNA for translation during normal and transcription-blocked development, *Biochim. Biophys. Acta*, 182, 203, 1969.
659. **Kedes, L. H. and Gross, P. R.,** Synthesis and function of messenger RNA during early embryonic development, *J. Mol. Biol.*, 42, 559, 1969.
660. **Slater, D. W. and Spiegelman, S.,** An estimation of genetic messages in the unfertilized echinoid egg, *Proc. Natl. Acad. Sci. U. S.*, 56, 164, 1966.
661. **Slater, D. W. and Spiegelman, S.,** Template capabilities and size distribution of echinoid RNA during early development, *Biochim. Biophys. Acta*, 166, 82, 1968.

662. **Glišin, V. R., Glišin, M. V., and Doty, P.,** The nature of messenger RNA in the early stages of sea urchin development, *Proc. Natl. Acad. Sci. U. S.,* 56, 285, 1966.
663. **Whiteley, A. H., McCarthy, B. J., and Whiteley, H. R.,** Changing population of messenger RNA during sea urchin development, *Proc. Natl. Acad. Sci. U. S.,* 55, 519, 1966.
664. **Liau, M. C., Hnilica, L. S., and Hurlbert, R. B.,** Regulation of RNA synthesis in isolated nucleoli by histones and nucleolar proteins, *Proc. Natl. Acad. Sci. U. S.,* 53, 626, 1965.
665. **Allfrey, V. G. and Mirsky, A. E.,** Role of histones in nuclear function, in *The Nucleohistones,* Bonner, J. and Ts'o, P. O. P., Eds., Holden-Day, San Fransisco, 1964, 267.
666. **Allfrey, V. G., Faulkner, R., and Mirsky, A. E.,** Acetylation and methylation of histones and their possible role in the regulation of RNA synthesis, *Proc. Natl. Acad. Sci. U. S.,* 51, 786, 1964.
667. **Vidali, G., Gershey, E. L., and Allfrey, V. G.,** Chemical studies of histone acetylation. The distribution of ϵ-N-methyllysine in calf thymus histones, *J. Biol. Chem.,* 243, 6361, 1968.
668. **Pogo, B. G. T., Pogo, A. O., Allfrey, V. G., and Mirsky, A. E.,** Changing patterns of histone acetylation and RNA synthesis in regeneration of the liver, *Proc. Natl. Acad. Sci. U. S.,* 59, 1337, 1968.
669. **Pogo, B. G. T., Pogo, A. O., and Allfrey, V. G.,** Histone acetylation and RNA synthesis in rat liver regeneration, *Genetics,* 61, Suppl. 1, 373, 1969.
670. **Nohara, H., Takahashi, T., and Ogata, K.,** Acetylation of histones by pigeon liver enzymes, *Biochim. Biophys. Acta,* 127, 282, 1966.
671. **Nohara, H., Takahashi, T., and Ogata, K.,** Enzymatic acetylation of histones and some chemical characters of their acetyl groups, *Biochim. Biophys. Acta,* 154, 529, 1968.
672. **Gallwitz, D. and Sekeris, C. E.,** The acetylation of histones of rat liver nuclei *in vitro* by acetyl CoA, *H. S. Z. Physiol. Chem.,* 350, 150, 1969.
673. **Allfrey, V. G.,**Functional and metabolic aspects of DNA-associated proteins, in *Histones and Nucleohistones,* Phillips, D. M. P., Ed., Plenum Press, London, 1971, 241.
674. **Allfrey, V. G., Pogo, B. G. T., Pogo, A. O., Kleinsmith, L. J., and Mirsky, A. E.,** The metabolic behaviour of chromatin, in *Histones,* DeReuck, A. V. S. and Knight, J., Eds., Churchill, London, 1966, 42.
675. **Allfrey, V. G.,** Structural modifications of histones and their possible role in the regulation of ribonucleic acid synthesis, *Proc. Can. Cancer Conf.,* 6, 313, 1965.
676. **Gallwitz, D.,** Enzymatic acetylation of HeLa cell histones in isolated nuclei *in vitro, H. S. Z. Physiol. Chem.,* 351, 1050, 1970.
677. **Libby, P. R.,** Histone acetylation by cell-free preparations from rat uterus: *in vitro* stimulation by estradiol-17β, *Biochem. Biophys. Res. Commun.,* 31, 59, 1968.
678. **Gallwitz, D.,** Acetylation of histones by a kinase from rat liver nuclei, *Biochem. Biophys. Res. Commun.,* 32, 117, 1968.
679. **Racey, L. A. and Byvoet, P.,** Histone acetyltransferase in chromatin, *Exp. Cell Res.,* 64, 366, 1971.
680. **Bondy, S. C., Roberts, S., and Morelos, B. S.,** Histone acetylating enzyme in brain, *Biochem. J.,* 119, 665, 1970.
681. **Paik, W. K., Pearson, D., Lee, H. W., and Kim, S.,** Nonenzymatic acetylation of histones with acetyl-CoA, *Biochim. Biophys. Acta,* 213, 513, 1970.
682. **Wilhelm, J. A. and McCarty, K. S.,** Partial characterization of the histones and histone acetylation in cell cultures, *Cancer Res.,* 30, 409, 1970.
683. **Wilhelm, J. A., Groves, C. M., and Hnilica, L. S.,** Nucleolar and extranucleolar chromatin of rat Novikoff hepatoma cells. A comparison of histone and nonhistone protein acetylation, *Biochim. Biophys. Acta,* 238, 295, 1971.
684. **Byvoet, P.,** Difference in turnover between histones and their acetyl-N- terminal groups, *Biochim. Biophys. Acta,* 160, 217, 1968.
685. **Pogo, B. G. T., Allfrey, V. G., and Mirsky, A. E.,** The effect of phytohemagglutinin on ribonucleic acid synthesis and histone acetylation in equine leukocyte, *J. Cell Biol.,* 35, 477, 1967.
686. **Ono, T., Terayama, H., Takaku, F., and Nakao, K.,** Hydrocortisone effect upon the phytohemagglutinin-stimulated acetylation of histones in human lymphocytes, *Biochim. Biophys. Acta,* 179, 214, 1969.
687. **Takaku, F., Nakao, K., Ono, T., and Terayama, H.,** Changes in histone acetylation and RNA synthesis in the spleen of polycythemic mouse after erythropoietin injection, *Biochim. Biophys. Acta,* 195, 396, 1969.
688. **Cross, M. E. and Ord, M. G.,** Changes in the phosphorylation and thiol content of histones in phytohemagglutinin-stimulated lymphocytes, *Biochem. J.,* 118, 191, 1970.
689. **Pogo, B. G. T., Allfrey, V. G., and Mirsky, A. E.,** RNA synthesis and histone acetylation during the course of gene activation in lymphocytes, *Proc. Natl. Acad. Sci. U. S.,* 55, 805, 1966.
690. **Byvoet, P. and Morris, H. P.,** N-acetylation of arginine-rich hepatoma histones, *Cancer Res.,* 31, 468, 1971.
691. **Wilhelm, J. A. and Hnilica, L. S.,** A comparison of histone and nonhistone protein acetylation between nucleolar and extranucleolar chromatin of normal and regenerating rat liver, *Biochim. Biophys. Acta,* in press.
692. **Racey, L. A. and Byvoet, P.,** Histone acetyltransferase from rat liver chromatin, *Fed. Proc.,* 30, 1083A, 1971.
693. **Gallwitz, D.,** Purification and properties of histone-specific acetyl-transferases from nuclei of different rat organs, *Fed. Proc.,* 30, 1083A, 1971.
694. **Marzluff, W. F. and McCarty, K. S.,** Two classes of histone acetylation in developing mouse mammary gland, *J. Biol. Chem.,* 245, 5635, 1970.

695. **Miller, D. M., Marzluff, W. F., and McCarty, K. S.,** Occurrence of ϵ-N-acetyl lysine in F3 and F2al histones, *Fed. Proc.*, 30, 1083A, 1971.
696. **Candido, E. P. M. and Dixon, G. H.,** Sites of *in vivo* acetylation in trout testis histone IV, *J. Biol. Chem.*, 246, 3182, 1971.
697. **Shepherd, G. R., Noland, B. J., and Hardin, J. M.,** Histone acetylation in synchronized mammalian cell cultures, *Biochim. Biophys. Acta*, 228, 544, 1971.
698. **Gallwitz, D.,** Extraction and partial purification of two histone-specific transacetylases from rat liver nuclei, *Biochem. Biophys. Res. Commun.*, 40, 236, 1970.
699. **Gallwitz, D.,** Organ specificity of histone acetyltransferases, *FEBS Lett.*, 13, 306, 1971.
700. **Noland, B. J., Hardin, J. M., and Shepherd, G. R.,** Histone acetyltransferase activity in synchronized mammalian cells, *Biochim. Biophys. Acta*, 246, 263, 1971.
701. **Inoue, A. and Fujimoto, D.,** Enzymatic deacetylation of histone, *Biochem. Biophys. Res. Commun.*, 36, 146, 1969.
702. **Inoue, A. and Fujimoto, D.,** Histone deacetylase from calf thymus, *Biochim. Biophys. Acta*, 220, 307, 1970.
703. **Libby, P. L.,** Activity of histone deacetylase in rat liver and Novikoff hepatoma, *Biochim. Biophys. Acta*, 213, 234, 1970.
704. **Racey, L. A. and Byvoet, P.,** *In vitro* acetylation of histones, *Proc. Am. Assoc. Cancer Res.*, 10, 71, 1969.
705. **Frenster, J. H., Allfrey, V. G., and Mirsky, A. E.,** Repressed and active chromatin isolated from interphase lymphocytes, *Proc. Natl. Acad. Sci. U. S.*, 50, 1026, 1963.
706. **Clever, U. and Ellgaard, E. G.,** Puffing and histone acetylation in polytene chromosomes, *Science*, 169, 373, 1970.
707. **Ellgaard, E. G.,** Gene activation without histone acetylation in *Drosophila melanogaster*, *Science*, 157, 1070, 1967.
708. **Clever, U.,** Control of chromosome puffing, in *Control of Nuclear Activity*, Goldstein, L., Ed., Prentice-Hall, New York, 1967.
709. **Allfrey, V. G.,** Changes in chromosomal proteins at times of gene activation, *Fed. Proc.*, 29, 1447, 1970.
710. **Ling, N. R.,** *Lymphocyte Stimulation*, North Holland, Amsterdam, 1968.
711. **Black, M. M. and Ansley, H. R.,** Antigen-induced changes in lymphoid cell histones. III. *In vitro* and in extract, *J. Cell Biol.*, 35, 619, 1967.
712. **Burton, D. W.,** Initial changes in the deoxyribonucleoprotein complexes of cangaroo lymphocytes stimulated with phytohaemagglutinin, *Exp. Cell Res.*, 49, 300, 1968.
713. **Zetterberg, A. and Auer, G.,** Early changes in the binding between DNA and histone in human leukocytes exposed to phytohemagglutinin, *Exp. Cell Res.*, 56, 122, 1968.
714. **Killander, D. and Rigler, R.,** Activation of deoxyribonucleoprotein in human leukocytes stimulated by phytohemagglutinin. I. Kinetics of the binding of Acridine orange to deoxyribonucleoprotein, *Exp. Cell Res.*, 54, 163, 1969.
715. **Rigler, R. and Killander, D.,** Activation of deoxyribonucleoprotein in human leukocytes stimulated by phytohemagglutinin. II. Structural changes of deoxyribonucleoprotein and synthesis of RNA, *Exp. Cell Res.*, 54, 171, 1969.
716. **Killander, D. and Rigler, R.,** Initial changes of deoxyribonucleoprotein and synthesis of nucleic acid in phytohemagglutinin-stimulated human leukocytes *in vitro*, *Exp. Cell Res.*, 39, 701, 1965.
717. **Ringertz, N. R., Darzynkiewicz, Z., and Bolund, L.,** Actinomycin binding properties of stimulated human lymphocytes, *Exp. Cell Res.*, 56, 411, 1969.
718. **Darzynkiewicz, Z., Bolund, L., and Ringertz, N. R.,** Nucleoprotein changes and initiation of RNA synthesis in PHA stimulated lymphocytes, *Exp. Cell Res.*, 56, 418, 1969.
719. **Mukherjee, A. B. and Cohen, M. M.,** Histone acetylation: cytological evidence in human lymphocytes, *Exp. Cell Res.*, 54, 257, 1969.
720. **Wilhelm, J. A. and McCarty, K. S.,** The uptake and turnover of acetate in HeLa cell histone fractions, *Cancer Res.*, 30, 418, 1970.
721. **Monjardino, J. P. P. V. and MacGillivray, A. J.,** RNA and histone metabolism in small lymphocytes stimulated by phytohemagglutinin, *Exp. Cell Res.*, 60, 1, 1970.
722. **Church, R. B. and McCarthy, B. J.,** Ribonucleic acid synthesis in regenerating and embryonic liver. I. The synthesis of new species of RNA during regeneration of mouse liver after partial hepatectomy, *J. Mol. Biol.*, 23, 459, 1967.
723. **Pogo, A. O., Littau, V. C., Allfrey, V. G., and Mirsky, A. E.,** Modification of ribonucleic acid synthesis in nuclei isolated from normal and regenerating liver: Some effects of salt and specific divalent cation, *Proc. Natl. Acad. Sci. U. S.*, 57, 743, 1967.
724. **Thaler, M. M. and Villee, C. A.,** Template activities in normal, regenerating and developing rat liver, *Proc. Natl. Acad. Sci. U. S.*, 58, 2055, 1967.
725. **Dahmus, M. F. and Bonner, J.,** Increased template activity of liver chromatin, a result of hydrocortisone administration, *Proc. Natl. Acad. Sci. U. S.*, 54, 1370, 1965.
726. **Beato, M., Homoki, J., Lukacs, I., and Sekeris, C. E.,** Increased template activity for RNA synthesis of rat liver nuclei incubated with cortisol *in vitro*, *H. S. Z. Physiol. Chem.*, 349, 1099, 1968.
727. **Korner, A.,** Ribonucleic acid and hormonal control of protein synthesis, *Progr. Biophys. Mol. Biol.*, 17, 61, 1967.
728. **Kenney, F. T. and Kull, F. J.,** Hydrocortisone-stimulated synthesis of nuclear RNA in enzyme induction, *Proc. Natl. Acad. Sci. U.S.*, 50, 493, 1963.

729. **Kidson, C. and Kirby, K. S.,** Selective alterations of mammalian messenger-RNA synthesis: Evidence for differential action of hormones on gene transcription, *Nature,* 203, 599, 1964.
730. **Kidson, C.,** Cortisol in the regulation of RNA and protein synthesis, *Nature,* 215, 779, 1967.
731. **Allfrey, V. G.,** Control mechanisms in ribonucleic acid synthesis, *Cancer Res.,* 26, 2026, 1966.
732. **Allfrey, V. G.,** Some observations on histone acetylation and its temporal relationship to gene activation, in *Regulatory Mechanisms for Protein Synthesis in Mammalian Cells,* San Pietro, A., Lamborg, M. R., and Kenney, F. T., Eds., Academic Press, New York, 1968, 65.
733. **Libby, P. R.,** Estrogen stimulation of histone acetylation in uterus, *Fed. Proc.,* 30, 361A, 1971.
734. **Anderson, N. O. and Gorski, J.,** The effect of estrogen on synthesis and acetylation of histones in rat uteri, *Fed. Proc.,* 30, 361A, 1971.
735. **Gallwitz, D. and Sekeris, C. E.,** Stimulation of RNA polymerase activity of rat liver nuclei by cortisol *in vitro* independent of effects on the acetylation and methylation of histones, *FEBS Lett.,* 3, 99, 1969.
736. **Pardue, M. L., Gerbi, S. A., Eckhardt, R. A., and Gall, J. G.,** Cytological localization of DNA complementary to ribosomal RNA in polytene chromosomes of *Diptera, Chromosoma,* 29, 268, 1970.
737. **Quagliarotti, G., Hidvegi, E., Wikman, J., and Busch, H.,** Structural analysis of nucleolar precursors of ribosomal ribonucleic acid. Comparative hybridizations of nucleolar and ribosomal ribonucleic acid with nucleolar deoxyribonucleic acid, *J. Biol. Chem.,* 245, 1962, 1970.
738. **Birnstiel, M. L.,** The nucleolus in cell metabolism, *Annu. Rev. Plant Physiol.,* 18, 25, 1967.
739. **Bekhor, I., Kung, G. M., and Bonner, J.,** Sequence-specific interaction of DNA and chromosomal protein, *J. Mol. Biol.,* 39, 351, 1969.
740. **Paul, J. and Gilmour, R. S.,** Organ-specific restriction of transcription in mammalian chromatin, *J. Mol. Biol.,* 34, 305, 1968.
741. **Tan, C. H. and Miyagi, M.,** Specificity of transcription of chromatin *in vitro, J. Mol. Biol.,* 50, 641, 1970.
742. **Bresnick, E.,** Regenerating liver: an experimental model for the study of growth, *Methods Cancer Res.,* 6, 347, 1971.
743. **Lindahl-Kiessling, K. and Peterson, R. D. A.,** The mechanism of phytohemagglutinin (PHA) action. II. The effect of certain enzymes and sugars, *Exp. Cell Res.,* 55, 81, 1969.
744. **Schildkraut, C. L. and Maio, J. J.,** Studies on the intracellular distribution and properties of mouse satellite DNA, *Biochim. Biophys. Acta,* 161, 76, 1968.
745. **Yasmineh, W. G. and Yunis, J. J.,** Satellite DNA in mouse autosomal heterochromatin, *Biochem. Biophys. Res. Commun.,* 35, 779, 1969.
746. **Yasmineh, W. G. and Yunis, J. J.,** Localization of mouse satellite DNA in constitutive heterochromatin, *Exp. Cell Res.,* 59, 69, 1970.
747. **Bernhard, W. and Granboulan, N.,** Electron microscopy of the nucleolus in vertebrate cells, in *Ultrastructure in Biological Systems,* Vol. 3, Dalton, A. J. and Haguenau, F., Eds., Academic Press, New York, 1968, 81.
748. **Wilhelm, J. A., Johnson, A. W., and Hnilica, L. S.,** Histone acetylation in differentiated and differentiating tissues, *Fed. Proc.,* 30, 1083A, 1971.
749. **Griswold, M. D. and Miller, G. L.,** Histones and their modifications during amphibian metamorphosis, *Comp. Biochem. Physiol.,* 39B, 445, 1971.
750. **Shepherd, G. R., Noland, B. J., and Roberts, C. N.,** Phosphorus in histones; *Biochim. Biophys. Acta,* 199, 265, 1970.
751. **Georgatsos, J. G., Rammos, G., Palavradzi, D., and Symeonidis, A.,** Phosphorylation of histones in mouse mammary tumors, *Eur. J. Cancer,* 4, 313, 1968.
752. **Stevely, W. S. and Stocken, L. A.,** Variations in the phosphate content of histone Fl in normal and irradiated tissues, *Biochem. J.,* 110, 187, 1968.
753. **Buckingham, R. H. and Stocken, L. A.,** Histone Fl, purification and phosphorus content, *Biochem. J.,* 117, 157, 1970.
754. **Schiltz, E. and Sekeris, C. E.,** Enzymatic phosphorylation of nuclear proteins by (γ-^{32}P)ATP in isolated rat liver nuclei, *H. S. Z. Physiol. Chem.,* 350, 317, 1969.
755. **Eltzina, N. V. and Veresotskaya, N. A.,** Phosphorylation of histones from liver and hepatoma ascites cells, *Biokhimiya,* 34, 921, 1969.
756. **Ord, M. G. and Stocken, L. A.,** Further studies on phosphorylation and the thiol/disulphide ratio of histones in growth and development, *Biochem. J.,* 112, 81, 1969.
757. **Buckingham, R. H. and Stocken, L. A.,** Histone Fl from rat thymus. Subfractionation and incorporation of (^{32}P) phosphate *in vitro, Biochem. J.,* 117, 509, 1970.
758. **Fitzgerald, P. J., Marsh, W. H., Ord, M. G., and Stocken, L. A.,** Histone changes during degeneration and regeneration of the pancreas, *Biochem. J.,* 117, 711, 1970.
759. **Klein, M. I. and Makman, M. H.,** Histone phosphorylation and cyclic AMP-dependent protein kinase in rat thymocytes, *Fed. Proc.,* 30, 220A, 1971.
760. **Balhorn, R., Rieke, W. O., and Chalkley, R.,** Rapid electrophoretic analysis for histone phosphorylation. A reinvestigation of phosphorylation of lysine-rich histone during rat liver regeneration, *Biochemistry,* 10, 3952, 1971.

761. **Kleinsmith, L. J., Allfrey, V. G., and Mirsky, A. E.,** Phosphorylation of nuclear protein early in the course of gene activation in lymphocytes, *Science,* 154, 780, 1966.
762. **Cross, M. E. and Ord, M. G.,** The transformation of lymphocytes by adenosine-3′:5′-cyclic monophosphate, and consequent changes in histone microstructure, *Biochem. J.,* 120, 21p, 1970.
763. **Adams, G. H. M., Vidali, G., and Neelin, J. M.,** Phosphate content of goose erythrocyte nuclei, *Can. J. Biochem.,* 48, 33, 1970.
764. **Seligy, V. L. and Neelin, J. M.,** Phosphorylation of histones in normal goose erythrocytes, *Can. J. Biochem.,* 49, 1062, 1971.
765. **Hayashi, T. and Iwai, K.,** Phosphate contents of calf thymus histone fractions, *J. Biochem.,* 68, 415, 1970.
766. **Langan, T. A., Rall, S. C., and Cole, R. D.,** Variation in primary structure at a phosphorylation site in lysine-rich histones, *J. Biol. Chem.,* 246, 1942, 1971.
767. **Jergil, B., Sung, M., and Dixon, G. H.,** Species- and tissue-specific patterns of phosphorylation of very lysine-rich histones, *J. Biol. Chem.,* 245, 5867, 1970.
768. **Marushige, K., Ling, V., and Dixon, G. H.,** Phosphorylation of chromosomal basic proteins in maturing trout testis, *J. Biol. Chem.,* 244, 5953, 1969.
769. **Stevely, W. S. and Stocken, L. A.,** Phosphorylation of rat-thymus histone, *Biochem. J.,* 100, 20c, 1966.
770. **Chae, C. B., Smith, M. C., and Irvin, J. L.,** Effect of histone phosphorylation on template activity of rat liver chromatin for RNA polymerase, *Fed. Proc.,* 30, 1084A, 1971.
771. **Adler, A. J., Schaffhausen, B., Langan, T. A., and Fasman, G. D.,** Altered conformational effects of phosphorylated lysine-rich histone (f-l) in f-l-deoxyribonucleic acid complexes. Circular dichroism and immunological studies, *Biochemistry,* 10, 909, 1971.
772. **Ord, M. G. and Stocken, L. A.,** Studies in synthesis of deoxyribonucleic acid. Radiobiochemical lesion in animal cells, *Nature,* 182, 1787, 1958.
773. **Pawse, A. P., Ord, M. G., and Stocken, L. A.,** Histone kinase and cell division, *Biochem. J.,* 122, 713, 1971.
774. **Gurley, L. R., Walters, R. A., and Enger, M. D.,** Isolation and characterization of histone Fl in ribosomes, *Biochem. Biophys. Res. Commun.,* 40, 428, 1970.
775. **Paik, W. K. and Lee, H. W.,** Enzymatic hydrolysis of histones in rat kidney microsomes, *Biochem. Biophys. Res. Commun.,* 38, 333, 1970.
776. **Church, R. B. and McCarthy, B. J.,** Ribonucleic acid synthesis in regenerating and embryonic liver. II. The synthesis of RNA during embryonic liver development and its relationship to regenerating liver, *J. Mol. Biol.,* 23, 477, 1967.
777. **Siebert, G., Ord, M. G., and Stocken, L. A.,** Histone phosphokinase activity in nuclear and cytoplasmic cell fractions from normal and regenerating rat livers, *Biochem. J.,* 122, 721, 1971.
778. **Cross, M. E. and Ord, M. G.,** Changes in histone phosphorylation and associated early metabolic events in pig lymphocyte cultures transformed by phytohaemagglutinin or 6-N,2′-O-dibutyryl adenosine 3′:5′-cyclic monophosphate, *Biochem. J.,* 124, 241, 1971.
779. **Shepherd, G. R., Noland, B. J., and Hardin, J. M.,** Histone phosphokinase levels in synchronized mammalian cells, *Exp. Cell Res.,* 67, 474, 1971.
780. **Yamamura, H., Takeda, M., Kumon, A., and Nishizuka, Y.,** Adenosine 3′,5′-cyclic phosphate-dependent and independent histone kinases from rat liver, *Biochem. Biophys. Res. Commun.,* 40, 675, 1970.
781. **Kuo, J. F. and Greengard, P.,** Cyclic nucleotide-dependent protein kinases. VII. Comparison of various histones as substrates for adenosine 3′,5′-monophosphate-dependent and guanosine 3′,5′-monophosphate-dependent protein kinases, *Biochim. Biophys. Acta,* 212, 434, 1970.
782. **Kaplowitz, P. B., Platz, R. D., and Kleinsmith, L. J.,** Nuclear phosphoproteins. III. Increase in phosphorylation during histone-phosphoprotein interaction, *Biochim. Biophys. Acta,* 229, 739, 1971.
783. **Schiltz, E. and Sekeris, C. E.,** Enzymatic phosphorylation of proteins of rat liver chromatin by (γ-^{32}P)ATP *in vitro, Experientia,* 27, 30, 1971.
784. **Walsh, D. A., Perkins, J. P., and Krebs, E. G.,** An adenosine 3′,5′-monophosphate-dependent protein kinase from rabbit skeletal muscle, *J. Biol. Chem.,* 243, 3763, 1968.
785. **Takeda, M., Yamamura, H., and Ohga, Y.,** Phosphoprotein kinases associated with rat liver chromatin, *Biochem. Biophys. Res. Commun.,* 42, 103, 1971.
786. **Langan, T. A.,** Phosphorylation of separate sites in lysine-rich histone by cyclic AMP-dependent and independent protein kinases, *Fed. Proc.,* 30, 1089A, 1971.
787. **Miyamoto, E., Petzold, G. L., Harris, J. S., and Greengard, P.,** Dissociation and concomitant activation of adenosine 3′,5′-monophosphate-dependent protein kinase by histone, *Biochem. Biophys. Res. Commun.,* 44, 305, 1971.
788. **Meisler, M. H. and Langan, T. A.,** Characterization of phosphatase specific for phosphorylated histones and protamine, *J. Biol. Chem.,* 244, 4961, 1969.
789. **Jergil, B. and Dixon, G. H.,** Protamine kinase from rainbow trout testis, *J. Biol. Chem.,* 245, 425, 1970.
790. **Benjamin, W. and Gellhorn, A.,** Acidic proteins of mammalian nuclei: isolation and characterization, *Proc. Natl. Acad. Sci. U. S.,* 59, 262, 1968.
791. **Platz, R. D., Kish, V. M., and Kleinsmith, L. J.,** Tissue specificity of nonhistone chromatin phosphoproteins, *FEBS Lett.,* 12, 38, 1970.

792. **Teng, C. S., Teng, C. T., and Allfrey, V. G.,** Studies on nuclear acidic proteins. Evidence for their phosphorylation, tissue specificity, selective binding to deoxyribonucleic acid, and stimulatory effects on transcription, *J. Biol. Chem.,* 246, 3597, 1971.
793. **Loeb, J. E. and Blat, C.,** Phosphorylation of some rat liver ribosomal proteins and its activation by cyclic AMP, *FEBS Lett.,* 10, 105, 1970.
794. **Kabat, D.,** Phosphorylation of ribosomal proteins in rabbit reticulocytes. Characterization and regulatory aspects, *Biochemistry,* 9, 4160, 1970.
795. **Kabat, D.,** Phosphorylation of ribosomal proteins in rabbit reticulocytes. A cell-free system with ribosomal protein kinase activity, *Biochemistry,* 10, 197, 1971.
796. **Martelo, O. J., Woo, S. L. C., Reinmann, E. M., and Davie, E. W.,** Effect of protein kinase on ribonucleic acid polymerase, *Biochemistry,* 9, 4807, 1970.
797. **Kim, S. and Paik, W. K.,** Studies on the origin of ϵ-N-methyl-L-lysine in protein, *J. Biol. Chem.,* 240, 4629, 1965.
798. **Hempel, K., Lange, H. W., and Birkofer, L.,** ϵ-N-Trimethyllysin, eine neue Aminosäure in Histonen, *Naturwissenschaften,* 55, 37, 1968.
799. **Hempel, K., Lange, H. W., and Birkofer, L.,** N^{ϵ}-Methylierte Lysine in Histonen aus Hühner-Erythrozyten, *H. S. Z. Physiol. Chem.,* 349, 603, 1968.
800. **Paik, W. K. and Kim, S.,** ω-N-methylarginine in histones, *Biochem. Biophys. Res. Commun.,* 40, 224, 1970.
801. **Paik, W. K. and Kim, S.,** ω-methylarginine in protein, *J. Biol. Chem.,* 245, 88, 1970.
802. **Friedman, M., Shull, K. H., and Farber, E.,** Highly selective *in vivo* ethylation of rat liver nuclear protein by ethionine, *Biochem. Biophys. Res. Commun.,* 34, 857, 1969.
803. **Gershey, E. L., Haslett, G. W., Vidali, G., and Allfrey, V. G.,** *J. Biol. Chem.,* 244, 4871, 1969.
804. **Byvoet, P.,** Uptake of label into methylated amino acids from rat tissue histones after *in vivo* administration of (Me-^{14}C) methionine, *Biochim. Biophys. Acta,* 238, 375a, 1971.
805. **Paik, W. K. and Kim, S.,** Protein methylation, *Science,* 174, 114, 1971.
806. **Patterson, B. D. and Davies, D. D.,** Specificity of the enzymatic methylation of pea histone, *Biochem. Biophys. Res. Commun.,* 34, 791, 1969.
807. **Sekeris, C. E., Sekeri, K. E., and Gallwitz, D.,** The methylation of histones of rat liver nuclei *in vitro, H. S. Z. Physiol. Chem.,* 348, 1660, 1967.
808. **Paik, W. K. and Kim, S.,** Enzymatic methylation of protein fractions from calf thymus nuclei, *Biochem. Biophys. Res. Commun.,* 29, 14, 1967.
809. **Burdon, R. H. and Garven, E. V.,** Enzymic modification of chromosomal molecules II. The formation of histone ϵ-N-trimethyl-L-lysine by a soluble chromatin methylase, *Biochim. Biophys. Acta,* 232, 371, 1971.
810. **Benjamin, W. B.,** Selective *in vitro* methylation of rat chromatin associated histone after partial hepatectomy, *Nature New Biol.,* 234, 18, 1971.
811. **Comb, D. G., Sarakar, N., and Pinzino, C. J.,** The methylation of lysine residues in proteins, *J. Biol. Chem.,* 241, 1857, 1966.
811a. **Gallwitz, D.,** Histone methylation. Partial purification of two histone-specific methyltransferases from rat thymus nuclei preferentially methylating histones F2a1 and F3, *Arch. Biochem. Biophys.,* 245, 650, 1971.
812. **Paik, W. K. and Kim, S.,** Solubilization and partial purification of protein methylase III from calf thymus nuclei, *J. Biol. Chem.,* 245, 6010, 1970.
813. **Paik, W. K. and Kim, S.,** Protein methylase. I. Purification and properties of the enzyme, *J. Biol. Chem.,* 243, 2108, 1968.
814. **Paik, W. K. and Kim, S.,** Enzymatic methylation of histones, *Arch. Biochem. Biophys.,* 134, 632, 1969.
815. **Kaye, A. M. and Sheratzky, D.,** Methylation of protein (histone) *in vitro:* Enzymic activity from the soluble fraction of rat organs, *Biochim. Biophys. Acta,* 190, 527, 1969.
816. **Kim, S. and Paik, W. K.,** Studies on the structural requirements of substrate protein for protein methylase II., *Biochemistry,* 10, 3141, 1971.
817. **Orenstein, J. M. and Marsh, W. H.,** Incorporation *in vivo* of methionine and ethionine into the methylation and ethylation of rat liver nuclear proteins, *Biochem. J.,* 109, 697, 1970.
818. **Turberville, C. and Craddock, V. M.,** Methylation of nuclear proteins by dimethylnitrosamine and by methionine in the rat *in vivo, Biochem. J.,* 124, 725, 1971.
819. **Burdon, R. H.,** Enzymic modification of chromosomal macromolecules. I. DNA and protein methylation in mouse tumor cell chromatin, *Biochim. Biophys. Acta,* 232, 359, 1971.
820. **Turner, G. and Hancock, R. L.,** Histone methylase activity of adult, embryonic and neoplastic liver tissues, *Life Sci.,* 9-II, 917, 1970.
821. **Burdon, R. H. and Pearce, C. A.,** Enzymic modification of chromosomal macromolecules. III. The effect of adenine nucleotides on histone modification in chromatin and possible influence of steroid hormones, *Biochim, Biophys. Acta,* 246, 561, 1971.
822. **Shepherd, G. R., Hardin, J. M., and Noland, B. J.,** Methylation of lysine residues of histone fractions in synchronized mammalian cells, *Arch. Biochem. Biophys.,* 143, 1, 1971.
823. **Borun, T. W., Pearson, D., and Paik, W. K.,** Histone methylation in HeLa S-3 cell cycle, *Fed. Proc.,* 30, 1083A, 1971.

824. **Dobozy, A., Hunyadi, J., and Simon, N.,** Demonstration of chloramphenicol hypersensitivity by measurement of histone methylation in lymphocyte cultures, *Clin. Exp. Immunol.,* 8, 917, 1971.
825. **Felix, K. and Rauch, H.,** The structure of the histone of the thymus gland IV., *H. S. Z. Physiol. Chem..* 200, 27, 1931.
826. **Jellum, E.,** Thiol content of calf thymus histone fractions, *Biochim. Biophys. Acta,* 115, 95, 1966.
827. **Jellum, E.,** Interaction of cystamine and cystamine derivatives with nucleic acids and nucleoproteins, *Int. J. Radiat. Biol.,* 9, 185, 1965.
828. **Jellum, E. and Eldjarn, L.,** Isolation of SH-containing DNA-nucleoproteins by chromatography on organomercurial polysaccharide, *Biochim. Biophys. Acta,* 100, 144, 1965.
829. **Deakin, H., Ord., M. G., and Stocken, L. A.,** "Glucose 6-phosphate dehydrogenase" activity and thiol content of thymus nuclei from control and X-irradiated rats, *Biochem. J.,* 89, 296, 1963.
830. **Greenstein, J. P.,** Nucleoproteins, *Adv. Protein Chem.,* 1, 209, 1944.
830a. **Sadgopal, A. and Bonner, J.,** Chromosomal proteins of interphase HeLa cells, *Biochim. Biophys. Acta,* 207, 206 1970.
831. **Sadgopal, A. and Bonner, J.,** Proteins of interphase and metaphase chromosomes compared, *Biochim. Biophys. Acta,* 207, 227, 1970.
832. **Hilton, J. and Stocken, L. A.,** The role of thiol groups in the modification of the template activity of histone-deoxyribonucleic acid complexes, *Biochem. J.,* 100, 21c, 1966.
833. **Stocken, L. A.,** Nature and function of nuclear thiol groups, in *Histones,* DeReuck, A. V. S. and Knight, J., Eds., Churchill, London, 1966, 62.
834. **Littau, V. C., Allfrey, V. G., Frenster, J. H., and Mirsky, A. E.,** Active and inactive regions of nuclear chromatin as revealed by electron microscope autoradiography, *Proc. Natl. Acad. Sci. U. S.,* 52, 93, 1964.
835. **Ord, M. G. and Stocken, L. A.,** Changes in the composition of acid extracts of purified *Echinus* nuclei during cleavage stages and from early embryos, *Biochem. J.,* 120, 671, 1970.
836. **Brachet, J.,** The role of sulfhydril groups in morphogenesis, *J. Exp. Zool.,* 142, 115, 1959.
837. **Chambon, P., Weill, J. D., Doly, J., Strosser, M. T., and Mandel, P.,** On the formation of a novel adenylic compound by enzymatic extracts of liver nuclei, *Biochem. Biophys. Res. Commun.,* 25, 638, 1966.
838. **Nishizuka, Y., Ueda, K., Nakazawa, K., and Hayaishi, O.,** Studies on the polymer of adenosine diphosphate ribose. I. Enzymic formation from nicotinamide adenine dinucleotide in mammalian nuclei, *J. Biol. Chem.,* 242, 3164, 1967.
839. **Sugimura, T., Fujimura, S., Hasegawa, S., and Kawamura, Y.,** Polymerization of the adenosine 5′-diphosphate ribose moiety of NAD by rat liver nuclear enzyme, *Biochim. Biophys. Acta,* 138, 438, 1967.
840. **Fujimura, S., Hasegawa, S., Shimizu, Y., and Sugimura, T.,** Polymerization of the adenosine 5′-diphosphate ribose moiety to nicotinamide-adenine dinucleotide by nuclear enzyme. I. Enzymatic reaction, *Biochim. Biophys. Acta,* 145, 247, 1967.
841. **Reeder, R. H., Ueda, K., Honjo, T., Nishizuka, Y., and Hayaishi, O.,** Studies on the polymer of adenosine diphosphate ribose. II. Characterization of the polymer, *J. Biol. Chem.,* 242, 3172, 1967.
842. **Hasegawa, S., Fujimura, S., Shimizu, Y., and Sugimura, T.,** The polymerization of adenosine 5′-diphosphate – ribose moiety of NAD by nuclear enzyme II. Properties of the reaction product, *Biochim. Biophys. Acta,* 149, 369, 1967.
843. **Ueda, K., Reeder, R. H., Honjo, T., Nishizuka, Y., and Hayaishi, O.,** Poly-adenosine diphosphate ribose synthesis associated with chromatin, *Biochem. Biophys. Res. Commun.,* 31, 379, 1968.
844. **Nishizuka, Y., Ueda, K., Honjo, T., and Hayaishi, O.,** Enzymic adenosine diphosphate ribosylation of histone and poly-adenosine diphosphate ribose synthesis in rat liver nuclei, *J. Biol. Chem.,* 243, 3765, 1968.
845. **Otake, H., Miwa, M., Fujimura, S., and Sugimura, T.,** Binding of ADP-ribose polymer with histones, *J. Biochem.,* 65, 145, 1969.
846. **Honjo, T., Nishizuka, Y., and Hayaishi, O.,** Diphteria toxin-dependent adenosine diphosphate ribosylation of aminoacyl transferase II and inhibition of protein synthesis, *J. Biol. Chem.,* 243, 3553, 1968.
847. **Burzio, L. and Koide, S. S.,** A functional role of polyADPR in DNA synthesis, *Biochem. Biophys. Res. Commun.,* 40, 1013, 1970.
848. **Burzio, L. and Koide, S. S.,** Mode of inhibition of DNA synthesis induced by adenosine diphosphoribosylation of nuclear proteins, *Biochem. Biophys. Res. Commun.,* 42, 1185, 1971.
849. **Srivastava, B. I. S.,** Hydroxyproline containing chromatin proteins from tobacco callus tissue, *Biochem. Biophys. Res. Commun.,* 41, 1357, 1970.
850. **Szybalski, W.,** Effects of elevated temperatures on DNA and on some polynucleotides.: Denaturation, renaturation and cleavage of glycosidic and phosphate ester bonds, in *Thermobiology,* Rose, A. H., Ed., Academic Press, New York, 1967, 73.
851. **Marmur, J., Rownd, R., and Schildkraut, C. L.,** Denaturation and renaturation of deoxyribonucleic acid, *Progr. Nucleic Acid Res.,* 1, 231, 1963.
852. **Chang, K. Y. and Carr, C. W.,** The binding of calcium with deoxyribonucleic acid and deoxyribonucleic acid complexes, *Biochim. Biophys. Acta,* 157, 127, 1968.
853. **Shapiro, J. T., Stannard, B. S., and Felsenfeld, G.,** The binding of small cations to deoxyribonucleic acid. Nucleotide specificity, *Biochemistry,* 8, 3233, 1969.
854. **Tabor, H. and Tabor, C. W.,** Spermidine, spermine, and related amines, *Pharmacol. Rev.,* 16, 245, 1964.

855. **Chevaillier, P.**, Étude au microscope électronique de l'interaction de l'acide désoxyribonucléique et des polyamines, *Exp. Cell Res.*, 58, 213, 1969.
856. **Heby, O. and Agrell, I.**, Observation on the affinity between polyamines and nucleic acids, *H. S. Z. Physiol. Chem.*, 352, 29, 1971.
857. **Agrell, I. and Heby, O.**, Interaction between spermine and histones at nucleoprotein formation, *H. S. Z. Physiol. Chem.*, 352, 39, 1971.
858. **Mahler, H. R., and Mehrotra, B. D.**, The interaction of nucleic acids with diamines, *Biochim. Biophys. Acta*, 68, 211, 1963.
859. **Liquori, A. M., Constantino, L., Crescenzi, V., Elia, V., Giglio, E., Puliti, R., De Santis Savino, M., and Vitagliano, V.**, Complexes between DNA and polyamines: a molecular model, *J. Mol. Biol.*, 24, 113, 1967.
860. **Dimicoli, J. L. and Hélène, C.**, Interactions entre acides aminés et acides nucléiques. III. Etude par absorption et résonance magnétique nucléaire de la formation de complexes entre le tryptophane et les constituants des acides nucléiques, *Biochimie*, 53, 331, 1971.
861. **Pieber, M., Romero, C., Tirapegui, C., and Tohá, J. C.**, Absence of hyperchromic effect in DNA heated above Tm, in the presence of tryptophan, *Z. Naturforsch.*, 24B, 508, 1969.
862. **Brown, P. E.**, The interaction of basic dipeptide methyl esters with DNA, *Biochim. Biophys. Acta*, 213, 282, 1970.
863. **Gabbay, E. J. and Kleinman, R.**, Topography of nucleic acid helices in solutions. V. The interactions of L-, D-, and DL-amino acid derivatives with nucleic acid helices. Demonstration of an asymetric surface, *J. Am. Chem. Soc.*, 89, 7123, 1967.
864. **Gabbay, E. J. and Kleinman, R. W.**, Topography of nucleic acid helices in solutions. The interaction specificities of optically active amino acid derivatives, *Biochem. J.*, 117, 247, 1970.
865. **Gabbay, E. J., Glasser, R., and Gaffney, B. L.**, Interaction specificity of nucleic acids, *Ann. N. Y. Acad. Sci.*, 171, 810, 1970.
866. **Matsuyama, A. and Nagata, C.**, Triplet-triplet energy transfer from histone to DNA in deoxyribonucleohistone, *Biochim. Biophys. Acta*, 224, 588, 1970.
867. **Crampton, C. F., Lipshitz, R., and Chargaff, E.**, Studies on nucleoproteins. II. Fractionation of deoxyribonucleic acid through fractional dissociation of their complexes with basic proteins, *J. Biol. Chem.*, 211, 125, 1954.
868. **Spitnik, P., Lipshitz, R., and Chargaff, E.**, Studies on nucleoproteins III. Deoxyribonucleic acid complexes with basic polyelectrolytes and their fractional extraction, *J. Biol. Chem.*, 215, 765, 1955.
869. **Lucy, J. A. and Butler, J. A. V.**, Fractionation of deoxyribonucleoprotein, *Biochim. Biophys. Acta*, 16, 431, 1955.
870. **Leng, M. and Felsenfeld, G.**, The preferential interactions of polylysine and polyarginine with specific base sequences in DNA, *Proc. Natl. Acad. Sci. U. S.*, 56, 1325, 1966.
871. **Shapiro, J. T., Leng, M., and Felsenfeld, G.**, Deoxyribonucleic acid-polylysine complexes. Structure and nucleotide specificity, *Biochemistry*, 8, 3219, 1969.
872. **Olins, D. E., Olins, A. L., and Von Hippel, P. H.**, Model nucleoprotein complexes: Studies on the interaction of cationic homopolypeptides with DNA, *J. Mol. Biol.*, 24, 157, 1967.
873. **Tsuboi, M., Matsuo, K., and Ts'o, P. O. P.**, Interaction of poly-L-lysine and nucleic acids, *J. Mol. Biol.*, 15, 256, 1966.
874. **Raukas, E.**, Melting point of DNA complexes with protamines, basic polypeptides, and polyethylene polyamine, *Biokhimiya*, 30, 1122, 1965.
875. **Miller, I. R. and Inbar, L.**, Interaction between DNA and the polybases polyarginine and polylysine in equivalent precipitates, *Biopolymers*, 7, 619, 1969.
876. **Yuki, A. and Fox, S. W.**, Selective formation of particles by binding of pyrimidine polyribonucleotides or purine polyribonucleotides with lysine-rich or arginine-rich proteinoids, *Biochem. Biophys. Res. Commun.*, 36, 657, 1969.
877. **Feughelman, M., Langridge, R., Seeds, W. E., Stockes, A. R., Wilson, H. R., Hooper, C. W., Wilkins, M. H. F., Barclay, R. K., and Hamilton, L. D.**, Molecular structure of deoxyribonucleic acid and nucleoprotein, *Nature*, 175, 834, 1955.
878. **Matsuo, K., Mitsui, Y., Iitaka, Y., and Tsuboi, M.**, Interaction of poly-l-lysine with nucleic acids III. An infrared and X-ray examination, *J. Mol. Biol.*, 38, 129, 1968.
879. **Matsuo, K., Fuke, M., Tsuboi, M., and Wada, A.**, Interaction of poly-l-lysine with nucleic acids. IV. A sedimentation and electron microscopic study, *Biochim. Biophys. Acta*, 179, 39, 1969.
880. **Olins, D. E., Olins, A. L., and Von Hippel, P. H.**, On the structure and stability of DNA-protamine and DNA-polypeptide complexes, *J. Mol. Biol.*, 33, 265, 1968.
881. **Ohba, Y.**, Structure of nucleohistone. II. Thermal denaturation, *Biochim. Biophys. Acta*, 123, 84, 1966.
882. **Shih, T. Y. and Bonner, J.**, Template properties of DNA-polypeptide complexes, *J. Mol. Biol.*, 50, 333, 1970.
883. **Wagner, K. G. and Arav, R.**, On the interaction of nucleotides with poly-l-lysine and poly-l-arginine. I. The influence of the nucleotide base on the binding behavior, *Biochemistry*, 7, 1771, 1968.
884. **Wagner, K. G.**, On the interaction of nucleotides with basic polyamino acids. Cooperative binding behavior, *Eur. J. Biochem.*, 10, 261, 1969.
885. **Chaudhuri, S. R. and Yang, J. T.**, Helix-coil transition of poly-l-ornithine in solution, *Biochemistry*, 7, 1379, 1968.
886. **Rifkind, J. M.**, Helix-coil transition of poly-l-arginine: A comparison with other basic polypeptides, *Biopolymers*, 8, 685, 1969.

887. **Davidson, B. and Fasman, G. D.,** The double stranded polyadenylic acid: Poly-l-lysine complex. A conformation study and characterization, *Arch. Biochem. Biophys.*, 144, 650, 1971.
888. **Rifkind, J. M. and Eichhorn, G.,** Specificity for the interaction of nucleotides with basic polypeptides, *Biochemistry*, 9, 1753, 1970.
889. **Lees, C. W. and Von Hippel, P. H.,** Hydrogen exchange studies of deoxyribonucleic acid-protein complexes. Development of a filtration method and application to the deoxyribonucleic acid-polylysine system, *Biochemistry*, 7, 2480, 1968.
890. **Cohen, P. and Kidson, C.,** Conformational analysis of DNA-poly-l-lysine complexes by optical rotatory dispersion, *J. Mol. Biol.*, 35, 241, 1968.
891. **Inoue, S. and Ando, T.,** Optical rotatory dispersion of nucleoclupeine and other model nucleoproteins, *Biochem. Biophys. Res. Commun.*, 32, 501, 1968.
892. **Inoue, S. and Ando, T.,** Interaction of clupeine with deoxyribonucleic acid I. Thermal melting and sedimentation studies, *Biochemistry*, 9, 395, 1970.
893. **Davidson, B. and Fasman, G. D.,** The single-stranded polyadenylic acid-poly-l-lysine complex. A conformational study and characterization, *Biochemistry*, 8, 4116, 1969.
894. **Haynes, M., Garrett, R. A., and Gratzer, W. A.,** Structure of nucleic acid-poly base complexes, *Biochemistry*, 9, 4410, 1970.
895. **Bonner, J., Dahmus, M. E., Fambrough, D., Huang, R. C. C., Marushige, K., and Tuan, D. Y. H.,** The biology of isolated chromatin, *Science*, 159, 47, 1968.
896. **Ilyin, Y. V. and Georgiev, G. P.,** Heterogeneity of deoxyribonucleoprotein particles as evidenced by utracentrifugation in cesium chloride density gradient, *J. Mol. Biol.*, 41, 299, 1969.
897. **Hancock, R.,** Separation by equilibrium centrifugation in CsCl gradients of density-labelled and normal deoxyribonucleoprotein from chromatin, *J. Mol. Biol.*, 48, 357, 1970.
898. **Huang, R. C. C., Bonner, J., and Murray, K.,** Physical and biological properties of soluble nucleohistones, *J. Mol. Biol.*, 8, 54, 1964.
899. **Hnilica, L. S. and Billen, D.,** The effect of DNA-histone interactions on the biosynthesis of DNA *in vitro, Biochim. Biophys. Acta,* 91, 271, 1964.
900. **Minyat, E. E., Borisova, O. F., Volkenstein, M. V., and Georgiev, G. P.,** On the deoxyribonucleoprotein structure I. Studies of base content of protein-free regions of DNP, *Mol. Biologia*, 4, 291, 1970.
901. **Olins, D. E., and Olins, A. L.,** Model nucleohistones: The interaction of F1 and F2al histones and native T7 DNA, *J. Mol. Biol.*, 57, 437, 1971.
902. **Shih, T. Y. and Bonner, J.,** Thermal denaturation and template properties of DNA complexes with purified histone fractions, *J. Mol. Biol.*, 48, 469, 1970.
903. **Ansevin, A. T. and Brown, B. W.,** Specificity in the association of histones with deoxyribonucleic acid. Evidence from derivative thermal denaturation profiles, *Biochemistry*, 10, 1133, 1971.
904. **Littau, V. C., Burdick, C. J., Allfrey, V. G., and Mirsky, A. E.,** The role of histones in the maintenance of chromatin structure, *Proc. Natl. Acad. Sci. U. S.*, 54, 1204, 1965.
905. **Mirsky, A. E., Burdick, C. J., Davidson, E. H., and Littau, V. C.,** The role of lysine-rich histone in the maintenance of chromatin structure in metaphase chromosomes, *Proc. Natl. Acad. Sci. U. S.*, 61, 592, 1968.
906. **Boublík, M., Šponar, J., and Šormová, Z.,** Interactions of deoxyribonucleic acids with basic proteins. I. The effect of lysine-rich histone on DNA macrostructure, *Collect. Czechosl. Chem. Commun.*, 32, 4319, 1967.
907. **Chalkley, G. R. and Jensen, R. H.,** A study of the structure of isolated chromatin, *Biochemistry*, 7, 4380, 1968.
908. **Jensen, R. H. and Chalkley, G. R.,** The physical state of nucleohistone under physiological ionic strength. The effect of interaction with free nucleic acids, *Biochemistry*, 7, 4388, 1968.
909. **Johns, E. W. and Butler, J. A. V.,** Specificity of the interactions between histones and deoxyribonucleic acid, *Nature*, 204, 853, 1964.
910. **Johns, E. W. and Forrester, S.,** Interactions between histone fractions and DNA, *Biochim. Biophys. Acta*, 209, 54, 1970.
911. **Touvet-Poliakow, M. C., Daune, M. P., and Champagne, M. H.,** Interactions entre l'acide déoxyribonucléique et les histones, *Eur. J. Biochem.*, 16, 414, 1970.
912. **Olins, D. E.,** Interaction of lysine-rich histones and DNA, *J. Mol. Biol.*, 43, 439, 1969.
913. **Wilkins, M. H. F., Zubay, G., and Wilson, H. R.,** X-ray diffraction studies of the molecular structure of nucleohistone and chromosomes, *J. Mol. Biol.*, 1, 179, 1959.
914. **Zubay, G. and Wilkins, M. H. F.,** An X-ray diffraction study of histone and protamine in isolation and in combination with DNA, *J. Mol. Biol.*, 4, 444, 1962.
915. **Zubay, G. and Wilkins, M. H. F.,** A note on reversible dissociation of deoxyribonucleohistone, *J. Mol. Biol.*, 9, 246, 1964.
916. **Zubay, G., Wilkins, M. H. F., and Blout, E. R.,** An X-ray diffraction study of a complex of DNA and a synthetic polypeptide, *J. Mol. Biol.*, 4, 69, 1962.
917. **Pardon, J. F., Wilkins, M. H. F., and Richards, B. M.,** Super-helical model for nucleohistone, *Nature*, 215, 508, 1967.

918. **Shih, T. Y. and Fasman, G. D.**, Circular dichroism studies of deoxyribonucleic acid complexes with arginine-rich histone IV (F2a1), *Biochemistry*, 10, 1675, 1971.
918a. **Matsuyama, A., Tagashira, Y., and Nagata, C.**, A circular dichroism study on the conformation of DNA in rat liver chromatin, *Biochim. Biophys. Acta*, 240, 184, 1971.
919. **Fasman, G. D., Schaffhausen, B., Goldsmith, L., and Adler, A.**, Conformational changes associated with F1 histone-deoxyribonucleic acid complexes. Circular dichroism studies, *Biochemistry*, 9, 2814, 1970.
920. **Adler, A. J. and Fasman, G. D.**, Circular dichroism studies of lysine-rich histone f-l-deoxyribonucleic acid complexes. Effect of salts and dioxane upon conformation, *J. Phys. Chem.*, 75, 1516, 1971.
921. **Šponar, J., Boublík, M., Frič, I., and Šormová, Z.**, Study of DNA conformation in native, partial and reconstituted nucleohistones, using hydrodynamic methods and optical rotatory dispersion, *Biochim. Biophys. Acta*, 209, 532, 1970.
922. **Frič, I. and Šponar, J.**, Circular dichroism of native and reconstituted nucleohistones, *Biopolymers*, 10, 1525, 1971.
923. **Fasman, G. D., Valenzuela, M. S., and Adler, A. J.**, Complexes of deoxyribonucleic acid with fragments of lysine-rich histone (f-1). Circular dichroism studies, *Biochemistry*, 10, 3795, 1971.
924. **Li, H. J. and Bonner, J.**, Interaction of histone half-molecules with deoxyribonucleic acid, *Biochemistry*, 10, 1461, 1971.
925. **Claude, A.**, Proteins, lipids, and nucleic acids in cell structure and functions, *Adv. Protein Chem.*, 5, 423, 1949.
926. **Mazen, A. and Champagne, M. H.**, Préparation et propriétés des nucléoprotéines de thymus de veau et d'érythrocytes de poulets, *Bull. Soc. Chim. Biol.*, 47, 1951, 1965.
927. **Welsh, R. S.**, Water-soluble, non-fibrous deoxyribonucleoprotein from calf thymus nuclei, *Nature*, 187, 943, 1960.
928. **Dingman, C. W. and Sporn, M. B.**, Studies on chromatin. I. Isolation and characterization of nuclear complexes of deoxyribonucleic acid, ribonucleic acid, and protein from embryonic and adult tissues of chicken, *J. Biol. Chem.*, 239, 3483, 1964.
929. **Messineo, L.**, The deoxyribonucleoproteins of sea urchin sperm. Extraction and analysis, *Arch. Biochem. Biophys.*, 96, 479, 1962.
930. **Marushige, K. and Bonner, J.**, Template properties of liver chromatin, *J. Mol. Biol.*, 15, 160, 1966.
931. **Alam, S. N., Alam, R. A., and Harbers, E.**, Studies on deoxyribonucleoproteins. VII. Purification of deoxyribonucleoproteins by carrier-free continuous electrophoresis, *Biochim. Biophys. Acta.*, 209, 550, 1970.
932. **Sonnenbichler, J. and Bornkamm, G. W.**, Zum Wirkungsmechanismus von Induktionstoffen I. Fraktionierung von Kalbsthymus-Chromatin und charakterisierung eines spezifischer DNA-Protein Komplexes, *H. S. Z. Physiol. Chem.*, 348, 1579, 1967.
933. **Sonnenbichler, J. and Nobis, P.**, Zum Wirkungsmechanismus von Induktionsstoffen II. Weitere Untersuchungen über einen Nukleinsäure-Protein Komplex aus Kalbsthymus-Chromatin, *H. S. Z. Physiol. Chem.*, 349, 1556, 1968.
934. **Sonnenbichler, J. and Nobis, P.**, The so-called non-histones from acid-treated calf thymus chromatin, *Eur. J. Biochem.*, 16, 60, 1970.
935. **Butler, J. A. V. and Davison, P. F.**, Deoxyribonucleoprotein: A genetic material, *Adv. Protein Chem.*, 18, 161, 1957.
936. **Srivastava, B. I. S.**, Studies on the chromatin of barley leaves during senescence, *Biochem. J.*, 110, 683, 1968.
937. **Harris, S. E. and Forrest, H. S.**, Template properties of embryo chromatin from eggs of the milkweed bug, *Oncopeltus fasciatus, Dev. Biol.*, 23, 324, 1970.
938. **Vendrely, R., Knobloch-Mazen A., and Vendrely, C.**, A comparative biochemical study of nucleohistones and nucleoprotamines in the cell nucleus, in *The Cell Nucleus*, Mitchell, J. S., Ed., Academic Press, New York, 1960, 200.
939. **Omata, S. and Ichii, S.**, Priming activity for RNA synthesis of chromatin from transplantable hepatomas and livers of tumor-bearing rats, *GANN*, 58, 55, 1967.
940. **Grunicke, H., Potter, V. R., and Morris, H. P.**, Comparative studies on nuclei and chromatin of hepatomas and rat liver, *Cancer Res.*, 30, 776, 1970.
941. **Smith, C. E. and Mora, P. T.**, Properties of chromatin from liver and from chemically induced minimal deviation hepatoma of the rat, *Biochim. Biophys. Acta*, 232, 643, 1971.
942. **Itzhaki, R. F.**, Structure and properties of rat thymus deoxyribonucleoprotein. I. Electric birefringence and centrifugation studies, *Proc. R. Soc.*, B164, 75, 1966.
943. **Davison, P. F. and Butler, J. A. V.**, The chemical composition of calf thymus nucleoprotein, *Biochim. Biophys. Acta*, 21, 568, 1956.
944. **Davison, P. F., Conway, B. E., and Butler, J. A. V.**, The nucleoprotein complex of the cell nucleus and its reactions, *Progr. Biophys. Biophys. Chem.*, 4, 148, 1954.
945. **Bartley, J. A. and Chalkley, R. G.**, The viscosity of nucleohistone in urea, *Biochim. Biophys. Acta*, 160, 224, 1968.
946. **Georgiev, G. P., Ilyin, Y. V., Tikhonenko, A. S., and Stelmashchuk, V. Y.**, The structure of chromosomal deoxyribonucleoproteins. I. The isolation of soluble DNP by urea treatment and its properties, *Mol. Biologia*, 4, 246, 1970.
947. **Dounce, A. L. and Hilgartner, C. A.**, A study of DNA nucleoprotein gels and the residual protein of isolated cell nuclei. Relationship to chromosomal structure, *Exp. Cell Res.*, 36, 228, 1964.
948. **Dounce, A. L.**, Nuclear gels and chromosomal structure, *Am. Sci.*, 59, 74, 1971.

949. **Fredericq, E. and Houssier, C.**, Physico-chemical properties of native and recombined calf thymus nucleohistones, *Eur. J. Biochem.*, 1, 51, 1967.
950. **Fredericq, E.**, Gel-forming properties of thymus nucleoproteins, *Biochim. Biophys. Acta*, 55, 300, 1962.
951. **Fredericq, E.**,Enzymatic degradation of thymus nucleoproteins, *Biochim. Biophys. Acta*, 68, 167, 1963.
952. **Fredericq, E.**, The chemical and physical properties of nucleohistones, in *Histones and Nucleohistones*, Phillips, D. M. P., Ed., Plenum Press, London, 1971, 135.
953. **Zubay, G.**, Nucleohistone structure and function, in *The Nucleohistones*, Bonner, J. and Ts'o, P. O. P., Eds., Holden-Day, San Francisco, 1964, 95.
954. **Itzhaki, R. F.**, Structure and properties of rat thymus deoxyribonucleoprotein. II. Effects of enzymes and other agents on electric birefrigence properties, *Proc. Royal Soc.*, B164, 411, 1966.
955. **Hilgartner, C. A.**, The binding of DNA to residual protein in mammalian nuclei, *Exp. Cell Res.*, 49, 520, 1968.
956. **Mackay, M., Hilgartner, C. A., and Dounce, L. A.**, Further studies of DNA-nucleoprotein gels and residual protein of isolated cell nuclei, *Exp. Cell Res.*, 49, 533, 1968.
957. **Kirby, K. S.**, The preparation of deoxyribonucleic acid by the p-amino-salicylate-phenol method, *Biochim. Biophys. Acta*, 36, 117, 1959.
958. **Balis, M. E., Salser, J. S., and Elder, A.**, A suggested role of amino-acids in deoxyribonucleic acid, *Nature*, 203, 1170, 1964.
959. **Frearson, P. M. and Kirby, K. S.**, Nucleoproteins 2. Fractionation of deoxyribonucleoproteins into acidic proteins and lysine-rich histones associated with deoxyribonucleic acid, *Biochem. J.*, 90, 578, 1964.
960. **Holoubek, V., Fanshier, L., Crocker, T. T., and Hnilica, L. S.**, Simultaneous increase in labeling of nuclear RNA and of DNA associated acidic nuclear proteins, *Life Sci.*, 5, 1691, 1966.
961. **Holoubek, V. and Crocker, T. T.**, DNA-associated acidic proteins, *Biochim. Biophys. Acta*, 157, 352, 1968.
962. **Salser, J. S. and Balis, M. E.**, Investigations on amino acids bound to DNA, *Biochim. Biophys. Acta*, 149, 220, 1967.
963. **Salser, J. S., Teller, M. N., and Balis, M. E.**, Changes in DNA-bound amino acids in experimental tumor transplants, *Cancer Res.*, 29, 1002, 1969.
964. **Bendich, A. and Rosenkranz, H. S.**, Some thoughts on the double-stranded model of deoxyribonucleic acid, *Progr. Nucleic Acid Res. Mol. Biol.*, 1, 219, 1963.
965. **Dumitrescu, C., Ghyka, G., and David, C.**, Modification of the macromolecular structure of the nucleoprotein induced by gel filtration on Sephadex, *Rev. Roum. Biochem.*, 7, 29, 1970.
966. **Ansevin, A. T., Hnilica, L. S., Spelsberg, T. C., and Kehm, S.**, Structure studies on chromatin and nucleohistones. Thermal denaturation profiles recorded in the presence of urea, *Biochemistry*, 10, 4793, 1971.
967. **Ris, H.**, The molecular organization of chromosomes, *Fron. Biol.*, 15, 237, 1969.
968. **Ris, H.**, Fine structure of chromosomes, *Proc. Royal Soc.*, B164, 246, 1966.
969. **Itzhaki, R. F. and Rowe, A. J.**, Structure and properties of rat thymus deoxyribonucleoprotein. Electron-microscope studies, *Biochim. Biophys. Acta*, 186, 158, 1969.
970. **Bram, S. and Ris, H.**, On the structure of nucleohistone, *J. Mol. Biol.*, 55, 325, 1971.
971. **Walker, I. O.**, Electrometric and spectrophotometric titration of histone and deoxyribonucleohistone, *J. Mol. Biol.*, 14, 381, 1965.
972. **Simpson, R. T.**, Modification of chromatin with acetic anhydride, *Biochemistry*, 10, 4466, 1971.
973. **Anderson, T.**, Technique for the preservation of three-dimensional structure in preparing specimens for the electron microscope, *Ann. N. Y. Acad. Sci.*, 13, 130, 1951.
974. **Yunis, J. J. and Yasmineh, W. G.**, Heterochromatin, satellite DNA, and cell function, *Science*, 174, 1200, 1971.
975. **Yunis, J. J. and Yasmineh, W. G.**, Satellite DNA in constitutive heterochromatin of the guinea pig, *Science*, 168, 263, 1970.
976. **Yasmineh, W. G. and Yunis, J. J.**, Repetitive DNA of *Microtus agrestis*, *Biochem. Biophys. Res. Commun.*, 43, 580, 1971.
977. **Mattoccia, E. and Comings, D. E.**, Buoyant density and satellite composition of DNA of mouse heterochromatin, *Nature New Biol.*, 229, 175, 1971.
978. **Eckhardt, R. A. and Gall, J. G.**, Satellite DNA associated with heterochromatin in *Rhynchosciara*, *Chromosoma*, 32, 407, 1971.
979. **Hsu, T. C. and Arrighi, F. E.**, Distribution of constitutive heterochromatin in mammalian chromosomes, *Chromosoma*, 34, 243, 1971.
980. **Jones, K. W.**, Chromosomal and nuclear location of mouse satellite DNA in individual cells, *Nature*, 225, 912, 1970.
981. **Pardue, M. L. and Gall, J. G.**, Chromosomal localization of mouse satellite DNA, *Science*, 168, 1356, 1970.
982. **Arrighi, F. E., Hsu, T. C., Saunders, P., and Saunders, G. F.**, Localization of repetitive DNA in the chromosomes of *Microtus agrestis* by means of *in situ* hybridization, *Chromosoma*, 32, 224, 1970.
983. **Brown, S. W.**, Heterochromatin, *Science*, 151, 417, 1966.
984. **Lima-de-Faria, A.**, DNA replication and gene amplification in heterochromatin, in *Handbook of Molecular Cytology*, Lima-de-Faria, A., Ed., North Holland Publ. Co., Amsterdam, 1969, 277.
985. **Frenster, J. H.**, Biochemistry and molecular biophysics of heterochromatin and euchromatin, in *Handbook of Molecular Cytology*, Lima-de-Faria, A., Ed., North Holland Publ. Co., Amsterdam, 1969, 251.
986. **Yunis, J. J. and Yasmineh, W. G.**, Satellite DNA in heterochromatin, *Adv. Cell Mol. Biol.*, 2, in press.

987. **Walker, P. M. B., Flamm, W. G., and McClaren, A.,** Highly repetitive DNA in rodents, in *Handbook of Molecular Cytology,* Lima-de-Faria, A., Ed., North Holland Publ. Co., Amsterdam, 1969, 52.

988. **Henning, W., Henning, I., and Stein, H.,** Repeated sequences in the DNA of *Drosophila* and their localization in giant chromosomes, *Chromosoma,* 32, 31, 1970.

989. **Jones, K. W. and Robertson, F. W.,** Localization of reiterated nucleotide sequences in *Drosophila* and mouse by *in situ* hybridization of complementary RNA, *Chromosoma,* 31, 331, 1970.

990. **Rae, P. M. M.,** Chromosomal distribution of rapidly reannealing DNA in *Drosophila melanogaster, Proc. Natl. Acad. Sci. U.S.,* 67, 1018, 1970.

991. **Gall, J. G., Cohen, E. H., and Polan, M. L.,** Repetitive DNA sequences in *Drosophila, Chromosoma,* 33, 319, 1971.

992. **Ris, H. and Kubai, D.,** Chromosome structure, *Annu. Rev. Genetics,* 4, 263, 1970.

993. **Duerksen, J. D. and McCarthy, B. J.,** Distribution of deoxyribonucleic acid sequences in fractionated chromatin, *Biochemistry,* 10, 1471, 1971.

993a. **De Bellis, R. H., Benjamin, W., and Gellhorn, A.,** Studies on *in vitro* RNA synthesis by chromatin fractions from rat liver and Novikoff hepatoma, *Biochem. Biophys. Res. Commun.,* 36, 166, 1969.

994. **Berlowitz, L.,** Analysis of histone *in situ* in developmentally inactivated chromatin, *Proc. Natl. Acad. Sci. U.S.,* 54, 476, 1965.

995. **Myśliwski, A.,** Reaction of histones with ammoniacal silver in chromatin in different states of condensation, *Acta Histochem.,* 39, 248, 1971.

996. **Pallotta, D., Berlowitz, L., and Rodriguez, L.,** Histones of genetically active and inactive chromatin in mealy bugs, *Exp. Cell Res.,* 60, 474, 1970.

997. **Comings, D. E.,** Histones of genetically active and inactive chromatin, *J. Cell Biol.,* 35, 699, 1967.

998. **Loewus, M. W.,** Analysis of chromatin in male and female mealy bugs, *Nature,* 218, 474, 1968.

999. **Frenster, J. H.,** Nuclear polyanions as de-repressors of synthesis of ribonucleic acid, *Nature,* 206, 680, 1965.

1000. **Gorovsky, M. A. and Woodward, J.,** Histone content of chromosomal loci active and inactive in RNA synthesis, *J. Cell Biol.,* 33, 723, 1967.

1001. **Harris, H.,** The reactivation of red cell nucleus, *J. Cell Sci.,* 2, 23, 1967.

1002. **Bolund, L., Ringertz, N. R., and Harris, H.,** Changes in the cytochemical properties of erythrocyte nuclei reactivated by cell fusion, *J. Cell Sci.,* 4, 71, 1969.

1003. **Gurdon, J. B. and Woodland, H. R.,** The cytoplasmic control of nuclear activity in animal development, *Biol. Rev.,* 43, 233, 1968.

1004. **Du Praw, E. J.,** Macromolecular organization of nuclei and chromosomes: a folded fibre model based on whole-mount electron microscopy, *Nature,* 206, 338, 1965.

1005. **Du Praw, E. J.,** Evidence for a "folded-fibre" organization in human chromosomes, *Nature,* 209, 577, 1966.

1006. **Du Praw, E. J.,** in *DNA and Chromosomes,* Holt, Rinehart, and Winston, New York, 1970, 132.

1007. **Huberman, J. A. and Riggs, A. D.,** Autoradiography of chromosomal DNA fibers from chinese hamster cells, *Proc. Natl. Acad. Sci. U.S.,* 55, 599, 1966.

1008. **Cantor, K. P. and Hearst, J. E.,** The structure of metaphase chromosomes. I. Electrometric titration, magnesium ion binding and circular dichroism, *J. Mol. Biol.,* 49, 213, 1969.

1009. **Chorazy, M., Bendich, A., Borenfreund, E., and Hutchison, D. J.,** Studies on the isolation of metaphase chromosomes, *J. Cell Biol.,* 19, 59, 1963.

1010. **Cantor, K. P. and Hearst, J. E.,** Isolation and partial characterization of metaphase chromosomes of mouse ascites tumor, *Proc. Natl. Acad. Sci. U. S.,* 55, 642, 1966.

1011. **Salzman, N. P., Moore, D. E., and Mendelsohn, J.,** Isolation and characterization of human metaphase chromosomes, *Proc. Natl. Acad. Sci. U. S.,* 56, 1449, 1966.

1012. **Huberman, J. A. and Attardi, G.,** Isolation of metaphase chromosomes from HeLa cells, *J. Cell Biol.,* 31, 95, 1966.

1013. **Huberman, J. A. and Attardi, G.,** Studies of fractionated HeLa cell metaphase chromosomes. I. The chromosomal distribution of DNA complementary to 28s and 18s ribosomal RNA and to cytoplasmic messenger RNA, *J. Mol. Biol.,* 29, 487, 1967.

1014. **Maio, J. J. and Schildkraut, C. L.,** Isolated mammalian metaphase chromosomes. I. General characteristics of nucleic acids and proteins, *J. Mol. Biol.,* 24, 29, 1967.

1015. **Lin, H. J., Karkas, J. D., and Chargaff, E.,** Template functions in the enzymic formation of polyribonucleotides. II. Metaphase chromosomes as templates in the enzymic synthesis of ribonucleic acids, *Proc. Natl. Acad. Sci. U. S.,* 56, 954, 1966.

1016. **Maio, J. J. and Schildkraut, C. L.,** Isolated mammalian metaphase chromosomes. II. Fractionated chromosomes of mouse and Chinese hamster cells, *J. Mol. Biol.,* 40, 203, 1969.

1017. **Mendelsohn, J., Moore, D. E., and Salzman, N. P.,** Separation of isolated Chinese hamster metaphase chromosomes into three size-groups, *J. Mol. Biol.,* 32, 101, 1968.

1018. **Wray, W. and Stubblefield, E.,** A new method for the rapid isolation of chromosomes, mitotic apparatus, or nuclei from mammalian fibroblasts at near neutral pH, *Exp. Cell Res.,* 59, 469, 1970.

1019. **Burkholder, G. D. and Mukherjee, B. B.,** Fractionation of isolated mammalian metaphase chromosomes, *Exp. Cell Res.,* 63, 213, 1970.

1020. **Zirkin, B. R. and Wolfe, S. L.**, The chemical composition of nuclei and chromosomes isolated by the Langmuir trough technique, *Chromosoma*, 32, 162, 1970.

1021. **Wolff, S.**, Strandedness of chromosomes, *Int. Rev. Cytol.*, 25, 279, 1969.

1022. **Prescott, D. M.**, The structure and replication of eucaryotic chromosomes, *Adv. Cell Mol. Biol.*, 1, 57, 1970.

1023. **Ris, H.**, Ultrastructure of the animal chromosome, in *Regulation of Nucleic Acid and Protein Biosynthesis*, Koningsberger, V. V. and Bosch, L., Eds., Elsevier, Amsterdam, 1967, 11.

1024. **Du Praw, E. J.**, The organization of nuclei and chromosomes in honeybee embryonic cells, *Proc. Natl. Acad. Sci. U. S.*, 53, 161, 1965.

1025. **Solari, A. J.**, The ultrastructure of chromatin fibers. I. The effect of spreading conditions, *Exp. Cell Res.*, 53, 553, 1968.

1026. **Solari, A. J.**, The ultrastructure of chromatin fibers. II. The ultrastructure of the loops from sea urchin sperm chromatin, *Exp. Cell Res.*, 53, 567, 1968.

1027. **Ohnuki, Y.**, Structure of chromosomes. I. Morphological studies of the spiral structure of human somatic chromosomes, *Chromosoma*, 25, 402, 1968.

1028. **Stubblefield, E. and Wray, W.**, Architecture of the Chinese hamster metaphase chromosome, *Chromosoma*, 32, 262, 1971.

1029. **Clarke, J. A., Rowland, G. F., and Salsbury, A. J.**, The surface and internal structure of metaphase chromatin, *Chromosoma*, 29, 74, 1970.

1030. **Bahr, G. F.**, Human chromosome fibers. Considerations of DNA-protein packing and of looping patterns, *Exp. Cell Res.*, 62, 39, 1970.

1031. **Garrett, R. A.**, Low angle X-ray diffraction from dilute nucleohistone gels, *Biochim. Biophys. Acta*, 246, 553, 1971.

1032. **Kobayashi, S., Maio, J. J., and Schildkraut, C. L.**, Histones of fractionated metaphase chromosomes of Chinese hamster cells, *Fed. Proc.*, 29, 730A, 1970.

1033. **John, P. C. L. and Jones, R. N.**, Molecular heterogeneity of soluble proteins and histones in relationship to the presence of B-chromosome in rye, *Exp. Cell Res.*, 63, 271, 1970.

1034. **Izawa, M., Allfrey, V. G., and Mirsky, A. E.**, Composition of the nucleus and chromosomes in the lampbrush stage of the newt oocyte, *Proc. Natl. Acad. Sci. U. S.*, 50, 811, 1963.

1035. **Izawa, M., Allfrey, V. G., and Mirsky, A. E.**, The relationships between RNA synthesis and loop structure in lampbrush chromosomes, *Proc. Natl. Acad. Sci. U.S.*, 49, 544, 1963.

1036. **Rae, P. M. M.**, Whole mount electron microscopy of *Drosophila* salivary chromosomes, *Nature*, 212, 139, 1966.

1037. **Berendes, H. D. and Meyer, G. F.**, A specific chromosome element, the telomere of *Drosophila* polytene chromosomes, *Chromosoma*, 25, 184, 1958.

1038. **Berendes, H. D.**, Polytene chromosome structure at submicroscopic level. I. A map of region X, 1-4 E of *Drosophila melanogaster, Chromosoma*, 29, 118, 1970.

1039. **Rao, M. V. N. and Ammermann, D. A.**, Polytene chromosomes and nucleic acid metabolism during macronuclear development in *Euplotes, Chromosoma*, 29, 246, 1970.

1040. **Nakanishi, Y. H., Kato, H., and Utsumi, S.**, Polytene chromosomes in silk gland cells of the silkworm, *Bombyx mori, Experientia*, 25, 384, 1969.

1041. **Dick, C. and Johns, E. W.**, Histone from *Drosophila melanogaster, Comp. Biochem. Physiol.*, 31, 529, 1969.

1042. **Robert, M. and Kroeger, H.**, Lokalisation zusätzlicher RNS-synthese in Trypsin-behandelten Riesenchromosomen von *Chironomus thummi, Experientia*, 21, 326, 1965.

1043. **Crampton, C. F., Lipshitz, R., and Chargaff, E.**, Studies on nucleoproteins. I. Dissociation and reassociation of the deoxyribonucleohistone of calf thymus, *J. Biol. Chem.*, 206, 499, 1954.

1044. **Shaw, L. M. J. and Huang, R. C. C.**, A description of two procedures which avoid the use of extreme pH conditions for resolution of components isolated from chromatin prepared from pig cerebellar and pituitary nuclei, *Biochemistry*, 9, 4530, 1970.

1045. **Henson, P. and Walker, I. O.**, The partial dissociation of nucleohistone by salts. Circular dichroism and denaturation studies, *Eur. J. Biochem.*, 16, 524, 1970.

1046. **Loeb, J. E.**, Gel filtration of deoxyribonucleoprotein on agarose, *Biochim. Biophys. Acta*, 157, 424, 1968.

1047. **Borchsenius, S. N., Turoverova, L. V., Vinokurova, T. I., Ilyina, I. D., and Vorobyev, V. I.**, Solubility and dissociation of chromosomal nucleoprotein complexes, *Mol. Biologia*, 3, 507, 1969.

1048. **Shih, T. Y. and Fasman, G. D.**, Conformation of deoxyribonucleic acid in chromatin: a circular dichroism study, *J. Mol. Biol.*, 52, 125, 1970.

1049. **Cattan, D., Bourgoin, D., and Joly, M.**, Change of conformation of DNA by association with proteins, *Eur. J. Biochem.*, 8, 541, 1969.

1050. **Simpson, R. T. and Sober, H. A.**, Circular dichroism of calf liver nucleohistone, *Biochemistry*, 9, 3103, 1970.

1051. **Wilhelm, F. X., Champagne, M. H., and Daune, M. P.**, Conformation du DNA dans la nucléoprotéine, *Eur. J. Biochem.*, 15, 321, 1970.

1052. **Wagner, T. and Spelsberg, T. C.**, Aspects of chromosomal structure. I. Circular dichroism studies, *Biochemistry*, 10, 2599, 1971.

1053. **Li, H. J., Isenberg, I., and Johnson, W. C.**, Absorption and circular dichroism studies on nucleohistone. IV. *Biochemistry*, 10, 2587, 1971.

1054. **Pemogorov, V. I., Debabov, V. G., Sladkova, I. A., and Rebentish, B. A.,** Structure of DNA and histones in the nucleohistone, *Biochim. Biophys. Acta,* 199, 556, 1970.
1055. **Henson, P. and Walker, I. O.,** The partial dissociation of nucleohistone by salts. Hydrodynamic and denaturation studies, *Eur. J. Biochem.,* 14, 345, 1970.
1056. **Inoue, S. and Ando, T.,** Interaction of clupeine with deoxyribonucleic acid. I. Thermal melting and sedimentation studies, *Biochemistry,* 9, 388, 1970.
1057. **Henson, P. and Walker, I. O.,** The structure of nucleohistone. Hydrodynamic behaviour at high ionic strength, *Eur. J. Biochem.,* 22, 1, 1971.
1058. **Smart, J. E. and Bonner, J.,** Studies on the role of histones in the structure of chromatin, *J. Mol. Biol.,* 58, 661, 1971.
1059. **Bram, S.,** The secondary structure of DNA in solution and in nucleohistone, *J. Mol. Biol.,* 58, 277, 1971.
1060. **Zama, M. and Ichimura, S.,** Difference between polylysine and polyarginine in changing DNA structure upon complex formation, *Biochem. Biophys. Res. Commun.,* 44, 936, 1971.
1061. **Giannoni, G., Peacocke, A. R., and Walker, I. O.,** Thymus deoxyribonucleoprotein. V. The effect of denaturation and sedimentation, *Biochim. Biophys. Acta,* 72, 469, 1963.
1062. **Kurtz, D. I. and Sinex, F. M.,** Age related differences in the association of brain DNA and nuclear protein, *Biochim. Biophys. Acta,* 145, 840, 1967.
1063. **Samis, H. V., Poccia, D. L., and Wulff, V. J.,** The effect of salt extraction and heat denaturation on the behavior of rat liver chromatin, *Biochim. Biophys. Acta,* 166, 410, 1968.
1064. **Brutlag, D., Schlehuber, C., and Bonner, J.,** Properties of formaldehyde-treated nucleohistone, *Biochemistry,* 8, 3214, 1969.
1065. **Ilyin, Y. V., Varshavsky, A. Y., and Georgiev, G. P.,** The structure of chromosomal deoxyribonucleoproteins. 3. DNP heterogeneity as evidenced by equilibrium centrifugation in CsCl density gradients, *Mol. Biologia,* 4, 821, 1970.
1066. **Klein, F. and Szirmai, J. A.,** Quantitative studies on the interaction of azure A with deoxyribonucleic acid and deoxyribonucleoprotein, *Biochim. Biophys. Acta,* 72, 48, 1963.
1067. **Miura, A. and Ohba, Y.,** Structure of nucleohistone. III. Interaction with toluidine blue, *Biochim. Biophys. Acta,* 145, 436, 1967.
1068. **Itzhaki, R. F.,** Studies on the accessibility of deoxyribonucleic acid in deoxyribonucleoprotein to cationic molecules, *Biochem. J.,* 122, 583, 1971.
1069. **Borisova, O. F. and Minyat, E. E.,** Complex of deoxyribonucleoprotein with acridine orange dye, *Mol. Biologia,* 3, 758, 1969.
1070. **Richards, B. M. and Pardon, J. F.,** The molecular structure of nucleohistone (DNH), *Exp. Cell Res.,* 62, 184, 1970.
1071. **Stollar, B. D.,** Immunochemical measurement of DNA in nucleoprotein with the use of anti-DNA antibodies from patients with systemic lupus erythromatosus, *Biochim. Biophys. Acta,* 209, 541, 1970.
1072. **Phillips, D. M. P.,** The binding of extra histone and protamine to deoxyribonucleoprotein, *Experientia,* 24, 668, 1968.
1073. **Combard-Vanderplancke, A. and Vendrely, R.,** Formation de complexes artificiels entre désoxyribonucléohistones et acide désoxyribonucléique on histone surajoutés, *Bull. Soc. Chim. Biol.,* 50, 427, 1968.
1074. **Ansevin, A. T., Getz, S., and Hnilica, L. S.,** The binding of additional histones to isolated chromatin, in preparation.
1075. **Itzhaki, R. F.,** Structure of deoxyribonucleoprotein as revealed by its binding to polylysine, *Biochem. Biophys. Res. Commun.,* 41, 25, 1970.
1076. **Clark, R. J. and Felsenfeld, G.,** Structure of chromatin, *Nature New Biol.,* 229, 101, 1971.
1077. **Britten, R. J. and Kohne, D. E.,** Repeated sequences in DNA, *Science,* 161, 529, 1968.
1078. **Itzhaki, R. F.,** The arrangement of proteins on the deoxyribonucleic acid in chromatin, *Biochem. J.,* 125, 221, 1971.
1079. **Mirsky, A. E.,** The structure of chromatin, *Proc. Natl. Acad. Sci. U.S.,* 68, 2945, 1971.
1079a. **Phillips, D. M. P.,** in preparation.
1080. **Busch, H., Steele, W. J., Hnilica, L. S., Taylor, C. W., and Mavioglu, H.,** Biochemistry of histones and the cell cycle, *J. Cell. Comp. Physiol.,* Suppl. 1, 62, 95, 1963.
1081. **Ursprung, H.,** Kernproteine und Genefunktion, *Naturwissenschaften,* 52, 375, 1965.
1082. **Frenster, J. H.,** A model of specific de-repression with interphase chromatin, *Nature,* 206, 1269, 1965.
1083. **Frenster, J. H.,** Mechanism of repression and de-repression within interphase chromatin, in *The Chromosome – Structural and Functional Aspects,* Dawe, C. J. and Yerganian, G., Eds., Williams & Wilkins, Baltimore, 1965, 78.
1084. **Frenster, J. H.,** Control of DNA strand separations during selective transcription and assynchronous replication, in *The Cell Nucleus – Metabolism and Radiosensitivity,* Klouwen, M. H., Ed., Taylor and Francis, Ltd., London, 1966, 27.
1085. **Nagumo, T.,** Complexes between acidic and basic proteins, *J. Biochem.,* 49, 379, 1961.
1086. **Haurowitz, F., Turner, J. E., and Bremer-Kennedy, M.,** Uber die gegenseitige Bindung von Eiweiss und Protamin, *H. S. Z. Physiol. Chem.,* 315, 127, 1959.
1087. **Hupka, S. and Thurzo, V.,** The interaction of calf thymus histone with human serum albumin and gamma globulin, *Neoplasma,* 8, 471, 1961.

1088. **Parsons, R. S. and Polya, J. B.**, Interaction of amides with proteins. VI. The action of protamine on proteins of human serum, *Enzymologia (Acta Biocatal.)*, 24, 360, 1962.
1089. **Agrell, I. P. S. and Christensson, E. G.**, Influence of polyvalent ions on histones isolated from different tissues, *Nature*, 191, 284, 1961.
1090. **Langan, T. A.**, A phosphoprotein preparation from liver nuclei and its effect on the inhibition of RNA synthesis by histones, in *The Regulation of Nucleic Acid and Protein Synthesis*, Koningsberger, V. V. and Bosch, L., Eds., Elsevier, Amsterdam, 1967, 233.
1091. **Wang, T. Y.**, Solubilization and characterization of the residual proteins of cell nucleus, *J. Biol. Chem.*, 241, 2913, 1966.
1092. **Wang, T. Y.**, The isolation, properties, and possible functions of chromatin acidic proteins, *J. Biol. Chem.*, 242, 1220, 1967.
1093. **Patel, G., Patel, V., Wang, T. Y., and Zobel, C. R.**, Studies on the nuclear residual proteins, *Arch. Biochem. Biophys.*, 128, 654, 1968.
1094. **Wang, T. Y. and Johns, E. W.**, Study of the chromatin acidic proteins of rat liver: Heterogeneity and complex formation with histones, *Arch. Biochem. Biophys.*, 124, 176, 1968.
1095. **Butler, J. A. V. and Johns, E. W.**, Interactions between histones and nucleic acids, *Biochem. J.*, 91, 15c, 1964.
1096. **Chipperfield, A. R.**, Effects of RNA and acidic protein on the interaction between DNA and histones, *Life Sci.*, 6, 2643, 1967.
1097. **Spelsberg, T. C. and Hnilica, L. S.**, The effects of acidic proteins and RNA on the histone inhibition of the DNA-dependent RNA synthesis *in vitro*, *Biochim. Biophys. Acta*, 195, 63, 1969.
1098. **Spelsberg, T. C., Kizer, P. E., and Hnilica, L. S.**, Dissociation of RNA-histone complexes by DNA, *Experientia*, 26, 136, 1970.
1099. **Sluyser, M. and Den Engelse, L.**, Removal of RNA from DNA preparations by the addition of lysine-rich histones, *Anal. Biochem.*, 25, 444, 1970.
1100. **Agrell, I. P. S.**, The stability of nucleoprotein complexes at varied ionic strength, *Life Sci.*, 10, part II, 349, 1971.
1101. **Agrell, I. P. S.**, Interactions between denatured DNA and RNA at the formation of nucleohistone, *Life Sci.*, 10, part II, 691, 1971.
1102. **Dastugue, B., Tichonicky, L., Hanoune, J., and Kruh, J.**, Liaison entre histones et RNA, *FEBS Lett.*, 8, 133, 1970.
1103. **Kruh, J., Tichonicky, L., and Wajcman, H.**, Action des protéines acides du noyau sur la synthèse acellulaire de l'hémoglobine et sur les RNA, *Biochim. Biophys. Acta*, 195, 549, 1969.
1104. **Wang, T. Y.**, Restoration of histone-inhibited DNA-dependent RNA synthesis by acidic chromatin proteins, *Exp. Cell Res.*, 53, 288, 1968.
1105. **Wang, T. Y.**, Diametric effects of histones and nonhistone proteins on DNA replication *in vitro*, *Exp. Cell Res.*, 57, 467, 1969.
1106. **Teng, C. S. and Hamilton, T. H.**, Role of chromatin in estrogen action in the uterus. II. Hormone induced synthesis of nonhistone acidic proteins which restore histone-inhibited DNA-dependent RNA synthesis, *Proc. Natl. Acad. Sci. U. S.*, 63, 465, 1969.
1107. **Wang, T. Y.**, Activation of transcription *in vitro* from chromatin by nonhistone proteins, *Exp. Cell Res.*, 61, 455, 1970.
1108. **Kamiyama, M. and Wang, T. Y.**, Activated transcription from liver chromatin by nonhistone proteins, *Biochim. Biophys. Acta*, 228, 563, 1971.
1109. **Kamiyama, M. and Dastugue, B.**, Rat liver nonhistone proteins: correlation between protein kinase activity and activation of RNA synthesis, *Biochem. Biophys. Res. Commun.*, 44, 29, 1971.
1110. **Kamiyama, M., Dastugue, B., and Kruh, J.**, Action of phosphoproteins and protein kinase from rat liver chromatin on RNA synthesis, *Biochem. Biophys. Res. Commun.*, 44, 1345, 1971.
1111. **Kraemer, R. J. and Coffey, D. S.**, The interaction of natural and synthetic polyanions with mammalian nuclei. I. DNA synthesis, *Biochim. Biophys. Acta*, 224, 553, 1970.
1112. **Kraemer, R. J. and Coffey, D. S.**, The interaction of natural and synthetic polyanions with mammalian nuclei. II. Nuclear swelling, *Biochim. Biophys. Acta*, 224, 568, 1970.
1113. **Brown, D. G. and Coffey, D. S.**, Release of nuclear DNA template restrictions by specific polyribonucleotides, *Science*, 171, 176, 1971.
1114. **Brown, D. G. and Coffey, D. S.**, The release of chromatin template restrictions by natural polyribonucleotides, *Biochem. Biophys. Res. Commun.*, 42, 326, 1971.
1115. **Getz, S., Ansevin, A. T., and Hnilica, L. S.**, The effects of various polyanions on thermal stability of chromatin, in preparation.
1116. **Kent, P. W., Hichens, M., and Ward, P. F. V.**, Fractionation of DNP by heparin and dextran sulphate, *Biochem. J.*, 68, 568, 1958.
1117. **Whitfield, J. F. and Perris, A. D.**, Dissolution of the condensed chromatin structures of isolated thymocyte nuclei and the disruption of deoxyribonucleoprotein by inorganic phosphate and phosphoprotein, *Exp. Cell Res.*, 49, 359, 1968.
1118. **Kinoshita, S.**, Heparin as possible initiator of genomic RNA synthesis in early development of sea urchin embryos, *Exp. Cell Res.*, 64, 403, 1971.

1119. **Miller, G., Berlowitz, L., and Regelson, W.,** Chromatin and histones in mealy bug cell explants: Activation and decondensation of facultative heterochromatin by synthetic polyanions, *Chromosoma,* 32, 251, 1971.
1120. **Meisler, M. H. and McCluer, R. H.,** Nucleohistone dissociation by ganglioside micelles, *Science,* 154, 896, 1966.
1121. **O'Malley, B. W., McGuire, W. L., Kohler, P. O., and Korenman, S. G.,** Studies on the mechanism of steroid hormone regulation of synthesis of specific proteins, *Recent Progr. Hormone Res.,* 25, 105, 1969.
1122. **Sekeris, C. E. and Lang, N.,** Bindung von ^{3}H-Cortison an Histone aus Rattenleber, *H. S. Z. Physiol. Chem.,* 340, 92, 1965.
1123. **Sluyser, M.,** Binding of hydrocortisone to rat liver histones, *J. Mol. Biol.,* 19, 591, 1966.
1124. **Amada, Y. and Yasuda, T.,** Changes in the basicity of histone fractions during callus induction, *Biochem. Biophys. Res. Commun.,* 43, 488, 1971.
1125. **Wilson, J. D. and Loeb, P. M.,** Estrogen and androgen control of cell biosynthesis in target organs, in *Developmental and Metabolic Control Mechanisms and Neoplasia,* Williams & Wilkins, Baltimore, 1965, 375.
1126. **Dingman, C. W. and Sporn, M. B.,** Actinomycin D and hydrocortisone: Intracellular binding in rat liver, *Science,* 149, 1251, 1965.
1127. **De Venuto, F., Kelleher, P. C., and Westphal, U.,** Interactions between corticosteroids and fractions of rat liver and muscle cells determined by "equilibrium fractionation" and equilibrium dialysis, *Biochim. Biophys. Acta,* 63, 434, 1962.
1128. **De Venuto, F. and Muldoon, T.,** Interactions between corticosteroids and fractions of mitochondria and nuclei from normal rat liver cells, *Exp. Cell Res.,* 50, 338, 1968.
1129. **Brunkhorst, W. K.,** Intracellular binding of corticosterone in thymus tissue, *Biochem. Biophys. Res. Commun.,* 35, 880, 1969.
1130. **Sluyser, M.,** Binding of testosterone and hydrocortisone to rat-tissue histones, *J. Mol. Biol.,* 22, 411, 1966.
1131. **Tsai, Y. H. and Hnilica, L. S.,** Interactions of histones with corticosteroid hormones, *Biochim. Biophys. Acta,* 238, 277, 1971.
1132. **Holoubek, V. and Driscoll, S. E.,** Cortisol binding fraction of rat liver chromatin, *Fed. Proc.,* 30, 1213A (1971).
1133. **Lavrinenko, I. A., Morozova, T. M., and Uschkova, L. F.,** Binding of 1,2-^{3}H-hydrocortisone by chromatin acidic proteins and histone *in vitro, Mol. Biologia,* 5, 17, 1971.
1134. **Dastugue, B., Defer, N., and Kruh, J.,** Binding of corticosterone to liver chromatin acidic proteins, *FEBS Lett.,* 16, 121, 1971.
1135. **Sunaga, K. and Koide, S. S.,** Interaction of calf thymus histones and DNA with steroids, *Steroids,* 9, 451, 1967.
1136. **Sunaga, K. and Koide, S. S.,** Factors influencing the interaction of steroids with calf thymus histones, *Arch. Biochem. Biophys.,* 122, 670, 1967.
1137. **Sunaga, K. and Koide, S. S.,** Structural specificity of the steroids interacting with calf thymus histones, *Biochem. Biophys. Res. Commun.,* 26, 342, 1967.
1138. **Tabachnick, M. and Giorgio, N. A.,** Interaction of thyroid hormone with histone from calf thymus nuclei, *Nature,* 212, 1610, 1966.
1139. **Sluyser, M.,** Effect of testosterone on the binding of prostate histone to DNA *in vitro, Biochem. Biophys. Res. Commun.,* 22, 336, 1966.
1140. **Sluyser, M.,** Interaction of steroid hormones with histones *in vitro, Biochim. Biophys. Acta,* 182, 235, 1969.
1141. **Monder, C.,** Stability of corticosteroids in aqueous solutions, *Endocrinology,* 82, 318, 1968.
1142. **Monder, C. and Walker, M. C.,** Interactions between corticosteriods and histones, *Biochemistry,* 9, 2489, 1970.
1143. **Ohtsuka, E. and Koide, S. S.,** *In vitro* stimulation of RNA synthesis in isolated rat liver nuclei by glucocorticoids, *Biochem. Biophys. Res. Commun.,* 35, 648, 1969.
1114. **Sunaga, K., Imamura, T., and Koide, S. S.,** Interaction of 21-dehydroprednisolone with amino acids, *Biochim. Biophys. Acta,* 210, 164, 1970.
1145. **Sunaga, K., Koide, S. S., and Tatum, A. H.,** Interaction of 21-dehydrocorticosteroids with peptides, *Biochim. Biophys. Acta,* 232, 347, 1971.
1146. **Madix, J. C. and Bresnick, E.,** Increased efficacy of liver chromatin as a template for RNA synthesis after administration of 3-methylcholanthrene, *Biochem. Biophys. Res. Commun.,* 28, 445, 1967.
1147. **Bresnick, E.,** Ribonucleic acid polymerase activity in liver nuclei from rats pretreated with 3-methylcholanthrene, *Mol. Pharmacol.,* 2, 406, 1966.
1148. **Bresnick, E. and Mossé, H.,** Activation of genetic transcription in rat liver chromatin by 3-methylcholanthrene, *Mol. Pharmacol.,* 5, 219, 1969.
1149. **Gelboin, H. V., Wortham, J. S., and Wilson, R. G.,** 3-methylcholanthrene and phenobarbital stimulation of rat liver RNA polymerase, *Nature,* 214, 281, 1967.
1150. **Piper, W. N. and Bousquet, W. F.,** Phenobarbital and methylcholanthrene stimulation of rat liver chromatin template activity, *Biochem. Biophys. Res. Commun.,* 33, 602, 1968.
1151. **Yee, M. and Bresnick, E.,** Effect of administration of 3-methylcholanthrene on the salt-extractable chromatin proteins of rat liver, *Mol. Pharmacol.,* 7, 191, 1971.
1152. **Sporn, M. B. and Dingman, C. W.,** Studies on chromatin. II. Effects of carcinogens and hormones on rat liver chromatin, *Cancer Res.,* 26, 2488, 1966.

1153. **Gelboin, H. V.**, Studies on the mechanism of methylcholanthrene induction of enzyme activities. II. Stimulation of microsomal and ribosomal amino acid incorporation: the effects of polyuridylic acid and actinomycin D, *Biochim. Biophys. Acta,* 91, 130, 1964.

1154. **Gelboin, H. V. and Blackburn, N. R.**, The stimulatory effect of 3-methylcholanthrene on benzpyrene hydroxylase activity in several rat tissues: Inhibition by actinomycin D and puromycin, *Cancer Res.*, 24, 356, 1964.

1155. **Fouts, J. R. and Rogers, L. A.**, Morphological changes in the liver accompanying stimulation of microsomal drug metabolizing enzyme activity by phenobarbital, chlordane, benzpyrene or methylcholanthrene in rats, *J. Pharmacol. Exp. Ther.*, 147, 112, 1965.

1156. **Conney, A. H., Miller, E. C., and Miller, J. A.**, The metabolsim of methylated aminoazo dyes. V. Evidence for induction of enzyme synthesis in the rat by 3-methylcholanthrene, *Cancer Res.*, 16, 450, 1956.

1157. **Conney, A. H., Miller, E. C., and Miller, J. A.**, Substrate-induced synthesis and other properties of benzpyrene hydroxylase in rat liver, *J. Biol. Chem.*, 228, 753, 1957.

1158. **Hey-Ferguson, A. and Bresnick, E.**, The binding of 3-methylcholanthrene to macromolecular components of rat liver preparations, *Mol. Pharmacol.*, 7, 183, 1971.

1159. **Furlong, N. B. and Thomann, A. J.**, Deoxyribonucleic acid polymerase activity during azo dye carcinogenesis, *Proc. Soc. Exp. Biol. Med.*, 115, 541, 1964.

1160. **Miller, J. A. and Miller, E. C.**, The metabolic activation of carcinogenic aromatic amines and amides, *Progr. Exp. Tumour Res.*, 11, 273, 1969.

1161. **Bates, R. R., Eaton, S. A., Morgan, D. L., and Yuspa, S. H.**, Replication of DNA after binding of the carcinogen 7,12-dimethylbenz-(α) anthracene, *J. Natl. Cancer Inst.*, 45, 1223, 1970.

1162. **Dingman, C. W. and Sporn, M. B.**, The binding of metabolites of amino azo dyes to rat liver DNA *in vivo, Cancer Res.*, 27, 938, 1967.

1163. **Arcos, J. C. and Argus, M. F.**, Molecular geometry and carcinogenic activity of aromatic compounds. New perspectives, *Adv. Cancer Res.*, 11, 305, 1968.

1164. **Brookes, P.**, On the interaction of carcinogens with DNA, *Biochem. Pharmacol.*, 20, 999, 1971.

1165. **Kapuler, A. M. and Michelson, A. M.**, The reaction of the carcinogen N-acetoxy-2-acetyl-amino fluorene with DNA and other polynucleotides and its stereochemical implications, *Biochim. Biophys. Acta*, 232, 436, 1971.

1166. **Grunberger, D. and Weinstein, I. B.**, Modifications of ribonucleic acid by chemical carcinogens. III. Template activity of polynucleotides modified by N-acetoxy-2-acetylamino flourene, *J. Biol. Chem.*, 246, 1123, 1971.

1167. **Lawson, T. A.**, The binding of 0-aminoazotoluene to deoxyribonucleic acid, ribonucleic acid and protein in the C 57 mouse, *Biochem. J.*, 109, 917, 1968.

1168. **Chan, E. W. and Ball, J. K.**, Polycyclic aromatic hydrocarbons. Covalent binding to DNA and effects on template function, *Biochim. Biophys. Acta*, 238, 46, 1971.

1169. **Sluyser, M.**, Interaction of carcinogens with histones *in vitro, Biochim. Biophys. Acta*, 154, 606, 1968.

1170. **Barry, E. J., Ovechka, C. A., and Gutman, H. R.**, Interaction of aromatic amines with rat liver proteins *in vitro.* II. Binding of N-2-fluorenylacetamide-9-C^{14} to nuclear proteins, *J. Biol. Chem.*, 243, 51, 1968.

1171. **Kodama, M., Tagashira, Y., and Nagata, C.**, Interaction of aromatic hydrocarbons with deoxyribonucleoprotein *in vitro, J. Biochem.*, 64, 81, 1968.

1172. **Lesko, S. A., Smith, A., Ts'o, P. O. P., and Umans, R. S.**, Interaction of nucleic acids IV. The physical binding of 3,4-benzpyrene to nucleosides, nucleotides, nucleic acids, and nucleoprotein, *Biochemistry*, 7, 434, 1968.

1173. **O'Brien, R. L., Stanton, R., and Craig, R. L.**, Chromatin binding of benzo(a) pyrene and 20-methylcholanthrene, *Biochim. Biophys. Acta*, 186, 414, 1969.

1174. **Lotlikar, P. D. and Paik, W. K.**, Binding of carcinogenic aromatic amines to rat liver nuclear acidic proteins *in vivo, Biochem. J.*, 124, 443, 1971.

1175. **Dijkstra, J. and Griggs, H. M.**, Binding of aminoazo dyes to serum albumin and to nuclear proteins of rat liver, *Br. J. Cancer*, 21, 205, 1967.

1176. **Bakay, B.**, Soluble proteins of liver nuclei of rats fed 3'-methyl-4-dimenthylaminoazobenzene, *Biochem. Pharmacol.*, 17, 689, 1968.

1177. **Schweppe, J. S., Kot, E., and Jungmann, R. A.**, Effects of gonadectomy on the uptake of polycyclic hydrocarbons by rat liver nuclei, DNA and histones, *Proc. Soc. Exp. Biol. Med.*, 138, 167, 1971.

1178. **Grover, P. L. and Sims, P.**, Interactions of the K-region epoxides of phenanthrene and dibenz(a,h)anthracene with nucleic acids and histone, *Biochem. Pharmacol.*, 19, 2251, 1970.

1179. **Sugimoto, T. and Terayama, H.**, Studies on carcinogen binding proteins I. Isolation and characterization of aminoazo dye-bound protein after administration of a single large dose of 3'-methyl-4-dimethylaminoazobenzene to rats, *Biochim. Biophys. Acta*, 214, 533, 1970.

1180. **Sorof, S., Young, E. M., and Ott, M. G.**, Soluble liver h proteins during hepatocarcinogenesis by aminoazo dyes and 2-acetylaminofluorene in the rat, *Cancer Res.*, 23, 864, 1963.

1181. **Barry, E. J. and Gutman, H. R.**, Interaction of aromatic amines with rat liver proteins *in vivo.* I. Preferential labeling of a cytoplasmic protein or proteins by N-2-fluorenyl-acetamide-9-^{14}C, *J. Biol. Chem.*, 241, 4600, 1966.

1182. **Litwack, G. and Morey, K. S.**, Cortisol metabolite binder. I. Identity with dimethylaminoazobenzene binding protein of liver cytosol, *Biochem. Biophys. Res. Commun.*, 38, 1141, 1970.

1183. **Ketterer, B., Ross-Mansell, P., and Whitehead, J. K.,** The isolation of carcinogen-binding protein from livers of rats given 4-dimethyl-aminoazobenzene, *Biochem. J.*, 103, 316, 1967.

1184. **Singer, S. and Litwack, G.,** Identity of corticosteroid binder I with the macromolecule binding 3-methylcholanthrene in liver cytosol *in vivo, Cancer Res.*, 31, 1364, 1971.

1185. **Bruchovsky, N. and Wilson, J. D.,** The conversion of testosterone to 5α-androstan-17βol-3-one by rat prostate *in vivo* and *in vitro, J. Biol. Chem.*, 243, 2012, 1968.

1186. **Rochefort, H. and Baulieu, E.,** New *in vitro* studies of estradiol binding in castrated rat uterus, *Endocrinology*, 84, 108, 1969.

1187. **Jensen, E. V., Numata, M., Smith, S., Suzuki, T., Brecher, P. I., and De Sombre, E. R.,** Estrogen-receptor interaction in target tissues, *Dev. Biol.*, Suppl. 3, 151, 1969.

1188. **De Sombre, E. R., Puca, G. A., and Jensen, E. V.,** Purification of an estrophilic protein from calf uterus, *Proc. Natl. Acad. Sci. U. S.*, 64, 148, 1969.

1189. **Shyamala, G. and Gorski, J.,** Estrogen receptors in the rat uterus. Studies on the interaction of cytosol and nuclear binding sites, *J. Biol. Chem.*, 244, 1097, 1969.

1190. **Beato, M., Biesewig, D., Braendle, W., and Sekeris, C. E.,** On the mechanism of hormone action. XV. Subcellular distribution and binding of 1,2-^{3}H cortisol in rat liver, *Biochim. Biophys. Acta*, 192, 494, 1969.

1191. **Beato, M., Seifart, K. H., and Sekeris, C. E.,** The effect of cortisol on the binding of actinomycin D to and on the template activity of isolated rat liver chromatin, *Arch. Biochem. Biophys.*, 138, 272, 1970.

1192. **Beato, M., Schmid, W., Braendle, W., and Sekeris, C. E.,** Partial purification of cortisol binding protein from rat liver cytosol, *Steroids*, 16, 207, 1970.

1193. **Matthysse, A. G.,** Organ specificity of hormone-receptor-chromatin interactions, *Biochim. Biophys. Acta*, 199, 519, 1970.

1194. **Morey, K. S. and Litwack, G.,** Isolation and properties of cortisol metabolite binding proteins of rat liver cytosol, *Biochemistry*, 8, 4813, 1969.

1195. **Brecher, P. I. and Wotiz, H. H.,** Stereospecificity of the uterine nuclear hormone receptors, *Proc. Soc. Exp. Biol. Med.*, 128, 470, 1968.

1196. **Swaneck, G. E., Chu, L. L. H., and Edelman, I. S.,** Stereospecific binding of aldosterone to renal chromatin, *J. Biol. Chem.*, 245, 5382, 1970.

1197. **Brecher, P. I. and Wotiz, H. H.,** Dissociation of estradiol from a uterine nuclear receptor, *Endocrinology*, 84, 718, 1969.

1198. **Chader, G. J., and Villee, C. A.,** Uptake of oestradiol by the rabbit hypothalamus. Specificity of binding by nuclei *in vitro, Biochem. J.*, 118, 93, 1970.

1199. **O'Malley, B. W., Sherman, M. R., and Toft, D. O.,** Progesterone "receptors" in the cytoplasm and nucleus of chick oviduct target tissue, *Proc. Natl. Acad. Sci. U. S.*, 67, 501, 1970.

1200. **O'Malley, B. W., Toft, D. O., and Sherman, M. R.,** Progesterone-binding components of chick oviduct, *J. Biol. Chem.*, 246, 117, 1971.

1201. **Steggles, A., Spelsberg, T. C., and O'Malley, B. W.,** Tissue specific binding *in vitro* of progesterone-receptor to chromatin of chick tissues, *Biochem. Biophys. Res. Commun.*, 43, 20, 1971.

1202. **Steggles, A. W., Spelsberg, T. C., Glasser, S. R., and O'Malley, B. W.,** Soluble complexes between steroid hormones and target tissue receptors bind specifically to target-tissue chromatin, *Proc. Natl. Acad. Sci. U. S.*, 68, 1479, 1971.

1203. **Spelsberg, T. C., Steggles, A. W., and O'Malley, B. W.,** Progesterone binding components of chick oviduct. III. Chromatin acceptor sites, *J. Biol. Chem.*, 246, 4188, 1971.

1204. **Gelboin, H. V.,** A microsome-dependent binding of benzo(a)pyrene to DNA, *Cancer Res.*, 29, 1272, 1969.

1205. **Grover, P. L. and Sims, P.,** Enzyme-catalyzed reactions of polycyclic hydrocarbons with deoxyribonucleic acid and protein *in vitro, Biochem. J.*, 110, 159, 1968.

1206. **Williams, D. J. and Rabin, B. R.,** Disruption by carcinogens of the hormone dependent association of membranes with polysomes, *Nature*, 232, 102, 1971.

1207. **Gilbert, W. and Müller-Hill, B.,** The *lac* operator is DNA, *Proc. Natl. Acad. Sci. U. S.*, 58, 2415, 1967.

1208. **Ptashne, M.,** Isolation of the λ phage repressor, *Proc. Natl. Acad. Sci. U. S.*, 57, 306, 1967.

1209. **Ptashne, M. and Hopkins, N.,** The operators controlled by the λ phage repressor, *Proc. Natl. Acad. Sci. U. S.*, 60, 1282, 1968.

1210. **Steinberg, R. A. and Ptashne, M.,** *In vitro* repression of RNA synthesis by purified λ phage repressor, *Nature New Biol.*, 230, 76, 1971.

1211. **Riggs, A. D., Reiness, G., and Zubay, G.,** Purification and DNA-binding properties of the catabolite gene activator protein, *Proc. Natl. Acad. Sci. U. S.*, 68, 1222, 1971.

1212. **Parks, J. S., Gottesman, M., Shimada, K., Weisberg, R. A., Perlman, R. L., and Pastan, I.,** Isolation of the *gal*-repressor, *Proc. Natl. Acad. Sci. U. S.*, 68, 1891, 1971.

1213. **Travers, A.,** Control of transcription in bacteria, *Nature New Biol.*, 229, 69, 1971.

1214. **Zubay, G., Gielow, L., and Englesberg, E.,** Cell-free studies on the regulation of the arabinose operon. Development of a cell-free system to synthesize l-ribulokinase coded by L-arabinose operon *ara*-OIBAD, *Nature New Biol.*, 233, 164, 1971.

1215. **Huang, R. C. C. and Bonner, J.**, Histone, a suppressor of chromosomal RNA synthesis, *Proc. Natl. Acad. Sci. U. S.*, 48, 1216, 1962.
1216. **Barr, G. C. and Butler, J. A. V.**, Histones and gene function, *Nature*, 199, 1170, 1963.
1217. **Johns, E. W. and Hoare, T. A.**, Histones and gene control, *Nature*, 226, 650, 1970.
1218. **Hindley, J.**, The relative ability of reconstituted nucleohistones to allow DNA-dependent RNA synthesis, *Biochem. Biophys. Res. Commun.*, 12, 175, 1963.
1219. **Skalka, A., Fowler, A. V., and Hurwitz, J.**, The effect of histones on the enzymatic synthesis of ribonucleic acid, *J. Biol. Chem.*, 241, 588, 1966.
1220. **Sonnenberg, B. P. and Zubay, G.**, Nucleohistone as a primer for RNA synthesis, *Proc. Natl. Acad. Sci. U. S.*, 54, 415, 1965.
1221. **Ray, A. K. and Zubay, G.**, RNA synthesis stimulated by sonicated nucleohistone, *Biochim. Biophys. Acta*, 129, 403, 1966.
1222. **Sluyser, M. and Snellen-Jurgens, N. H.**, Interaction of histones and nucleic acids *in vitro*, *Biochim. Biophys. Acta*, 199, 490, 1970.
1223. **Hoare, T. A. and Johns, E. W.**, The physical state of deoxyribonucleoprotein and its ability to direct ribonucleic acid synthesis *in vitro*, *Biochem. J.*, 119, 931, 1970.
1224. **Bonner, J. and Huang, R. C. C.**, Methodology for the study of the template activity of chromosomal nucleohistone, *Biochem. Biophys. Res. Commun.*, 22, 211, 1966.
1225. **Clark, P. R. and Byvoet, P.**, Inhibition of RNA-polymerase by arginine-rich histone fractions, *Experientia*, 26, 725, 1970.
1226. **Moskowitz, G., Ogawa, Y., Starbuck, W. C., and Busch, H.**, Comparative effects on RNA polymerase of the whole GAR histone and its peptides containing clusters of basic amino acids, *Biochem. Biophys. Res. Commun.*, 35, 741, 1969.
1226a. **Moskowitz, G. J., Wilson, R. K., Starbuck, W. C., and Busch, H.**, Comparative effects of the GAR histone and its cyanogen bromide cleavage products on RNA polymerase, *Physiol. Chem. Phys.*, 2, 217, 1970.
1227. **Fox, C. F., Gumport, R. I., and Weiss, S. B.**, The enzymatic synthesis of ribonucleic acid. V. The interaction of ribonucleic acid polymerase with nucleic acids, *J. Biol. Chem.*, 240, 2101, 1965.
1228. **Goldstein, J.**, Resistance of RNA to thermal denaturation in the presence of polyamines, *Biochim. Biophys Acta*, 123, 620, 1966.
1229. **Suwalsky, M., Traub, W., Schmueli, U., and Subirana, J. A.**, An X-ray study of the interaction of DNA with spermine, *J. Mol. Biol.*, 42, 363, 1969.
1230. **Konishi, G. and Koide, S. S.**, Stimulation of RNA polymerase activity by histones, polyamino acids, and polypeptide hormones, *Experientia*, 27, 262, 1971.
1231. **Spelsberg, T. C. and Hnilica, L. S.**, Studies on the RNA polymerase-histone complexes, *Biochim. Biophys. Acta*, 195, 55, 1969.
1232. **Spelsberg, T. C., Tankersley, S., and Hnilica, L. S.**, Inhibition of RNA polymerase activity by arginine-rich histones, *Experientia*, 25, 129, 1969.
1233. **Spelsberg, T. C., Tankersley, S., and Hnilica, L. S.**, The interaction of RNA polymerase with histones, *Proc. Natl. Acad. Sci. U. S.*, 62, 1218, 1969.
1234. **Bazill, G. W. and Philpot, J. S. L.**, Studies on the assay of primer DNA in the presence of histone and nucleoprotein and in isolated nuclei, *Biochim. Biophys. Acta*, 76, 223, 1963.
1235. **Billen, D. and Hnilica, L. S.**, Some aspects of inhibition of DNA synthesis *in vitro* by histone preparations, *J. Cell Biol.*, 19, 7a, 1963.
1236. **Sluyser, M., Thung, P. J., and Emmelot, P.**, Inhibition of deoxyribonucleic acid synthesis in regenerating rat liver by the administration of histone *in vivo*, *Biochim. Biophys. Acta*, 108, 249, 1965.
1237. **Gurley, L. R., Irvin, J. L., and Holbrook, D. J.**, Inhibition of DNA polymerase by histones, *Biochem. Biophys. Res. Commun.*, 14, 527, 1964.
1238. **Wood, W. G., Irvin, J. L., and Holbrook, D. J.**, Inhibition of calf thymus deoxyribonucleic acid polymerase by histones, *Biochemistry*, 7, 2256, 1968.
1239. **Schwimmer, S. and Bonner, J.**, Nucleohistone as template for the replication of DNA, *Biochim. Biophys. Acta*, 108, 67, 1965.
1240. **Chambon, P., Ramuz, M., and Doly, J.**, Relationships between soluble DNA-dependent RNA polymerase and "aggregate" RNA polymerase, *Biochem. Biophys. Res. Commun.*, 21, 156, 1965.
1241. **Chambon, P., Karon, H., Ramuz, M., and Mandel, P.**, The influence of ionic strength and polyanion on the transcription *in vitro*. II. Effects on the template efficiency of rat liver chromatin for a purified bacterial RNA polymerase, *Biochim. Biophys. Acta*, 157, 520, 1968.
1242. **Chambon, P., Ramuz, M., Mandel, P., and Doly, J.**, The influence of ionic strength and a polyanion on transcription *in vitro*. I. Stimulation of the aggregate RNA polymerase from rat liver nuclei, *Biochim. Biophys. Acta*, 157, 504, 1968.
1243. **Goldberg, I. H.**, Ribonucleic acid synthesis in nuclear extracts of mammalian cells grown in suspension cultures; effect of ionic strength and surface-active agents, *Biochim. Biophys. Acta*, 51, 201, 1961.
1244. **Koslov, Y. V. and Georgiev, G. P.**, Mechanism of inhibitory action of histones on DNA template activity *in vitro*, *Nature*, 228, 245, 1970.

1245. **Smart, J. E. and Bonner, J.,** Studies on the role of histones in relation to the template activity and precipitability of chromatin at physiological ionic strength, *J. Mol. Biol.,* 58, 675, 1971.
1246. **Hnilica, L. S., Ansevin, A. T., and Getz, S.,** Thermal denaturation of partially dehistonized chromatin preparations, in preparation.
1247. **Brasch, K., Seligy, V. L., and Setterfield, G.,** Effects of low salt concentrations on structural organization and template activity of chromatin in chicken erythrocyte nuclei, *Exp. Cell Res.,* 65, 61, 1971.
1248. **Varshavsky, A. Y., Ilyin, Y. V., Kladyckov, V. I., and Senchenkov, E. P.,** Collapse of extended deoxyribonucleoprotein molecules upon increase of the ionic strength of solution, *Biochim. Biophys. Acta,* 246, 583, 1971.
1249. **Chevaillier, P.,** Problémes morphologiques liés a l'étude biochimique des différentes fractions de chromatine extraites du noyou du foie de souris, *Exp. Cell Res.,* 67, 466, 1971.
1250. **Lezzi, M.,** Differential effects of sodium and potassium ions on the template activity of rat liver chromatin, *Physiol. Chem. Phys.,* 1, 447, 1969.
1251. **Kurashina, Y., Ohba, Y., and Mizuno, D.,** Template activity of partially dehistonized nucleohistones for DNA dependent RNA polymerase, *J. Biochem.,* 67, 661, 1970.
1252. **O'Meara, A. R. and Herrmann, R. L.,** A biphasic activation of template by removal of protein from chromatin, *Fed. Proc.,* 29, 913a, 1970.
1253. **Gilmour, R. S. and Paul, J.,** RNA transcription from reconstituted nucleoprotein is similar to natural RNA, *J. Mol. Biol.,* 40, 137, 1969.
1254. **Pogo, A. O., Allfrey, V. G., and Mirsky, A. E.,** Evidence for increased DNA template activity in regenerating liver nuclei, *Proc. Natl. Acad. Sci. U.S.,* 56, 550, 1966.
1255. **Bannai, S. and Terayama, H.,** Regulation of RNA synthesis in chromatin (aggregate enzyme) from rat liver, *Biochim. Biophys. Acta,* 142, 410, 1967.
1256. **Bannai, S. and Terayama, H.,** Template activity of chromatins isolated from regenerating rat liver, *J. Biochem.,* 66, 289, 1969.
1257. **Farber, J., Rovera, G., and Baserga, R.,** Template activity for chromatin during stimulation of cellular proliferation in human diploid fibroblasts, *Biochem. J.,* 122, 189, 1971.
1258. **Flickinger, R. A., Greene, R., Kohl, D. M., and Miyagi, M.,** Patterns of synthesis of DNA-like RNA in parts of developing frog embryo, *Proc. Natl. Acad. Sci. U.S.,* 56, 1712, 1966.
1259. **Flickinger, R. A., Coward, S. J., Miyagi, M., Moser, C., and Rollins, F.,** The ability of DNA and chromatin of developing frog embryos to prime for RNA polymerase-dependent RNA synthesis, *Proc. Natl. Acad. Sci. U.S.,* 53, 783, 1965.
1260. **Coward, S. J., Rollins, E., and Flickinger, R. A.,** The equivalence of DNA from developing frog embryos, *Exp. Cell Res.,* 42, 42, 1966.
1261. **Pyhtilä, M. J. and Sherman, F. G.,** Age-associated studies on thermal stability and template effectiveness of DNA and nucleoprotein from beef thymus, *Biochem. Biophys. Res. Commun.,* 31, 340, 1968.
1262. **Von Hahn, H. P.,** Structural and functional changes in nucleoprotein during the ageing of the cell, *Gerontologia,* 16, 116, 1970.
1263. **Seligy, V. and Miyagy, M.,** Studies of template activity of chromatin isolated from metabolically active and inactive cells, *Exp. Cell Res.,* 58, 27, 1969.
1264. **Kernell, A. M., Bolund, L., and Ringertz, N. R.,** Chromatin changes during erythropoiesis, *Exp. Cell Res.,* 65, 1, 1971.
1265. **Holoubek, V.,** Repressed chromatin from virus infected and from differentiated cells, *Experientia,* 21, 437, 1965.
1266. **Steinhart, W. L.,** Alteration of template activity of chromatin from livers of *Pneumococcus*-infected mice, *Proc. Soc. Exp. Biol. Med.,* 137, 619, 1971.
1267. **Kim, K. H. and Cohen, P. P.,** Modification of tadpole liver chromatin by thyroxine treatment, *Proc. Natl. Acad. Sci. U.S.,* 55, 1251, 1966.
1268. **Barker, K. L. and Warren, J. C.,** Template activity of uterine chromatin: control by estradiol, *Proc. Natl. Acad. Sci. U.S.,* 56, 1298, 1966.
1269. **Breuer, C. B. and Florini, J. R.,** Effects of ammonium sulfate, growth hormone, and testosterone propionate on ribonucleic acid polymerase and chromatin activities in rat skeletal muscle, *Biochemistry,* 5, 3857, 1966.
1270. **Liao, S., Barton, R. W., and Lin, A. H.,** Differential synthesis of ribonucleic acid in prostatic nuclei: Evidence for selective gene transcription induced by androgens, *Proc. Natl. Acad. Sci. U.S.,* 55, 1593, 1966.
1271. **Stackhouse, H. L., Chetsanga, C. J., and Tan, C. H.,** The effect of cortisol on genetic transcription in rat liver chromatin, *Biochim. Biophys. Acta,* 155, 159, 1968.
1272. **Hamilton, T. H.,** Control by estrogen of genetic transcription and translation, *Science,* 161, 649, 1968.
1273. **Teng, C. S. and Hamilton, T. H.,** The role of chromatin in estrogen action in the uterus. I. The control of template capacity and chemical composition and the binding of H^3-estradiol-17β, *Proc. Natl. Acad. Sci. U.S.,* 60, 1410, 1968.
1274. **Morgan, C. R. and Bonner, J.,** Template activity of liver chromatin increased by *in vivo* administration of insulin, *Proc. Natl. Acad. Sci. U.S.,* 65, 1077, 1970.
1275. **Venis, M. A.,** Stimulation of RNA transcription from pea and corn DNA by protein retained on Sepharose coupled to 2,4-dichlorophenoxyacetic acid, *Proc. Natl. Acad. Sci. U.S.,* 68, 1824, 1971.
1276. **Dati, F. A. and Maurer, H. R.,** Estradiol mediated template activity of mouse and rat uterine chromatin: An unsolved problem, *Biochim. Biophys. Acta,* 246, 589, 1971.

1277. **Hallick, R. B. and De Luca, H. F.**, Vitamin D_3-stimulated template activity of chromatin from rat intestine, *Proc. Natl. Acad. Sci. U.S.*, 63, 528, 1969.

1278. **Steinhart, W. L.**, Diurnal rhythmicity in template activity of mouse liver chromatin, *Biochim. Biophys. Acta*, 228, 301, 1971.

1279. **Kurtz, D. I. and Sinex, F. M.**, Age related differences in the association of brain DNA and nuclear proteins, *Biochim. Biophys. Acta*, 145, 840, 1967.

1280. **Von Hahn, H. P.**, Age-related alterations in the structure of DNA. II. The role of histones, *Gerontologia*, 10, 174, 1964.

1281. **Von Hahn, H. P.**, A model of "regulatory" ageing of the cell at the gene level, *J. Gerontol.*, 21, 291, 1966.

1282. **Von Hahn, H. P. and Fritz, E.**, Age-related alterations in the structure of DNA. III. Thermal stability of rat liver DNA, related to age, histone content and ionic strength, *Gerontologia*, 12, 237, 1966.

1283. **Pyhtilä, M. J. and Sherman, F. G.**, Studies on the histones and DNA of calf and old cow thymus, *J. Gerontol.*, 23, 450, 1968.

1284. **Pyhtilä, M. J. and Sherman, F. G.**, Influence of age on rat liver and kidney chromatin, *Gerontologia*, 15, 321, 1969.

1285. **Von Hahn, H. P., Miller, J., and Eichhorn, G. L.**, Age-related alterations in the structure of nucleoprotein. IV. Changes in the composition of whole histone from rat liver, *Gerontologia*, 15, 293, 1969.

1286. **Auer, G., Zetterberg, A., and Killander, D.**, Changes in binding between DNA and arginine residues in histones induced by cell crowding, *Exp. Cell Res.*, 62, 32, 1970.

1287. **Urban, V. and Schvarcenberg, H.**, Impact of whole γ-irradiation on nuclear histones from rat spleen and liver, *Strahlentherapie*, 132, 4, 1967.

1288. **Bauer, R. D., Johanson, R., and Thomasson, W. A.**, Alterations of DNA-histone interactions following irradiation, *Int. J. Radiat. Biol.*, 16, 575, 1969.

1289. **Olkowski, Z. and Manocha, S. L.**, Quantitative studies on histones and DNA in the motoneurons of spinal cord of mice after total-body gamma-irradiation, *Strahlentherapie*, 142, 365, 1971.

1290. **Robinson, M. G., Weiss, J. J., and Wheeler, C. M.**, Irradiation of deoxyribonucleohistone solutions. II. Labilization of the DNA-histone linkage, *Biochim. Biophys. Acta*, 124, 181, 1966.

1291. **Lloyd, P. H. and Peacocke, A. R.**, The dissociation of histone from deoxyribonucleohistone by γ-irradiation, *Biochem. J.*, 109, 341, 1968.

1292. **Skalka, M. and Matyášová, J.**, Post-irradiation changes in the deoxyribonucleoprotein complex in animal tissues revealed by action of heparin, *Nature*, 212, 1253, 1967.

1293. **Skalka, M. and Matyášová, J.**, The increased sensitivity of the deoxyribonucleoprotein complex of irradiated tissue to the action of heparin, *J. Radiat. Biol.*, 11, 193, 1966.

1294. **Kolíbalová, A.**, Changes in DNP histones after irradiation and contact with heparin, *Folia Biol.*, 16, 356, 1970.

1295. **Čejková, M., Skalka, M., and Matyášová, J.**, The effect of irradiation on deoxyribonucleoproteins in animal tissues. IV. Changes in the histone component of deoxyribonucleoprotein isolated from tissues of irradiated mice and of DNP fraction after interactions with heparin, *Folia Biol.*, 17, 189, 1971.

1296. **Matyášová, J., Skalka, M., Čejková, M., and Blažíček, G.**, The effect of radiation on deoxyribonucleoprotein of mouse spleen and thymus after irradiation, *Folia Biol.*, 17, 328, 1971.

1297. **Peacocke, A. R. and Preston, B. N.**, Effect of gamma-rays on thymus nucleoprotein in solution, *Nature*, 192, 228, 1961.

1298. **Robinson, M. G., Weiss, J. J., and Wheeler, C. M.**, Irradiation of deoxyribonucleohistone solutions. I. Amino acid destruction, *Biochim. Biophys. Acta*, 124, 176, 1966.

1299. **Lloyd, P. H. and Peacocke, A. R.**, The action of γ-rays on deoxyribonucleohistone in solution, *Proc. Royal Soc.*, B, 164, 40, 1966.

1300. **Clever, U.**, Regulation of chromosome function, *Annu. Rev. Genetics*, 2, 11, 1968.

1301. **Grossbach, U.**, Chromosomen-Aktivität und biochemische Zelldifferenzierung in den Speicheldrüsen von *Camptochironomus*, *Chromosoma*, 28, 136, 1969.

1302. **Clever, U. and Romball, C. G.**, RNA and protein synthesis in the cellular response to a hormone ecdysone, *Proc. Natl. Acad. Sci. U.S.*, 56, 1470, 1966.

1303. **Berendes, H. D.**, Factors involved in the expression of gene activity in polytene chromosomes, *Chromosoma*, 24, 418, 1968.

1304. **Berendes, H. D.**, Amino acid incorporation into giant chromosomes of *Drosophila hydei*, *Drosophila Inf. Ser.*, 42, 102, 1970.

1305. **Holt, T. K. H.**, Local protein accumulation during gene activation. I. Quantitative measurements on dye binding capacity at subsequent stages of puff formation in *Drosophila hydei*, *Chromosoma*, 32, 64, 1970.

1306. **Beerman, W.**, Control of differentiation at the chromosomal level, *J. Exp. Zool.*, 157, 49, 1964.

1307. **Beerman, W. and Bahr, G. F.**, The submicroscopic structure of the Balbiani ring, *Exp. Cell Res.*, 6, 195, 1954.

1308. **Stevens, B. J. and Swift, H.**, RNA transport from nucleus to cytoplasm in *Chironomus* salivary glands, *J. Cell Biol.*, 31, 55, 1966.

1309. **Liao, S., Lin, A. H., and Barton, R. W.**, Selective stimulation of ribonucleic acid synthesis in prostatic nuclei by testosterone, *J. Biol. Chem.*, 241, 3869, 1966.

1310. **Drews, J. and Brawerman, G.,** Alterations in the nature of ribonucleic acid synthesized in rat liver during regeneration and after cortisol administration, *J. Biol. Chem.,* 242, 801, 1967.
1311. **Liao, S. and Lin, A. H.,** Prostatic nuclear chromatin: an effect of testosterone on the synthesis of ribonucleic acid in cytidyl (3′,5′) guanosine, *Proc. Natl. Acad. Sci. U.S.,* 57, 379, 1967.
1312. **Trachewsky, D. and Segal, S. J.,** Differential synthesis of ribonucleic acid in uterine nuclei: evidence for selective gene transcription induced by estrogens, *Eur. J. Biochem.,* 4, 279, 1968.
1313. **O'Malley, B. W. and McGuire, W. L.,** Studies on the mechanism of estrogen-mediated tissue differentiation: regulation of nuclear transcription and induction of new RNA species, *Proc. Natl. Acad. Sci. U.S.,* 60, 1527, 1968.
1314. **Yu, F. L. and Feigelson, P.,** The sequential stimulation of uracil-rich and guanine-rich RNA species during cortisone induction of hepatic enzymes, *Biochem. Biophys. Res. Commun.,* 35, 499, 1969.
1315. **Hahn, W. E., Schjeide, O. A., and Gorbman, A.,** Organ-specific estrogen-induced RNA synthesis resolved by DNA-RNA hybridization in the domestic fowl, *Proc. Natl. Acad. Sci. U.S.,* 62, 112, 1969.
1316. **Doenecke, D. and Sekeris, C. E.,** Effect of cortisol on RNA synthesis as detected by hybridization with differentially renaturing DNA species, *FEBS Lett.,* 8, 61, 1970.
1317. **Oravec, M. and Korner, A.,** Stimulation of synthesis of DNA-like and ribosomal RNA by growth hormone, *J. Mol. Biol.,* 58, 489, 1971.
1318. **Thompson, E. B., Tomkins, G. M., and Curran, J. F.,** Induction of tyrosine α-ketoglutarate transaminase by steroid hormones in a newly established tissue culture cell line, *Proc. Natl. Acad. Sci. U.S.,* 56, 296, 1966.
1319. **Notides, A. and Gorski, J.,** Estrogen-induced synthesis of a specific uterine protein, *Proc. Natl. Acad. Sci. U.S.,* 56, 230, 1966.
1320. **O'Malley, B. W. and Kohler, P. O.,** Hormonal induction of specific proteins in chick oviduct cell cultures, *Biochem. Biophys. Res. Commun.,* 28, 1, 1967.
1321. **Sokoloff, L., Roberts, P. A., Januska, M. M., and Kline, J. E.,** Mechanisms of stimulation of protein synthesis by thyroid hormones *in vivo, Proc. Natl. Acad. Sci. U.S.,* 60, 652, 1968.
1322. **De Angelo, A. B. and Gorski, J.,** Role of RNA synthesis in the estrogen induction of a specific uterine protein, *Proc. Natl. Acad. Sci. U.S.,* 66, 693, 1970.
1323. **Shelton, K. R. and Allfrey, V. G.,** Selective synthesis of nuclear acidic proteins in liver cells stimulated by cortisol, *Nature,* 228, 132, 1970.
1324. **Teng, C. S. and Hamilton, T. H.,** Regulation by estrogen of organ-specific synthesis of a nuclear acidic protein, *Biochem. Biophys. Res. Commun.,* 40, 1231, 1970.
1325. **Britten, R. J. and Davidson, E. H.,** Gene regulation for higher cells: A theory, *Science,* 165, 349, 1969.
1326. **Spelsberg, T. C. and Sarkissian, I. V.,** Isolation and analysis of the proteins of plant nuclei from two stages of differentiation of *Phaseolus vulgaris, Phytochemistry,* 9, 1385, 1970.
1327. **Cohen, S. N., Spicer, S. S., and Yielding, K. L.,** Observations on the histone/DNA ratio of the immature rabbit uterus following administration of estrogen, *Biochim. Biophys. Acta,* 87, 511, 1964.
1328. **Hanoune, J. and Feigelson, P.,** The absence of cortisone effect on the synthesis of specific histones and ribosomal protein subunits in liver, *Biochem. Biophys. Res. Commun.,* 34, 215, 1969.
1329. **Yasuda, T. and Yamada, Y.,** Complex formation by 2, 4-dichlorophenoxyacetic acid with histones during callus induction, *Biochem. Biophys. Res. Commun.,* 40, 649, 1970.
1330. **Barker, K. L.,** Estrogen induced synthesis of histones and specific nonhistone protein in the uterus, *Biochemistry,* 10, 284, 1971.
1331. **Mueller, G. C., Gorski, J., and Aizawa, Y.,** The role of protein synthesis in early estrogen action, *Proc Natl. Acad. Sci. U.S.,* 47, 164, 1961.
1332. **Salganik, R. I., Morozova, I. M., and Zakharov, M. A.,** The changes in the state of rat liver chromatin under hydrocortisone induction, *Biochim. Biophys. Acta,* 174, 755, 1969.
1333. **Katz, J., Levitz, M., Gorstein, F., and Troll, W.,** Influence of trypsin on *in vitro* RNA synthesis by disrupted nuclei of rat and human uteri in different hormonal states, *Endocrinology,* 87, 294, 1970.
1334. **Gillespie, D. and Spiegelman, S.,** A quantitative assay for DNA-RNA hybrids with DNA immobilized on a membrane, *J. Mol. Biol.,* 12, 829, 1965.
1335. **McCarthy, B. J. and Church, R. B.,** The specificity of molecular hybridization reactions, *Annu. Rev. Biochem.,* 39, 131, 1970.
1336. **Smith, K. D., Church, R. B., and McCarthy, B. J.,** Template specificity of isolated chromatin, *Biochemistry,* 8, 4271, 1969.
1337. **Paul, J. and Gilmour, R. S.,** Restriction of deoxyribonucleic acid template activity in chromatin is organ specific, *Nature,* 210, 992, 1966.
1338. **Paul, J. and Gilmour, R. S.,** Template activity of DNA is restricted in chromatin, *J. Mol. Biol.,* 16, 242, 1966.
1339. **Huang, R. C. C. and Huang, P. C.,** Effect of protein-bound RNA associated with chick embryo chromatin on template specificity of the chromatin, *J. Mol. Biol.,* 39, 365, 1969.
1340. **Spelsberg, T. C. and Hnilica, L. S.,** Proteins of chromatin in template restriction. II. Specificity of RNA synthesis, *Biochim. Biophys. Acta,* 228, 212, 1971.

1341. **Spelsberg, T. C. and Hnilica, L. S.**, Deoxyribonucleoproteins and the tissue specific restriction of the DNA in chromatin, *Biochem. J.*, 120, 435, 1970.
1342. **Huang, R. C. C.**, Dihydrouridylic acid containing RNA from chicken embryo chromatin, *Fed. Proc.*, 26, 603, 1967.
1343. **Bekhor, I., Bonner, J., and Dahmus, G. K.**, Hybridization of chromosomal RNA to native DNA, *Proc. Natl. Acad. Sci. U.S.*, 62, 271, 1969.
1344. **Gilmour, R. S. and Paul, J.**, Role of non-histone components in determining organ specificity of rabbit chromatins, *FEBS Lett.*, 9, 242, 1970.
1345. **Nomura, M., Traub, P., Guthrie, G., and Nashimoto, H.**, The assembly of ribosomes, *J. Cell Physiol.*, 74, Suppl. 1, 241, 1969.
1346. **Melli, M., Whitfield, C., Rao, K. V., Richardson, M., and Bishop, J. O.**, DNA-RNA hybridization in vast DNA excess, *Nature New Biol.*, 231, 8, 1971.
1347. **Brown, I. R. and Church, R. B.**, RNA transcription from nonrepetitive DNA in the mouse, *Biochem. Biophys, Res. Commun.*, 42, 850, 1971.
1348. **Spelsberg, T. C., Wilhelm, J. A., and Hnilica, L. S.**, Nuclear proteins in genetic restriction. II. The nonhistone proteins in chromatin, *Sub-Cell Biochem.*, 1, 107, 1972.
1349. **Baserga, R. and Stein, G.**, Nuclear acidic proteins and cell proliferation, *Fed. Proc.*, 30, 1752, 1971.
1350. **Dastugue, B., Tichonicky, L., Penit-Soria, J., and Kruh, J.**, Comparison des protéines de la chromatine du noyau et du cytoplasme et des RNA des trois types de cellules différenciées, *Bull. Soc. Chim. Biol.*, 52, 391, 1970.
1351. **Loeb, J. E. and Creuzet, C.**, Comparison des propriétés électrophorétiques des protéines nucléaires de différent tissus, *Bull. Soc. Chim. Biol.*, 52, 1007, 1970.
1352. **Leveson, J. E. and Peacocke, A. R.**, Complexes of non-histone protein in calf thymus chromatin, *Biochim. Biophys. Acta*, 149, 311, 1967.
1353. **Steele, W. J. and Busch, H.**, Studies on acidic nuclear proteins of the Walker tumor and liver, *Cancer Res.*, 23, 1153, 1963.
1354. **Umansky, S. R., Tokarskaya, V. I., Zotova, R. N., and Migushina, V. L.**, Isolation and heterogeneity of nonhistone proteins of rat liver chromatin, *Mol. Biol.*, 5, 270, 1971.
1355. **MacGillivray, A. J., Carroll, D., and Paul, J.**, The heterogeneity of the nonhistone chromatin proteins from mouse tissues, *FEBS Lett.*, 13, 204, 1971.
1356. **Kleinsmith, L. J., Heidema, J., and Carroll, A.**, Specific binding of rat liver nuclear proteins to DNA, *Nature*, 226, 1025, 1970.
1357. **Wilhelm, J. A., Ansevin, A. T., Johnson, A. W., and Hnilica, L. S.**, Proteins of chromatin in genetic restriction. IV. Comparison of histone and nonhistone proteins of rat liver nucleolar and extranucleolar chromatin, *Biochim. Biophys. Acta*, 272, 220, 1972.
1358. **Gronow, M. and Griffiths, G.**, Rapid isolation and separation of the nonhistone proteins of rat liver nuclei, *FEBS Lett.*, 15, 340, 1971.
1359. **Marushige, K., Brutlag, D., and Bonner, J.**, Properties of chromosomal nonhistone protein of rat liver, *Biochemistry*, 7, 3149, 1968.
1360. **Shelton, K. R. and Neelin, J. M.**, Nuclear residual proteins from goose erythroid cells and liver, *Biochemistry*, 10, 2342, 1971.
1361. **Dravid, A. R. and Burdman, J. A.**, Acidic proteins in rat brain nuclei: disc electrophoresis, *J. Neurochem.*, 15, 25, 1968.
1362. **Elgin, S. C. R. and Bonner, J.**, Partial fractionation and chemical characterization of the major nonhistone chromosomal proteins, *Biochemistry*, 11, 772, 1972.
1363. **Sperti, S., Lorini, M., Pinna, L. A., and Moret, V.**, Phosphorylserine sequences in phosphoproteins from Ehrlich ascites cells, *Biochim. Biophys. Acta*, 82, 476, 1964.
1364. **Kleinsmith, L. J. and Allfrey, V. G.**, Nuclear phosphoproteins. I. Isolation and characterization of a phosphoprotein fraction from calf thymus nuclei, *Biochim. Biophys. Acta*, 175, 123, 1969.
1365. **Kleinsmith, L. J. and Allfrey, V. G.**, Nuclear phosphoproteins. II. Metabolism of exogenous phosphoprotein by intact nuclei, *Biochim. Biophys. Acta*, 175, 136, 1969.
1366. **Le Stourgeon, W. M. and Rusch, H. P.**, Nuclear acidic protein changes during differentiation in *Physarum polycephalum, Science*, 174, 1233, 1971.
1367. **McClure, M. E., Groves, C. M., and Hnilica, L. S.**, Nonhistone protein synthesis in developing sea urchin embryos, in preparation.
1368. **Alberts, B. M. and Frey, L.**, T-4 Bacteriophage gene-32: A structural protein in the replication and recombination of DNA, *Nature*, 227, 1313, 1970.
1369. **Alberts, B. M., Amodio, F. J., Jenkins, M., Gutmann, E. D., and Ferris, F. L.**, Studies with DNA-cellulose chromatography. I. DNA-binding proteins from *E. coli, Cold Spring Harbor Symp. Quant. Biol.*, 33, 289, 1968.
1370. **Wakabayashi, K. and Hnilica, L. S.**, DNA-binding nonhistone proteins from rat chromatin, in preparation.
1371. **Salas, J. and Green, H.**, Proteins binding to DNA and their relation to growth in cultured mammalian cells, *Nature New Biol.*, 229, 165, 1971.
1372. **Crippa, M.**, Regulatory factor for the transcription of the ribosomal genes in amphibian oocytes, *Nature*, 227, 1138, 1970.

1373. **Poonian, M. S., Schlabach, A. J., and Weissbach, A.,** Covalent attachment of nucleic acids to agarose for affinity chromatography, *Biochemistry*, 10, 420, 1971.

1374. **Cavalieri, L. F. and Carroll, E.,** A DNA-acrylamide gel column for analyzing proteins that bind to DNA. I. DNA polymerase, *Proc. Natl. Acad. Sci. USA*, 67, 807, 1970.

1375. **Davis, R. H., Copenhaver, J. H., and Carver, M. J.,** Characterization of acidic proteins in cell nuclei from rat brain by high-resolution acrylamide gel electrophoresis, *J. Neurochem.*, 19, 473, 1972.

1376. **Wang, T. Y.,** Tissue specificity of nonhistone chromosomal proteins, *Exp. Cell. Res.*, 69, 217, 1971.

1377. **Wilhelm, J. A., Groves, C. M., and Hnilica, L. S.,** Lack of major cytoplasmic protein contamination of rat liver nuclear chromatin, *Experientia,* in press.

1378. **Chytil, F. and Spelsberg, T. C.,** Tissue differences in antigenic properties of non-histone protein-DNA complexes, *Nature New Biol.,* 233, 215, 1971.

1379. **Henning, N., Frenger, W., Scheiffarth, F., and Assaf, A.,** Immunologische Studien über die Antigenität von Zellkernsubstanzen und über Zellkern-Antikörper, *Z. Rheumaforsch.,* 21, 13, 1962.

1380. **Spelsberg, T. C., Steggles, A. W., Chytil, F., and O'Malley, B. W.,** Progesterone-binding components of chick oviduct. V. Exchange of progesterone-binding capacity from target to nontarget tissue chromatins, *J. Biol. Chem.,* 247, 1368, 1972.

1381. **Chytil, F., Glasser, S. R., and Spelsberg, T. C.,** Nuclear acidic protein-DNA complexes during rat liver development, *Fed. Proc.,* 31, 418, 1972.

1382. **Wakabayashi, K., Groves, C. M., and Hnilica, L. S.,** Antigenic properties of chromatins from rat liver tissues, in preparation.

1383. **McClure, M. E. and Hnilica, L. S.,** Nuclear proteins in genetic restriction. III. The cell cycle, *Sub-Cell. Biochem.,* in press.

1384. **Buck, M. D. and Schauder, P.,** *In vivo* stimulation of ^{14}C-amino acid incorporation into nonhistone proteins in rat liver chromatin induced by insulin and cortisol, *Biochim. Biophys. Acta,* 224, 644, 1970.

1385. **Ruddon, R. W. and Rainey, C. H.,** Stimulation of nuclear protein synthesis in rat liver after phenobarbital administration, *Biochem. Biophys. Res. Commun.,* 40, 152, 1970.

1386. **Hamilton, T. H., Teng, C. S., and Means, A. R.,** Early estrogen action: nuclear synthesis and accumulation of protein correlated with enhancement of two DNA-dependent RNA polymerase activities, *Proc. Natl. Acad. Sci. USA,* 59, 1265, 1968.

1387. **Malpoix, P. J.,** Stimulation by erythropoietin of histone and nonhistone chromatin protein synthesis in disaggregated foetal mouse liver, *Exp. Cell Res.,* 65, 393, 1971.

1388. **Mukherjee, A. B.,** Relationship between nucleic acids, histone and nonhistone protein synthesis in human lymphocytes stimulated with phytohemagglutinin, *Can. J. Genet. Cytol.,* 12, 151, 1970.

1389. **McClure, M. E. and Hnilica, L. S.,** Temporal studies of the biosynthesis of chromatin in synchronized mammalian cell cultures, in *Oncology 1970,* Vol. 1, Clark, R. L., Cumley, R. W., McCay, J. E., and Copeland, M. M., Eds., Year Book Medical Publ., Chicago, 1971, 494.

1390. **Jockusch, B. M., Brown, D. F., and Rusch, H. P.,** Synthesis of a nuclear protein in G_2-phase, *Biochem. Biophys. Res. Commun.,* 38, 279, 1970.

1391. **Stein, G. S. and Borun, T. W.,** The synthesis of acidic chromosomal proteins during the cell cycle of HeLa S-3 cells. I. The accelerated accumulation of acidic residual nuclear protein before the initiation of DNA replication, *J. Cell Biol.,* 52, 292, 1972.

1392. **Borun, T. W. and Stein, G. S.,** The synthesis of acidic chromosomal proteins during the cell cycle of HeLa S-3 cells. II. The kinetics of residual protein synthesis and transport, *J. Cell Biol.,* 52, 308, 1972.

1393. **Ahmed, K. and Ishida, H.,** Effect of testosterone on nuclear phosphoproteins of rat ventral prostate, *Mol. Pharmacol.,* 7, 323, 1971.

1394. **Ahmed, K.,** Studies on nuclear phosphoproteins of rat ventral prostate: incorporation of ^{32}P from [γ-^{32}P] ATP, *Biochim. Biophys. Acta,* 243, 38, 1971.

1395. **Benjamin, W. B. and Goodman, R. M.,** Phosphorylation of dipteran chromosomes and rat liver nuclei, *Science,* 166, 629, 1969.

1396. **Helmsing, P. J. and Berendes, H. D.,** Induced accumulation of nonhistone proteins in polytene nuclei of *Drosophila hydei, J. Cell Biol.,* 50, 893, 1971.

1397. **Travers, A. A.,** Positive control of transcription by a bacteriophage Sigma factor, *Nature,* 225, 1009, 1970.

1398. **Stein, G. S. and Baserga, R.,** Cytoplasmic synthesis of acidic chromosomal proteins, *Biochem. Biophys. Res. Commun.,* 44, 218, 1971.

1399. **Wu, R. S. and Warner, J. R.,** Cytoplasmic synthesis of nuclear proteins in various cell fractions after brief pulses, *J. Cell Biol.,* 51, 643, 1971.

1400. **Craig, N. and Perry, R. P.,** Persistent cytoplasmic synthesis of ribosomal proteins during the selective inhibition of ribosomal RNA synthesis, *Nature New Biol.,* 229, 75, 1971.

1401. **Monod, J., Changeux, J. P., and Jacob, F.,** Allosteric proteins and cellular control systems, *J. Mol. Biol.,* 6, 306, 1963.

1402. **Monod, J., Cohen-Bazire, G., and Cohn, M.,** Biosynthèse de la β-galactosidase (lactase) chez *E. coli.* La specificité de l'induction, *Biochim. Biophys. Acta,* 7, 585, 1951.

1403. **Kepes, A.**, Kinetics of induced enzyme synthesis. Determination of mean life of galactosidase specific messenger RNA, *Biochim. Biophys. Acta*, 76, 293, 1963.
1404. **Jacob, F. and Monod, J.**, Genetic repression, allosteric inhibition, and cellular differentiation, in *Cytodifferentiation and Macromolecular Synthesis, 21st Growth Symposium*, Locke, M., Ed., Academic Press, New York, 1963, 30.
1405. **Jacob, F. and Monod, J.**, Genetic mapping of the elements of the lactose region in *Escherichia coli*, *Biochem. Biophys. Res. Commun.*, 18, 693, 1965.
1406. **Wilhelm, J. A., Spelsberg, T. C., and Hnilica, L. S.**, Nuclear proteins in genetic restriction. I. The histones, *Sub-Cell. Biochem.*, 1, 39, 1971.
1407. **Kozlov, Y. and Georgiev, G. P.**, Control of the biosynthesis of informational RNA in animal cells. 3. The mechanism of RNA synthesis inhibition by chromatin proteins, *Mol. Biol.*, 5, 789, 1971.
1408. **Frenster, J. H.**, Ultrastructural probes of chromatin within human lymphocytes, *Nature New Biol.*, 236, 176, 1972.
1409. **Huang, R. C. C. and Bonner, J.**, Histone-bound RNA, a component of native nucleohistone, *Proc. Natl. Acad. Sci. USA*, 54, 960, 1965.
1410. **Bonner, J. and Huang, R. C. C.**, Histones as specific repressors of chromosomal RNA synthesis, in *Histones*, De Reuck, A. V. S. and Knight, J., Eds., Churchill, London, 1966, 18.
1411. **Jacobson, R. A. and Bonner, J.**, Studies of the chromosomal RNA and of the chromosomal RNA-binding protein of higher organisms, *Arch. Biochem. Biophys.*, 146, 557, 1971.
1412. **Jacobson, R. A. and Bonner, J.**, The occurrence of dihydroribothymidine in chromosomal RNA, *Biochem. Biophys. Res. Commun.*, 33, 716, 1968.
1413. **Bonner, J. and Widholm, J.**, Molecular complementarity between nuclear DNA and organ-specific chromosomal RNA, *Proc. Natl. Acad. Sci. USA*, 57, 1379, 1967.
1414. **Dahmus, M. and McConnell, D.**, Chromosomal ribonucleic acid of rat ascites cells, *Biochemistry*, 8, 1524, 1969.
1415. **Shih, T. Y. and Bonner, J.**, Chromosomal RNA of calf thymus chromatin, *Biochim. Biophys. Acta*, 182, 30, 1969.
1416. **Mayfield, J. E. and Bonner, J.**, Tissue differences in rat chromosomal RNA, *Proc. Natl. Acad. Sci. USA*, in press.
1417. **Holmes, D., Mayfield, J. E., Murthy, L., and Bonner, J.**, Rapidly-labeled, high molecular weight, nuclear RNA and its relationship to chromosomal RNA, *Biochemistry*, in press.
1418. **Church, R. B. and McCarthy, B. J.**, Changes in nuclear and cytoplasmic RNA in regenerating mouse liver, *Proc. Natl. Acad. Sci. USA*, 58, 1548, 1967.
1419. **Mayfield, J. E. and Bonner, J.**, A partial sequence of nuclear events in regenerating rat liver, *Proc. Natl. Acad. Sci. USA*, 69, 7, 1972.
1420. **Sypherd, P. S. and Strauss, N.**, The role of RNA in repression of enzyme synthesis, *Proc. Natl. Acad. Sci. USA*, 50, 1059, 1963.
1421. **Commerford, S. L. and Delihas, N.**, Examination of the nucleohistone from mouse liver and intestine for RNA covalently linked to histone, *Proc. Natl. Acad. Sci. USA*, 56, 1759, 1966.
1422. **Von Heyden, H. W. and Zachau, H. G.**, Characterization of RNA in fractions of calf thymus chromatin, *Biochim. Biophys. Acta*, 232, 651, 1971.
1423. **Artman, M. and Roth, J. S.**, Chromosomal RNA: an artifact of preparation? *J. Mol. Biol.*, 60, 291, 1971.
1424. **Szeszák, F. and Pihl, A.**, Heterogeneity of chromosomal RNA, *FEBS Lett.*, 20, 177, 1972.
1425. **Bekhor, I. and Bavetta, L. A.**, An unusual property of chromatin isolated from mammalian salivary glands, *Biochem. Biophys. Res. Commun.*, 42, 615, 1971.
1426. **Pogo, A. O.**, Modification of ribonucleic acid synthesis in isolated rat liver nuclei by low salt concentrations and specific divalent cations, *Biochim. Biophys. Acta*, 182, 57, 1969.
1427. **Kroeger, H.**, Hormones, ion balances and gene activity in dipteran chromosomes, *Mem. Soc. Endocrinol.*, 15, 55, 1967.
1428. **Moulé, J.**, The effects of magnesium and manganese on the purine ratio of RNA transcribed *in vitro* by rat liver nuclei, *Eur. J. Biochem.*, 17, 544, 1970.
1429. **Burgess, R. R.**, RNA polymerase, *Annu. Rev. Biochem.*, 40, 711, 1971.
1430. **Butterworth, P. H., Cox, R. F., and Chesterton, C. J.**, Transcription of mammalian chromatin by mammalian DNA-dependent RNA polymerases, *Eur. J. Biochem.*, 23, 229, 1971.
1431. **Reeder, R. H. and Brown, D. D.**, Transcription of the ribosomal RNA genes of an amphibian by the RNA polymerase of a bacterium, *J. Mol. Biol.*, 51, 361, 1970.
1432. **Warner, J. R., Soeiro, R., Birnboim, H. C., Girard, M., and Darnell, J. E.**, Rapidly labeled HeLa cell nuclear RNA. I. Identification by zone sedimentation of a heterogeneous fraction separate from ribosomal precursor RNA, *J. Mol. Biol.*, 19, 349, 1966.
1433. **Attardi, G., Parnas, H., Hwang, M. I. H., and Attardi, B.**, Giant-size rapidly labeled nuclear ribonucleic acid and cytoplasmic messenger ribonucleic acid in immature duck erythrocytes, *J. Mol. Biol.*, 20, 145, 1966.
1434. **Soeiro, R., Birnboim, H. C., and Darnell, J. E.**, Rapidly labeled HeLa cell nuclear RNA. II. Base composition and cellular localization of a heterogeneous RNA fraction, *J. Mol. Biol.*, 19, 326, 1966.
1435. **Darnell, J. E.**, Ribonucleic acids from animal cells, *Bacteriol. Rev.*, 32, 262, 1968.
1436. **Soeiro, R., Vaughan, M. H., Warner, J. R., and Darnell, J. E.**, The turnover of nuclear DNA-like RNA in HeLa cells, *J. Cell Biol.*, 39, 112, 1968.

1437. **Georgiev, G. P., Samarina, O. P., Lerman, M. I., Smirnov, M. N., and Severtzov, A. N.**, Biosynthesis of messenger and ribosomal ribonucleic acids in the nucleolochromosomal apparatus of animal cells, *Nature*, 200, 1291, 1963.
1438. **Penman, S.**, RNA metabolism in the HeLa cell nucleus, *J. Mol. Biol.*, 17, 117, 1966.
1439. **Shearer, R. W. and McCarthy, B. J.**, Evidence for ribonucleic acid molecules restricted to the cell nucleus, *Biochemistry*, 6, 283, 1967.
1440. **Pagoulatos, G. N. and Darnell, J. E.**, Fractionation of heterogeneous nuclear RNA: Rates of hybridization and chromosomal distribution of reiterated sequences, *J. Mol. Biol.*, 54, 517, 1970.
1441. **Shearer, R. W. and McCarthy, B. J.**, Characterization of RNA molecules restricted to the nucleus in mouse L-cells, *J. Cell. Physiol.*, 75, 97, 1970.
1442. **Shearer, R. W. and McCarthy, B. J.**, Related base sequences in the DNA of simple and complex organisms. IV. Evolutionary divergence of base sequences in mouse L-cell cytoplasmic and nucleus-restricted RNA, *Biochem. Genet.*, 4, 395, 1970.
1443. **Church, R. B., Luther, S. W., and McCarthy, B. J.**, RNA synthesis in Taper hepatoma and mouse liver cells, *Biochim. Biophys. Acta*, 190, 30, 1969.
1444. **Birnboim, H. C., Pène, J. J., and Darnell, J. E.**, Studies on HeLa cell nuclear DNA-like RNA by RNA-DNA hybridization, *Proc. Natl. Acad. Sci. USA*, 58, 320, 1967.
1445. **Davidson, E. H. and Britten, R. J.**, Note on the control of gene expression during development, *J. Theor. Biol.*, 32, 123, 1971.
1446. **Britten, R. J. and Davidson, E. H.**, Repetitive and non-repetitive DNA sequences and a speculation on the origins of evolutionary novelty, *Q. Rev. Biol.*, 46, 111, 1971.
1447. **Georgiev, G. P.**, On the structural organization of operon and the regulation of RNA synthesis in animal cells, *J. Theor. Biol.*, 25, 473, 1969.
1448. **Tsanev, R. and Sendov, B.**, Possible molecular mechanism for cell differentiation in multicellular organisms, *J. Theor. Biol.*, 30, 337, 1971.
1449. **Crick, F.**, General model for the chromosomes of higher organisms, *Nature*, 234, 25, 1971.
1450. **Gurdon, J. B. and Woodland, H. R.**, On the long-term control of nuclear activity during cell differentiation, *Curr. Top. Dev. Biol.*, 5, 39, 1970.

Acknowledgments

Sincere appreciation for permission to reproduce published material is extended to the following: Dr. J. Bonner and the Academic Press for Figures 2 and 38; Dr. J. Sonnenbichler and the Elsevier Publishing Co. for Figure 6; Dr. G. R. Chalkley and the American Society of Biological Chemists for Figures 14, 15, 16, 17, and 18; Dr. B. R. Zirkin and the Academic Press for Figure 22; Dr. E. M. Bradbury and the Plenum Press for Figures 23, 24, 25, 26, and 27; Dr. G. H. Dixon, the Academic Press and the National Research Council of Canada for Figures 21, 28, and 30; Dr. A. Orengo and the Academic Press for Figure 29; Dr. P. O. P. Ts'o and the Academic Press for Figure 31; Dr. H. Ris, the Elsevier Publishing Co., and the Academic Press for Figures 32, 33a, and 37; Drs. B. R. Brinkley and M. E. McClure for Figure 33B; Dr. G. P. Georgiev for Figures 34 and 35; Dr. E. J. DuPraw and the Holt, Rinehart and Winston Co. for Figure 36; Dr. A. T. Ansevin and the American Chemical Society for Figures 39, 41, 42, 43, 44, 45, 46, 47, and 48; Dr. T. C. Spelsberg, the Elsevier Publishing Co., and the National Academy of Sciences for Figures 49, 50, 51, 57, 58, 59, and 60; Dr. G. J. McCarthy and the American Chemical Society for Figures 52, 53, and 54; Dr. J. Paul and the Academic Press for Figures 55 and 56; Dr. M. Gronow and the North-Holland Publishing Co. for Figure 61; Miss C. M. Groves and Ms. J. Thomson for Figures 62, 63, and 64; Dr. R. T. Britten and the American Association for the Advancement of Science for Figure 65; and Dr. G. P. Georgiev and the Academic Press for Figure 66.

ADDENDUM

Recently, at the Gordon Research Conference on Nuclear Proteins, Chromatin Structure, and Gene Regulation (Beaver Dam, Wisc., July 3 – 7, 1972) a unified nomenclature for histone fractions was accepted by all participants. The amino acid composition of each histone is used for its description, i.e., one letter abbreviations of the names of the three amino acids most abundant or most characteristic for the fraction determines its name. Since one letter abbreviations K, A, R, P, L, E, G, and S stand for lysine, alanine, arginine, proline, leucine, glutamic acid, glycine, and serine, respectively, the new symbols for the common histones are as follows:

Old	New
F1 (I)	KAP
F2b (IIb2)	KAS
F2a2 (IIb1)	LAK
F2a1 (IV)	GRK
F3 (III)	ARE

Additional description of each histone, such as its source (animal or tissue), electrophoretic heterogeneity (mainly F1 or KAP histones), chemical modifications resulting in electrophoretic heterogeneity (acetylation of F2a1 or GRK, F3 or ARE histones, phosphorylation of F1 or KAP histones), etc. will be included in the subscript. Hence, the chicken erythrocyte-specific histone F2c (V) is $KSA_{chicken}$ and the trout testis histone fraction T is $AKP_{trout\ testis}$. Details concerning the new histone nomenclature can be obtained from Professor James Bonner, Division of Biology, California Institute of Technology, Pasadena, California 91109.